REFRACTIONAL INFLUENCES IN ASTROMETRY AND GEODESY

INTERNATIONAL ASTRONOMICAL UNION
UNION ASTRONOMIQUE INTERNATIONALE

SYMPOSIUM No. 89
ORGANIZED BY IAU IN COOPERATION WITH IAG AND IUGG
HELD IN UPPSALA, SWEDEN, 1–5 AUGUST 1978

REFRACTIONAL INFLUENCES IN ASTROMETRY AND GEODESY

EDITED BY

ERIK TENGSTRÖM
Geodetic Institute, Uppsala University, Sweden

and

GEORGE TELEKI
Astronomical Observatory, Beograd, Yugoslavia

with the cooperation of
I. OHLSSON

D. REIDEL PUBLISHING COMPANY
DORDRECHT : HOLLAND / BOSTON : U.S.A. / LONDON : ENGLAND

Library of Congress Cataloging in Publication Data

Main entry under title:

Refractional Influences in Astrometry and Geodesy

(IAU Symposium; No. 89)
Includes index.
1. Refraction, Astronomical—Congresses. 2. Astrometry—Congresses.
3. Geodesy—Congresses. I. Tengström, Erik. II. Teleki, G. III. Ohlsson, Ingegerd. IV. International Astronomical Union. V. International Association of Geodesy. VI. International Union of Geodesy and Geophysics. VII. Series: International Astronomical Union. Symposium; No. 89.
QB155.R33   522'.9   79-21466
ISBN 90-277-1037-6
ISBN 90-277-1038-4 pbk.

Published on behalf of
the International Astronomical Union
by
D. Reidel Publishing Company, P.O. Box 17, Dordrecht, Holland

All Rights Reserved
Copyright © 1979 by the International Astronomical Union

Sold and distributed in the U.S.A., Canada, and Mexico
by D. Reidel Publishing Company, Inc.
Lincoln Building, 160 Old Derby Street, Hingham,
Mass. 02043, U.S.A.

No part of the material protected by this copyright notice may be reproduced or utilized in any form or by any means, electronic or mechanical, including photocopying, recording or by any informational storage and retrieval system, without written permission from the publisher

Printed in The Netherlands

A curious fellow named Snell
Found a Law we all think is swell
Observers all buy it
But when they apply it
Their answers are all shot to hell.

(J.A. Hughes 1978)

TABLE OF CONTENTS

PREFACE	xi
LIST OF PARTICIPANTS	xiii
OPENING ADDRESS / M. Holmdahl	xv
ADDRESS / T.J. Kukkamäki	xvii
ADDRESS / G. Teleki	xix
ADDRESS / E. Tengström	xxi
G. TELEKI / Research into astronomical refraction - today and tomorrow	1
W.J. ALTENHOFF / Refraction in radioastronomy	11
J.A. HUGHES / Environmental systematics and astronomical refraction II	13
H. YASUDA and R. FUKAYA / An origin of the variations of astronomical refraction	27
V.I. SERGIENKO / On astronomical refraction for a three-dimensional model atmosphere	35
J. DOMMANGET / Les effets de la réfraction atmosphérique sur les coordonnées tangentielles en astrométrie	39
V.I. IVANOV / Atmospheric turbulence effects on astrometric determinations	67
J. SAASTAMOINEN / On the calculation of refraction in model atmospheres	73
S. MIKKOLA / Calculating astronomical refraction by means of continued fractions	95
C. SUGAWA and N. KIKUCHI / On the characteristics of astronomical refraction in the northern hemisphere	103
S. TAKAGI and Y. GOTO / Astronomical refraction at the Mizusawa Latitude Observatory	119
B.K. BAGILDINSKY, S.P. PULIAEV and E.G. ZHILINSKY / Chromatic refraction in the vertical circle observation	125
DISCUSSION ABOUT REFRACTION CORRECTIONS IN STAR CATALOGUE WORK AND GEODETIC ASTRONOMY / Chairman: J.A. Hughes	129

D.G. CURRIE / Two color refractometry, precision stellar catalogs, and the role of anomalous refraction — 131

A.J. ANDERSON / Path length variations due to changes in tropospheric refraction — 157

W.J. ALTENHOFF / Diurnal changes of radio refraction — 163

P.V. ANGUS-LEPPAN / Use of meteorological measurements for computing refractional effects - a review — 165

M.K. SZACHERSKA and Z. WIŚNIEWSKI / Statistics applied to the estimation of the influence of the environment on results of observations — 179

L. HRADILEK / Evaluation of refraction by geodetic measurements — 191

R. BLAŽEK and L. HRADILEK / Investigation on refraction in trigonometrical leveling traverses — 195

K. RAMSAYER / The accuracy of the determination of terrestrial refraction from reciprocal zenith angles — 203

H. KAHMEN / Numerical filtering of refraction coefficients — 213

F.K. BRUNNER / Vertical refraction angle derived from the variance of the angle-of-arrival fluctuations — 227

D.C. WILLIAMS / Results from an absolute test of the NPL dispersometer over 4 km — 239

S-G. MÅRTENSSON / Experiences from IDM measurements at the test base of the Geodetic Institute of Uppsala University — 241

J. MILEWSKI / Possibilities of increasing the accuracy in the determination of refractional angles with Tengström's IDM — 249

R. PAZUS / Application of refractional effects in geodetic framework of Iraq by continuous trilateration — 267

J. KAKKURI and O. OJANEN / Computing parallactic refraction for stellar triangulation — 275

DISCUSSION ABOUT COMPUTING PARALLACTIC REFRACTION IN MODEL ATMOSPHERES / Chairman: T.J. Kukkamäki — 281

EDM PANEL ON INSTRUMENTS AND ATMOSPHERICAL CORRECTIONS / Chairman: I. Brook — 285

T.J. KUKKAMÄKI / Levelling refraction research, its present state and future possibilities — 293

TABLE OF CONTENTS

O. REMMER / The direct experimental detection of the
systematic error of precise levelling   301

S.R. HOLDAHL / Removal of refraction errors in geodetic
leveling   305

J. LARSSON / Refractional effects in photogrammetry   321

J.C. DE MUNCK / The use of multi- or single wave methods
to eliminate terrestrial refraction from geodetic measurements   327

ROUND TABLE DISCUSSION ABOUT THE USE OF DISPERSION METHODS
FOR DETERMINING REFRACTIONAL EFFECTS IN ASTRONOMY AND GEODESY /
Chairman: E. Tengström   331

G.H. LILJEQUIST / Notes on some abnormal refraction phenomena
in the atmosphere   343

COOPERATION BETWEEN ASTRONOMERS, METEOROLOGISTS AND GEODESISTS
FOR SOLVING REFRACTIONAL PROBLEMS / Chairman: G. Teleki   351

WORKSHOP ON WAVE PROPAGATION / Chairman: H. Kahmen   355

F.K. BRUNNER / On precursors of electromagnetic wave propagation   357

WORKSHOP ON LOCAL GEODYNAMICS FROM REFRACTION CORRECTED
MEASUREMENTS / Chairman: A.H. Dodson   371

A.H. DODSON / The role of refraction in the measurement of
three-dimensional movements by geodetic methods   373

CLOSING SESSION   381

RESOLUTIONS   385

JOINT MEETING OF THE WORKING GROUP ON ASTRONOMICAL REFRACTION,
IAU COMMISSION 8, AND SSG I.42 ON "ELECTROMAGNETIC WAVE PROPAGATION
AND REFRACTION IN THE EARTH'S ATMOSPHERE" OF IAG / Chairman:
G. Teleki   389

INDEX   391

PREFACE

E. Tengström
G. Teleki

The IAU symposium no 89, 'Refractional Influences in Astrometry and Geodesy', was held at the University of Uppsala, Sweden, during August 1 through August 5, 1978. It was a joint meeting of IAU and IAG, cosponsored by IAG and IUGG. Considerable financial support was also given by the Swedish Natural Science Research Council, the Wallenberg Foundation in Stockholm and the University of Uppsala.

The Scientific Organizing Committee included G. Teleki (chairman), W. Altenhoff (Bonn), P. Bender (Boulder), L. Hradilek (Prague), J.A. Hughes (Washington D.C.), J. Kakkuri (Helsinki), A.J. Nefedieva (Kazan), J. Saastamoinen (Ottawa), C. Sugawa (Mizusawa) and E. Tengström (Uppsala). The Local Organizing Committee included E. Tengström (chairman and convenor), I. Ohlsson (adm. secretary and treasurer), P. Hodacs, S-G. Mårtensson, S. Eklund and M. O'Shaughnessy, all from Uppsala.

This symposium was the first international meeting of astronomers and geodesists connected with refraction phenomena.

The proceedings of this symposium contain all papers accepted by its Scientific Organizing Committee, and all discussions around them together with all contributions during special workshops and round-table discussion programs.

The material for the account of this joint IAU/IAG symposium was given as camera-ready manuscripts demanded from the publisher, and prepared from manuscripts submitted by the responsible participants. Almost all contributions have been contained, either in their complete form or as abbreviations. The discussions, which were recorded on tape during the sessions, thanks to efforts of Mr Hodacs, have been followed in detail by the editors Teleki and Tengström, who have carefully listened at the tapes from various sessions. A first draft of the discussions was then sent out to all participants involved. As soon as their corrections were made available, the definitive text has been written in the Reidel format by I. Ohlsson. If no response to our inquiry for eventual changes has reached us in time, the contributor's original wordings have been accepted as printable.

The editors are extremely grateful for the very efficient help in editing the proceedings of this first joint meeting between IAU and IAG on Refraction, given by the administrative secretary of the Local Organizing Committee, Miss Ingegerd Ohlsson, Uppsala.

At last, the editors like to stress the fact, that this joint meeting on Refraction, according to its 8th resolution, will be able to speed up the effort for creating an interunion body on this subject (IAU/IUGG). The immediate effect of this resolution must certainly be, that the Executive Committees of IAU and IAG must take it into consideration in their most urgent plans for common future activities.

LIST OF PARTICIPANTS

Altenhoff, W.J., Bonn, BRD
Anderson, A.J., Uppsala, Sweden
Anderson, E.G., New Brunswick, Canada
Angus-Leppan, P.V., Kensington, Australia
Bauch, A., München, BRD
Brook, I., Gävle, Sweden
Chaudhary, D.R., Jaipur, India
Currie, D.G., Maryland, USA
Dodson, A.H., Nottingham, U.K.
Dommanget, J., Bruxelles, Belgium
Ehlert, D., Frankfurt, BRD
Eklund, S., Uppsala, Sweden
Garfinkel, B., New Haven, USA
Gächter, B., Heerbrugg, Switzerland
Hodacs, P., Uppsala, Sweden
Holdahl, S.R., Rockville, USA
Hradilek, L., Praha, Czechoslovakia
Hughes, J.A., Washington D.C., USA
Israelsson, S., Uppsala, Sweden
Johansson, S., Gävle, Sweden
Kahmen, H., Karlsruhe, BRD
Kakkuri, J., Helsinki, Finland
Kaniuth, K., München, BRD
Kautzleben, H., Potsdam, DDR
Kukkamäki, T.J., Helsinki, Finland
Larsson, J., Stockholm, Sweden
Liljequist, G., Uppsala, Sweden
Milewski, J., Koszalin, Poland
de Munck, J.C., Delft, The Netherlands
Mårtensson, S-G., Uppsala, Sweden
Oja, T., Uppsala, Sweden
Oterma, L., Turku, Finland
Pelzer, H., Hannover, BRD
Poder, K., Charlottenlund, Denmark
Ramsayer, K., Stuttgart, BRD
Remmer, O., Charlottenlund, Denmark
Saastamoinen, J., Ottawa, Canada
Sugawa, C., Mizusawa, Japan
Szacherska, M., Warszawa, Poland
Teleki, G., Beograd, Yugoslavia
Tengström, E., Uppsala, Sweden
Williams, D.C., Teddington, U.K.
Griffiths, A., Teddington, U.K.

OPENING ADDRESS BY:

M. Holmdahl
Rector Magnificus of Uppsala University
Sweden

Ladies and Gentlemen,
On behalf of Uppsala University it is a great pleasure for me, as my first official duty as rector of this 501 years old university, to extend a most hearty welcome to all of you participants in this symposium on Refractional Influences in Astrometry and Geodesy. I understand that this symposium is a joint meeting sponsored by the International Astronomical Union and the International Association of Geodesy at the invitation of the Swedish National Committee for Astronomy and the Swedish National Committee for Geodesy and Geophysics. I welcome especially the president of the IAG, professor T.J. Kukkamäki, and the presidents of special study groups in IAU and IAG, dr G. Teleki and our own professor E. Tengström, who also have acted as presidents of the Scientific Organizing Committee and the Local Organizing Committee of this symposium, respectively.

I understand that the topic of this symposium is of utmost importance to solve common problems in geodesy and astronomy with main emphasis on the use of dispersion methods for determining refractional effects in astronomy and geodesy. In the field of astronomical refraction I have been told that scientists in Japan have made very important contributions and it is therefore with great pleasure we welcome the presence of professor C. Sugawa at this symposium.

Uppsala University last year celebrated its 500 years anniversary under the leadership of our great past president, professor Torgny Segerstedt, a scientific year with a series of symposia and congresses in different fields with participation of 6000 scientists from all over the world. We are glad that these activities continue and that you have chosen Uppsala as the place for your symposium. A university never can be proud of age alone. Scientific traditions must be matched by front-line activities in many fields to prove the value and youth of a university in the international scientific community. In your field of interference between Geodesy, Meteorology and Astronomy our own university institutions have made important new contributions especially through professor Tengström and professor Liljequist and their coworkers. The presence of so many young active foreign scientists here at this symposium I take as a proof of the international acceptance of the quality of their work in your respective fields here in Uppsala.

We thank our guests, foreign and inland, for coming to our university. Though 500 years old with your astronomical measures measured, it is comparatively young - especially in spirit we hope.

With the rapid expansion of knowledge in all fields of basic and applied research it is not surprising that in all countries the university student number has increased rapidly during the last two decades. It immatriculated at this university nearly the same number of students as was immatriculated during the previous 475 years of this university's existence. I do not need to stress that quantity may create problems for quality. That is why we are so thankful for the necessary international cooperation. International contacts stress the need for quality in higher education and research and demonstrate that quality in these important fields never can be judged by national but only by international standard.

Finally, at this occasion, let me wish you all a prosperous scientific meeting and rewarding days here in Uppsala. If these will be memorable for you, your host university will feel amply rewarded. If the conference will give you new ideas and encouragement for future work, I am sure the organizers will feel equally rewarded. I take this opportunity on behalf of the university to thank them. You name them yourselves but I include all members of the Local and Scientific Organizing Committees for their work.

Thank you and very welcome!

ADDRESS BY:

T.J. Kukkamäki
President of the International Association of Geodesy (IAG)
Finland

In the last century the limits of accuracy in geodesy were set by incompletenesses in instruments, while ambient factors had minor influence. Since then progress in instrument building has been so great that improvements in theodolites were no longer able to increase horizontal accuracy significantly, and the movements of the instrument stands and, especially, lateral refraction now have a greater effect on accuracy than incompletenesses in the instruments. This state of affairs was still more evident for trigonometrical height determination as vertical refraction markedly surpasses the defects of instruments.

Conditions in levelling have also reached this state. The invar rods and levelling instruments were completed to the extent that refraction now sets the limits of accuracy possible to be achieved.

Under these conditions, it was more fruitful to try to develope research on the effects of ambient factors than on the observation instruments themselves.

The situation changed essentially when electronic methods and instruments entered geodesy. These new instruments and methods soon reached accuracies that are, even some decades, better than those obtained by classical means. The problems of ambient factors, I mean here the refraction mostly, got characters completely different from the earlier.

Along a side vertical and lateral refraction, longitudinal refraction entered the picture, especially in electronic distance measurement, EDM.

Research into refraction in the old sense is still going on. Research in the new sense consists of vertical, lateral and longitudinal effects. Longitudinal refraction is considered to be belonging to the EDM techniques and methods, and thus is to be treated in those connections.

Vertical and lateral refraction are the subjects of this symposium. This question is very important for improving the accuracy of classical geodetic measurements. As the classical data in most disciplines form the fundamental basis for both classical and modern measurements, these studies are very important for geodetic science as a whole.

The refraction in astrometry goes far beyond my knowledge. I do know however that it is very important for astronomical geodesy in a vertical and a lateral sense, and for space geodesy in a vertical, a lateral and a longitudinal sense.

The International Association of Geodesy and the International Astronomical Union are co-sponsoring this symposium, which is expected to give valuable results to both for these sciences, geodesy and astrometry.

On behalf of the IAG, I would like to thank our Swedish hosts, especially professor Tengström, for the invitation to this meeting. I wish its work the best success.

ADDRESS BY:

> G. Teleki
> Chairman of the Scientific Organizing Committee,
> Chairman of the Working Group on Astronomical Refraction of
> Commission 8 of the International Astronomical Union (IAU)
> Yugoslavia

Dear Colleagues, Ladies and Gentlemen,

I take great pleasure in having the honor to greet you in the name of the Scientific Organizing Committee of this symposium and to wish you successful work.

Permit me to greet, in the name of all participants, our distinguished guests who honored us by their presence. We have with us: professor Holmdahl, the Rector Magnificus of the Uppsala University and professor Kukkamäki, the president of the International Association of Geodesy, who is also an active participant of this meeting. I am also happy to greet the members of the Scientific Organizing Committee, of the IAU Commission 8 Working Group on Astronomical Refraction, of the IAG Special Study Group I.42 on Electromagnetic Wave Propagation and Refraction in the Earth's Atmosphere, and all invited speakers.

Another pleasant duty has fallen to me. In the name of all of you I would like to greet professor Tengström, the convenor and chairman of the Local Organizing Committee and to thank him for the work performed by him in the organizing of this meeting. He not only kindly accepted the organization but gave an immense contribution to the shaping of the profile of this meeting.

Professor Tengström was greatly aided in his work by: miss Ohlsson, secretary and treasurer, and the members of the Organizing Committee: messrs. Hodacs, Mårtensson, Eklund and O'Shaughnessy. To all of them we owe our sincere gratitude.

In planning of this symposium, we had two tasks in view: first, to summarize the results achieved in different fields and to stimulate thereby further researches and, second, by the exchange of views and experiences in the fields concerned to promote the knowledge of the refraction as an unique phenomenon. Thus, our task has been and continues to be: the analysis and the synthesis of the refractional influences. Even if the times were different it would be quite normal to proceed in this way.

However, at a time like ours, these questions are getting particular importance. For once, both astrometry and geodesy are at a turning point of their development. By the introduction of a new, extra-atmospheric observing technique the accuracy of our measurements is being, or might substantially be, enhanced. This is not to say that the corresponding ground-based observations, in which refractional influences are an important factor, are to be, once and for all, discarded, but an increased accuracy of them is what is strongly demanded. We have reached the point where a better knowledge of the refractional influences is an imperative, a basic requirement if ground-based observations are to be, if not quite competetive, then at least complemental to the extra-atmospheric observations.

I believe I am not exaggerating by my emphasizing the importance of the knowledge of refractional, more exactly, of the atmospheric influences on the future development of geodesy and astrometry, as one may quite safely assume that ground-based geodesy and astrometry are not going to disappear at least in the near future. To what extent atmospheric influences will affect their results largely depends upon us. It is therefore our obligation to contribute, by our, as exact as possible, accounting for, reducing or eliminating those influences, to the increase of the accuracy of our measurements. It would be a good thing if no quotation would be made any longer of the wellknown Newcomb's statement about astronomical refraction from 1906: "There is no branch of practical astronomy on which so much has been written as on astronomical refraction, and which is still in so unsatisfactory state".

I certainly do not expect the aspired turn to be brought about exactly at this session. But I am certain that we are going, by our exchange of experiences, to contribute to the progress of these researches. I therefore sincerely regret, and I believe that all of you share this statement, that we do not have today with us some of the prominent researchers in this field. However, I am hopeful that through the proceedings of this symposium the contact with all interested will be maintained and that in this way the development of the investigations into refraction will be speeded up.

We are going to listen here on various refractional influences. It would be correct, however, to treat them as informations on the refraction as an unique phenomenon, bringing thereby the required synthesis into effect. I am confident that, in doing so, we are furthering our investigations. It is even of interest to examine the possibility of uniting corresponding working, e.g. study groups of refraction of the IAU and IAG into one sole body as an inter-union commission for researches into refraction.

It is my heartful wish that all of us, on closing this symposium, should be content with its work and that we will have clear picture as to the future of our investigations. Hoping for this, let me welcome you once again and wish you, besides a successful work, a pleasant stay in Uppsala. To our hosts once again our best thanks.

ADDRESS BY:

>E. Tengström
>Convener of the symposium,
>Chairman of the Special Study Group I.42 of the International
>Association of Geodesy (IAG)
>Sweden

Your Magnificence, Mr Teleki, Mr President of IAG, Ladies and Gentlemen,

It means a great honor for me that I have been entrusted to convene you to this symposium, the topic of which deals with a most important problem, especially in Geodesy and Astrometry, namely the problem of eliminating refractional influences from observations of various electromagnetic wave signals.

I am especially indebted to Dr Teleki, who is the official representative of the Astronomical Union here. He is also president of the Working Group on Astronomical Refraction in Commission 8 of the International Astronomical Union. He early suggested a joint symposium to be held between the International Astronomical Union (IAU) and the International Association of Geodesy (IAG) for the purpose of establishing a close cooperation in studying refractional phenomena, connected with the propagation of wave-trains toward an observer, transmitted by terrestrial or extraterrestrial signal sources.

Dr Teleki's intention was originally to arrange a symposium, with same title as the one you are now attending, in his home-country Yugoslavia, at its wonderful capital Belgrade. Due to some insurmountable difficulties, Teleki's idea of holding the symposium at Belgrade could not be realized. When he asked me for advice, I immediately told him that Uppsala University would probably help to arrange the symposium, if he and IAU together with IAG and the International Union of Geodesy and Geophysics (IUGG) liked to accept an invitation from the Swedish National Committee for Astronomy and from the Swedish National Committee for Geodesy and Geophysics to a symposium on the aforesaid topics at Uppsala University under my convenor-ship.

Thanks the very positive attitude to the proposal from aforementioned organizations, shown by the General Secretary of IAU, Mrs E. Müller, her Assistant General Secretary, Mr P.A. Wayman, and the General Secretary of IAG, Mr M. Louis, the symposium was approved and called IAU Symposium No 89 on "Refractional Influences in Astrometry and Geodesy", cosponsored by IAG. The IAU support to this conference has been twofold, namely, a substantial travelling support for invited participants and the printing of the proceedings of the symposium by the D. Reidel Publishing Co., Holland. IAG has contributed with money for the organization of the symposium and IUGG with grants for travelling to invited geodesists. The Swedish Natural Science Research Council (NFR), the Wallenberg foundation and the Rector of Uppsala University have also pecuniarily helped the Geodetic Institute to be able to carry through the symposium. I thank them all for their contributions.

Last but not least, - from the depth of my heart - I have to express my gratitude to all members of our small Local Organizing Committee, who have devoted every minute of their time to solving the organization problems. Especially I like to thank my secretary Miss Ingegerd Ohlsson for her efficient way of attacking every problem appearing during the preparation of this meeting.

The official guests of this symposium are, besides Dr Teleki, who represents the International Astronomical Union, the Rector Magnificus of the University of Uppsala, professor Holmdahl, and the President of the International Association of Geodesy, professor Kukkamäki.

Professor Holmdahl has kindly invited you to a reception in the University building this night at 7 p.m. Uppsala University has always been supporting international cooperation in science. I am glad, that also this international event has been observed by the highest authority of this university through professor Holmdahl's eagerness to make your acquaintance and to be informed of your work during the reception he will host today.

My old friend, professor Kukkamäki is not only an important official representative being the president of IAG. He as well as other people at his famous Geodetic Institute in Helsinki is a good friend of my small institute at Hällby since long time ago. His successor as director of Geodeettinen Laitos, professor Kakkuri, is also here to demonstrate our close cooperation in various fields of geodesy, not least in refraction.

Kukkamäki is also the great old man in the field of nivellitic refraction, a subject which is included as being very essential in the programme of the symposium. Not only that. He will actively participate in our work in this area by presenting a review paper on the subject.

I thank professor Holmdahl and professor Kukkamäki for their willingness to be here with us to demonstrate their interest in our work.

As chairman of the IAG Special Study Group I.42 on "Electromagnetic Wave Propagation and Refraction in the Earth's Atmosphere" I am glad to welcome very important members of this group, who liked to contribute to our discussions. Scientists as K. Poder, I. Brook, L. Hradilek, K. Ramsayer, J. Saastamoinen, P.V. Angus-Leppan, H. Kahmen, D.C. Williams,

J.C. de Munck, are all wellknown and their presence here convinces me that with them, the symposium must be a good one.

I am sorry, that we could not have among us this time a lot of geodesists, who have worked in the field of refraction as members of SSG I.42, e.g. T. Glissmann, M.T. Prilepin, M.C. Thompson, H. Hopfield, E. Livieratos, T. Parm, A.R. Robbins. We send them all our best regards, hoping that we may meet during a planned symposium on refraction in Canberra 1979, that is during the next general assembly of IUGG.

Signs of life from aforementioned members of SSG I.42 have, however, reached me. Several of these members have sent important papers to be discussed here. Some have only demonstrated their unhappiness because of impossibilities to attend the symposium for various reasons. In any case 42% of the members of the group are present here and they will all actively contribute to our work.

Dr Teleki's Working Group on "Astronomical Refraction" comprises 11 members. Of them, 5 are present here as active contributors. This means almost 50% participation, which result I like to emphasize with congratulations to the chairman of the group.

Hoping that this small but - at least as regards the number of attendants - representative symposium will be able to throw new light on the problem of wave propagation and refraction, of importance to all of us in astronomy and geodesy, I wish to welcome you all to a fruitful meeting.

I have not yet mentioned the role of meteorological expertness for our work. But it is clear, that meteorological information can never be avoided in our work, even if we believe that astronomical and geodetic refraction studies might perhaps give more to the meteorologists than they can give to us. The difficult theoretical investigations of atmospherical turbulence must be supported by observations of wave propagation and refraction, as was clearly understood by the famous meteorologist Brocks long time ago. Brocks also formulated the title of research "Geodetic Meteorology", a title which I like to reverse to "Meteorological Geodesy".

Some of our guests during this symposium are pure meteorologists. Professor Liljequist, chief of the meteorological department of the university of Uppsala, will join us at certain sessions and give us valuable information of his experience of anomalous refraction phenomena in the Antarctic, eventually confirming the experience obtained by Angus-Leppan.

Liljequist's assistant Dr Israelsson will also give us a lot of valuable informations as to the wave propagation in the turbulent atmosphere. I am very grateful for his contribution. His field is sound wave propagation. But the main principles of treating propagation questions are still the same for geodesists working in the spectral regions of optics and radiowaves.

At the end of this talk, I like to tell you, that I have written to professor Tatarski in Moscow, asking him to review the situation as regards turbulence effects on wave propagation, if possible during our symposium. I got an answer from him, that the Soviet regulations did not permit him to attend at such a short notice (June 1978). But he was apparently extremely interested in our work. The books by Tatarski will, of course, constitute an important background for our discussions of wave propagation in turbulent atmospheres. Certain problems, connected with Tatarski's theory of such atmospheres will be touched by Kahmen and Ivanov.

I hope that you will feel well during this small symposium at the university of Uppsala, and that you will go home full of new ideas for solving this essential problem of refractional effects on wave-train information, important for Astronomy and Geodesy and through these sciences also of great value for studying turbulence in our atmosphere, which is of extreme importance for meteorologists.

So, Uppsala University and the Geodetic Institute at Hällby are both wishing that your participation in this symposium will give you something of value to carry home.

RESEARCH INTO ASTRONOMICAL REFRACTION - TODAY AND TOMORROW
(A Survey Paper)

G. Teleki
Astronomical Observatory, Belgrade, Yugoslavia

SUMMARY: Proceeding from the present-day and future needs of astrometry, current researches into astronomical refraction as well as its prospective development are analysed. Particular attention is dedicated to the constructing of new refraction tables, determination of the refractional influences by way of experiments, site selection of the observing stations and to the preventive measures to be taken at them. Need and usefullness are emphasized of the cooperation of experts specialized in astronomical, terrestrial, photogrammetric, radio and other kinds of refractional influences.

1. INTRODUCTION

Members of the Working Group on Astronomical Refraction of the IAU Commission 8 have laid down in a collection of papers (Teleki, 1974a) their views on the present state and future of astronomical refraction investigations. The following conclusion is found in the introductory article (Teleki, 1974b) of this paper collection: "Refractional investigations are needed especially at this moment when we have new astrometrical methods, fundamentally different from the classical ones, and we must compare the old and the new data. At this comparison the possible refractional influences must be known with higher accuracy than nowadays, because there are some unconfirmed beliefs and prejudices related to that". This statement remains in force at present as well. If any change has taken place in the last 4-5 years, it consisted in the tendencies in astrometry becoming prominent toward adopting new methods and instruments, their being intensively worked upon, but the impression is not to remove that the researches into astronomical refraction do not follow this trend.

Besides classical instruments already in existence, in the astrometry of the future new ground-based optical instruments, artificial satellites with astrometric missions, radio astrometry (very-long-baseline interferometry) as well as laser techniques must be reckoned with. To illustrate what new types of instruments are capable of offering let

the example of "Space Astrometry Hipparcos" satellite (ESA, 1978) be
pointed out: an accuracy of 0''.0015 is expected in the position determination of stars brighter than 11th magnitude, that of parallax up to 0''.002
and annual proper motion to 0''.002. Classical ground-based instruments, whose
internal errors are characterised by the amounts 0''.2 - 0''.4, whereas
external errors might even be larger, cannot compete with the new
techniques. To achieve this, new ground-based instruments should be created
and placed into favourable astro-climatic conditions. Høg (1975) puts the
estimate of the productivity of the present-day and future meridian
circles as follows: if the annual weight of the present-day meridian
circles is taken as unit, the weight of the photo-electrical circles
amounts to 15, that of horizontal circles to 40, and, if the latter are
placed in exceptionally favourable conditions, a weight up to 800 is
attainable. But even if the latter came true, great though its
significance might be, it is nevertheless reasonable to expect these
results to be at least one order of magnitude inferior to those
obtainable by Hipparcos Satellite. Still this ground-based technique
might be a complementary one to the new techniques. Failing this, the
reason of existence of these instruments and methods is lost - as will
soon be the case of the classical methods of pursuing the elements of
the Earth's rotation.

It is self-evident that in speaking about astro-climatically favourable
location, total effect of the atmosphere on the astrometric results is
meant (Teleki, 1978). A location is favourable provided it renders
possible a reduction, elimination or reliable calculation of individual
effects. By appropriate protection of the instruments it becomes to
considerably reduce, even to eliminate atmospheric effects on them and
on the equipment attached to them. Thus, refractional influences
constitute the basic problem. They can be reduced or, given favourable
conditions, even calculated, which is otherwise impossible at locations
with complex atmospheric structure (on this question see 4.1.).

What about the present-day accuracy of the determination of refractional
influences? The accuracy of the calculated refraction value within the
optical range is of the order of a few hundreths of an arc second, the
uncertainty increasing with zenith distances (Teleki, 1974b).
A somewhat better position prevails in the calculation of the refraction
of radio-waves, but it is even there not satisfactory (Bean, Teleki,
1974; Altenhoff, 1974). The calculation of the influence of the local
atmosphere constitutes central problem in both cases.

It is therefore evident that a better knowledge of the refractional
influences, in parallel with the existence of more perfect instruments,
provides one of the keys to the increase in the accuracy of the ground-
-based astrometric results. To give accurate values of refraction is as
difficult a task, if not even more difficult, than the construction of
a new high quality astrometric instrument. This in itself speaks of the
importance of the study of refraction for the future of the ground-
-based astrometric measurements.

In this review we are going to confine ourselves to the part of the optical refraction called astronomical refraction, decomposing it into pure and anomalous refraction (Teleki, 1974b). Pure refraction is defined as the component of the astronomical refraction in the vertical plane of the observer's place, assuming the atmosphere is a medium with characteristics of an ideal gas in hydrostatic equilibrium, in a Newtonian gravitational field and with spherically symmetric distribution of density. As for anomalous refraction, its definition is: the difference between the true and the pure refraction. Pure refraction is calculated by the classical integral of refraction, which serves for the forming of various refraction tables. One may therefore say that the anomalies are corrections to the tabulated refraction values.

2. PURE REFRACTION, REFRACTION TABLES

2.1. In solving classical integrals of refraction not only formulae, as simple as only possible, for the calculation of the influence are required, but the possibility is of prime importance of their application at greater zenith distances. Out of more recent expressions that of Saastamoinen (1974) is expedient up to $86°$ zenith distance, while that of Garfinkel (1967) is applicable to all zenith distances from $0°$ through $180°$. These are, to be sure, rather mathematical than physical and meteorological, possibilities of the refraction calculation. It is realistic to assume the refraction tables as being capable of securing 98% of the true value of that influence for the zenith distances up to $60°-70°$. At zenith distances beyond this limit the deviations may be excessive (Kolchinskij, 1976), thus astrometric measurements, requiring high precision, are not to be made as far off zenith. This is confirmed by Vasilenko and Kharitonova (1977) who, by applying the method of statistical orthogonal expansion of refractive index, obtained more accurate values of the influences at greater zenith distances ($80°-88°$); it is estimated that the total calculated error amounts to about $3\rlap{.}''9$ (i.e. 0.3%).

2.2. In the current astronomical praxis most in use are Pulkovo Refraction Tables, Fourth Edition (1956). The investigations of Teleki (1967), Nefed'eva (1973) and others have revealed the deficiencies of these Tables, of which we are going to mention: inadequate value of the refraction constant and the correction for the air temperature.

It became clear that more up-to-date refraction tables were necessary. Two activities were therefore started: first, proceeding from her researches, Nefed'eva (1973)compiled new tables, and, second, the need was pointed out of the forming of international tables, to be standard in astrometry.

2.3. The constructing of the international tables of the kind indicated has been urged by the resolution of the IAU Commission 8 in 1973. As can be seen from detailed plan of their forming (Teleki, 1977), the goal set were tabulated values closer to reality. This means that instead of the refraction values obtained for some idealized conditions, such values are given as correspond closer to the real physical and meteorological circumstances.

Basic question is, doubtless, that of choosing the model of atmosphere most corresponding to reality. The work on this is going on, but early researches of Sugawa and Kikuchi (1974, 1975) already let us comprehend the importance and need of these analyses.

The tables are to be formed in such a way that the refractional influences may be calculated directly in terms of the refractive index in the surface air layer. The possibility is thereby presented to make use of data obtained by the refractometer.

In connection with the forming of these modern tables or of some others on the same lines, the question arises whether they are necessary at all at the present time. As an alternative, the calculation is suggested on the basis of aerological measurements, carried out simultaneously or approximately at the same time with the astrometric observations, the use of local refraction tables etc. In handling questions of this type the principle is to be followed: such method should be applied which under circumstances is capable of offering optimal results. It may be said that a need for accurate tables continues to be felt. One of the reasons is their being capable of providing the first frame of the refractional influences, indispensable for the comparison of the data obtained at various stations.

3. EXPERIMENTAL METHODS

Calculated values of refractional effects on the basis of tables - obtained for some averaged and idealized conditions - are long since insufficient to meet astrometric needs. Hence many efforts have been made to improve them. This is being performed either by determination of the anomalous refraction (as corrections to the tables), or by determination of the total refractional effect (indenpendent of tabular values) for a given place and time. This is usually made by way of experiments.

In seeking true values, emphasis is to be placed on the following: not only values of the anomalies or the total influences in the vertical plane are required (as in the case of pure refraction), but values of the horizontal components are necessary too. Thus, full space characteristics of the effects must be ascertained.

In view of this and by meteorological considerations we are led to infer that local effects are most difficult to know - those not contained in the tables (except effects of the average temperature, humidity and pressure), but essential for obtaining true values (Teleki, 1967; 1969; 1974c). To be more explicit here is an information:

by exploring the atmospheric turbulence by means of radio sondes up to 20 km altitude, Barletti et al. (1977) came to the conclusion that once the local turbulence (ground convection and orographic disturbances) is eliminated, all sites show the same average atmospheric properties.

It is therefore understandable that such methods are to be considered as most convenient which take into account just these local factors. Ideal methods, at the same time most difficult also, are those intended to determine the influences in their totality and not only parts of them.

3.1. The oldest method of determining refraction values is that based directly on the results of astrometric measurements. The first refraction tables owe their origin to this procedure. This method requires, quite clearly, a definite mathematical model of the refraction influences which makes possible the singling out of separate components. As a result of their analysis, Teleki and Shevarlich (1971) have stated the following factors, as impending the shaping of a satisfactory model: variable nature of meteorological elements, irregular formation of the air layers around the instruments, the variability of the instrumental characteristics (systematical and accidental) and errors in star coordinates. If these reasons stand - and they do stand - then it is practically impossible to devise an universal method for singling out of refractional influences from astrometric measurements. The authors came to the conclusion that a rigorous method must fulfil the following conditions: a) the shortest possible time for the measurement of the observational values required, b) the use of high quality instruments, whose characteristics and constants are practically unchangeable, c) the elimination of errors in star coordinates, and d) the analysis is to be made within narrow zones according to zenith distances. The meaning of these conditions is the suppression of the interpolation and extrapolation alike in time as in space. It is just upon these foundations that the authors proposed a method of the determination of anomalies in the circum-zenithal zone, which however has not as yet been employed (due in the first place to technical difficulties).

Unfortunately, the methods proposed do not, in their majority, comply with the above stated principles. By habit, universal methods are offered, implying the adopted models as being general: these methods are based on models, expressed, in the last instance, by $tg\ z$ or $sec^2 z$ (where z is the apparent zenith distance). In consequence, the nature of the values derived becomes very problematical. As an example take the determination of the so called constant of refraction from observations of one and the same star at upper and lower culminations (Teleki, 1977).

It may therefore stated that the methods intended at the singling out of refractional influence from astrometric measurements are without prospects in the future.

3.2. In view of what is presented in 3.1., attempts have been made and will be made to devise rigorous methods by which atmospheric parameters could be determined and on that basis refractional influences calculated. In brief, the procedure should consist not in singling out effects from astrometric results but should be quite indenpendent of them.

Aerological measurements have permitted a better knowledge of the free atmosphere, with all its variations in time and space. The calculations of the refractional influences directly from aerological measurements have already began. It was considered that in doing so instantaneous state of the local atmosphere has been fully accounted for. But this is only partially so. The main objection rests on the fact that these calculations are made on the basis of one sounding, as a matter of fact in the vertical direction, implying thereby the air layers to be spherically and the law of tg z to be valid. But it is evidently not true (Teleki, 1974c).

Aerological as well as analogous measurements by means of artificial satellites (Teleki, 1974d) yield useful informations about the upper atmosphere layers, but leave open the question of the exploration the atmosphere in the vicinity of the astrometric instruments. There, unfortunately no advance has been recorded. The determination of the direction and velocity of winds and temperature measurements on a single or several points were not sufficient to provide necessary elements for the calculation of the meteorological model of the local atmosphere.

What might be expected in the future? More rigorous methods should be applied in the research into the lower atmosphere - one of the methods is based on the acoustic sounding techniques (Bean, Teleki, 1974). However, as the lower atmosphere is a very turbulent and complex medium, one cannot expect this method to yield complete and precise information on the refractional influences in the near future. There is still a strong reason for the hope that the meteorology of boundary layer will develop in the years to come, which will stimulate research into refraction. New analysis may therefore be expected of the density field around the observational pavilion (Kakuta et al., 1974), of the influence of the urban heat islands (Hughes, 1974), various characteristics of the local atmosphere (Goto, 1978) and others.

3.3. In recent time much is worked in geodesy upon multiple-wavelength (dispersion) method for the elimination of the refractional influences - a survey of these investigations is given in Prilepin (1974). The basic principle (Joshi, Mueller, 1974) is described by the following: if several frequencies operating simultaneously to measure range between two points were assumed to follow the same straight line between end points (although there will be little deviation), then the only source of error in each of measurements is the retardation due to refraction of the atmosphere. The idea was put forward to apply this method in astrometry (Tengström, 1968). The method is not without its deficiencies and cannot simply be used even in geodesy, still less in astrometry. The striving of E.Tengström at the Geodetic Institute at the Uppsala University and of D. Currie at the University of Maryland

Physics Department to apply the dispersion method in astrometry therefore be acclaimed. The results may be expected to be interesting as such methods are just looked for by which determination or elimination of the total amount and not of part of refraction is possible.

3.4. The artificial satellites with astrometric missions can in the future render great services in providing information on the refractional influences. It will be possible to determine via these satellites stellar coordinates with high accuracy, accordingly it will be possible to eliminate the hindrance under c) in 3.1. Of particular usefulness for discerning instantaneous effects will be simultaneous observation from the satellite and with ground-based instruments. But this possibility may be reckoned with only by the end the next decade.

4. PREVENTION

The most promising "combating" the unsufficiently accurate refractional influences is prevention: the best possible location of the instrument, most adequate pavilion and observing method.

4.1. The majority of the astrometric instruments is unadequately located (proximity of cities and buildings; nonadequate pavilions; intricate air density field around the instruments; to many cloudy nights; unsteady image, etc.). Consequently the accuracy of the results of observations is considerably lowered. What prospects are offered by a convenient location can be seen from Høg's estimate (1975), referred to already in the Introduction. In any case the inference should run on the following lines: before deciding upon the location for an astrometric instrument, meteorologic exploration should carefully be carried out, while the station and the pavilion should be constructed in such a fashion as to secure the best possible stability and at the same time the density field as simple as possible (Teleki, 1974b). The instrument housing should be aired by ventilator.

These preventive measures should be taken not only on account of the refraction of the light ray but also in order to reduce the amount of the total atmospheric influence on the instrument and its accessories. The importance of the instrument protection is a fact sufficiently pointed out. Less emphasised but equally important is the protection of accessories. As a highly positive example we mention the introduction of vacuum meridian marks (Mitich, 1974; Mitich, Pakvor, 1977), by means of which the pursuing of the azimuth variations of the instrument is performed with an accuracy bordering to the theoretical one (mean error of a single setting on the meridian mark is $\pm 0\overset{s}{.}005$ while that of ordinary meridian mark is $\pm 0\overset{s}{.}013$ at best).

4.2. Much can be gained by the correct choice of observing methods of determining certain quantities (latitude, declination, etc.) enabling the reduction or even elimination of the refractional influences.

Differential methods - such as Talcott's method of latitude determination, astro-photographic methods of position determination, etc. - are capable of eliminating the better part of these influences, but not all, as anomalous refraction is usually left out, affecting to a considerable degree the final accuracy (Teleki, 1968).

It is difficult to devise such observing methods which would enable refractional influences to be completely eliminated. This is a consequence of the difficulties associated with the choice of a proper model through which this elimination would be possible. The same applies to the classical methods (see 3.1.) as well as to the more recent dispersion method (see 3.3.). Thus, the procedures suggested are to be looked upon rather critically. All to often we are confronted with some mathematical solution of the problem which is lacking sufficient physical and meteorological backing.

4.3. With the observing methods currently in use, where no possibility of eliminating refractional influences is given, care should be taken that observations are not made at too large zenith distances, certainly not beyond $60°-70°$. This limitation is imposed alike to absolute as well as to differential methods of observations.

4.4. One should be careful in taking meteorological data on the surface layer, necessary for the computing of the pure refraction. By this we mean the measurement of these quantities at a given time (not the averaged ones), made with sufficiently precise equipment and always close to the object-glass of the instrument. These measurements should preferably be made by refractometers, which yield refractive indices by a whole order of magnitude more precise than those by the formulae (Teleki, 1974b; 1974c).

5. CONCLUSIONS

At the end, the following conclusions might be drawn:

5.1. Astrometry needs more accurately known refractional influences than we are nowadays able to offer;

5.2. Refraction tables, giving values for ideal and average conditions, must be brought closer to the reality;

5.3. Not even improved refraction tables are capable of meeting present-day needs, which necessitates special investigation of refractional effects at a given place and for a given time;

5.4. Preventive measures should be strictly observed; careful site selection, adequate pavilions, instruments, accessories and observing methods. Observations should not be made at zenith distances over

$60°-70°$, and meteorological data should cautiously be collected. Refraction, though a unique phenomenon, has hitherto been investigated piecemeal as astronomical, terrestrial, photogrammetric, radio or some other. A synthesis of these partial knowledges recommends itself as well as a continual cooperation of the experts in these fields. Astrometry would immensely benefit by such cooperation as more information would be acquired, in the first place that related to the surface air layers, which in other fields is considerable more widely investigated.

REFERENCES

1. Altenhoff, W.J.: 1974, Publ. Obs. Astron. Belgrade 18, pp. 17-20.
2. Barletti, R., Cepatelli, G., Paterno, L., Righini, A., Speroni,N.: 1977, Astron. Astrophys. 54, pp. 649-659.
3. Bean, B.R., Teleki, G.: 1974, Publ. Obs. Astron. Belgrade 18, pp. 21-44.
4. ESA (European Space Agency) document DP/PS (78) 13, 1978.
5. Garfinkel, B.: 1967, Astr. Journal 72, pp. 235-254.
6. Goto, T.: 1978, Publ. Int. Lat. Obs. Mizusawa 12 (in print).
7. Høg, E.: 1975, ESRO (European Space Research Organization) document SP-108, p. 63.
8. Hughes, J.A.: 1974, Publ. Obs. Astron. Belgrade 18, pp. 63-81.
9. Joshi, C.H., Mueller, I.I.: 1974, Publ. Obs. Astron. Belgrade 18, p. 149.
10. Kakuta, C., Teleki, G., Goto, T.: 1974, Publ. Obs. Astron. Belgrade 18, pp. 159-180.
11. Kolchinskij, I.G.: 1976, Astrometrija i astrofizika, Kiev 28, pp. 52-65.
12. Mitich, L.A.: 1974, Trans. IAU, p. 31.
13. Mitich, L.A., Pakvor, I.: 1977, Bull. Obs. Astron. Belgrade 128, pp. 11-15.
14. Nefed'eva, A.I.: 1973, Izv. Astr. Obs. A. P. Engelgardta 40, pp. 3-69.
15. Prilepin, M.T.: 1974, Proc. Int. Symp. on Terrestrial ADM and Atm. Effects on Angular Measurements, Stockholm 5, 3.
16. Saastamoinen, J.: 1974, ibid, 9.
17. Sugawa, C., Kikuchi, N.: 1974, Proc. Int. Lat. Obs. Mizusawa 14, pp. 145-162.
18. Sugawa, C., Kikuchi, N.: 1975, Proc. Int. Lat. Obs. Mizusawa 15, pp. 1-16.
19. Tengström, E.: 1968, Bull. Géod. 87, p. 30.
20. Teleki, G.: 1967, Publ. Obs. Astron. Belgrade 13, p. 68.
21. Teleki, G.: 1968, Acta Geodaet., Geophys. et Montanist., Budapest 3, pp. 237-240.
22. Teleki, G.: 1969, Studii si cerc. de astronomie, Bucurest 14, pp. 3-7.
23. Teleki, G. (editor): 1974a, Publ. Obs. Astron. Belgrade 18, pp. 1-234.
24. Teleki, G.: 1974b, ibid, p. 9.
25. Teleki, G.: 1974c, ibid, p. 226.

26 Teleki, G.: 1974d, Proc. Int. Symp. on Terrestrial ADM and Atm. Effects on Angular Measurements, Stockholm 5, 1.
27 Teleki, G.: 1977, Bull. Obs. Astron. Belgrade 128, pp. 7-10.
28 Teleki, G.: 1978, "Atmospheric Influences in the Fundamental Astrometry", Bull. Obs. Astron. Belgrade 129 (in print)
29 Teleki, G., Shevarlich, B.: 1971, Publ. Dept. Astron. Belgrade 3, 5.
30 Vasilenko, N.A., Kharitonova. T.N.: 1977, Astrometrija i Astrofizika, Kiev 31, pp. 38-41.

REFRACTION IN RADIOASTRONOMY

Wilhelm J. Altenhoff
Max-Planck-Institut für Radioastronomie
Auf dem Hügel 69
5300 Bonn, West Germany

ABSTRACT

In the last few years there was no paper in the astronomical literature reporting on new measurements of radioastronomical refraction. But there are some important review papers on different aspects of this topic, e.g. on the ionosphere by M.M. Komesaroff (1960) and T. Hagfors (1976), on the prediction of tropospheric refraction with ground based meteorological data by B.R. Bean (1962), B.R. Bean and G. Teleki (1974), and R.K. Crane (1976), on range measurements, predicted again by surface weather data, by H.S. Hopfield (1971).

For astrometry only interferometric observations give the necessary positional accuracy. The result of the interferometer is not affected, if the delay by the atmosphere is the same for each antenna, therefore this technique is not sensitive to normal refraction.

Differential delays can originate in the uneven distribution of water vapor in the atmosphere. Wesseling et al (1974) measured these differential delays with an infrared hygrometer; their attempts to correlate the water vapor with the observed interferometer phase were only partially successful.

Hinder (1970) has investigated the differential delays as function of baseline and of season. One can conclude that improvements of positional accuracy either by going to longer baselines or to shorter wavelengths depend on the success to predict the differential delays either by infrared hygrometers or similar means.

REFERENCES

Bean, B.R.: 1962, Proc. I.R.E., 50, 260.

Bean, B.R., Teleki, G.: 1974, in Publ. Obs. Astron. Belgrade, Vol. 18, edited by G. Teleki.

Crane, R.K.: 1976, in Methods of Experimental Physics, Vol. 12B, edited by M.L. Meeks, New York.

Hagfors, T.: 1974, in Methods of Experimental Physics, Vol. 12B, edited by M.L. Meeks, New York.

Hinder, R.A.: 1970, Nature, 225, 614.

Hopfield, H.S.: 1971, Radioscience, 6, 357.

Komesaroff, M.M.: 1960, Australian J. Phys., 13, 153.

Wesseling, K.H., Basart, J.P., Nance, J.L.: 1974, Radioscience, 9, 349.

ENVIRONMENTAL SYSTEMATICS AND ASTRONOMICAL REFRACTION II.

James A. Hughes

U. S. Naval Observatory
Washington, D. C., USA

ABSTRACT

The role of urban heat islands in producing systematic isopycnic tilts is explored in more detail, and with greater rigor, than in Part I of this series. (Perth, 1974).

Specifically, a three dimensional integration is carried out, and light rays are, in effect, "traced" through the resulting perturbation field by evaluating the integral of anomalous refraction. This is done for various values of the parameters, viz., wind direction and observatory location relative to the heat island, strength of the central perturbation, zenith distance of the observed object, etc.

It is stressed that heat islands are not the only source of such systematic effects.

Finally, a brief discussion of some possible methods of determining observationally the effects here treated theoretically, as well as other site dependent effects, is appended.

INTRODUCTION

The effects which an urban heat island can have upon astronomical refraction were briefly discussed in an essentially qualitative way in a previous paper by the author (1). That paper will be referred to as I.

In the present work a more quantitative approach is taken. The new elements which are introduced to make this possible are: (1) a three dimensional heat island is used, (2) the perturbation field is used to calculate the isopycnic tilt, which in turn is used to evaluate the integral of anomalous refraction, and (3) the anomalous refraction is evaluated for non-zenith observations.

The point of view taken here is that of an astrometrist engaged in fundamental meridian observations which ultimately lead to fundamental reference systems such as the FK4 catalog.

Certainly no one engaged in such work would locate, initially, an astrometric observatory within the sphere of influence of such a perturbing factor as an urban heat island. However, such islands have a way of growing, and in the process they often engulf long established observatories. Among several possible examples of such cases, two are: the Cerro Calan Observatory in Santiago, Chile, and my own institution, the U. S. Naval Observatory in Washington, D. C. When the latter was established at its present location in 1895 it was remarked that it would be, "one-hundred years", before any deleterious effects could be expected from the capital city. The Observatory is located approximately 6 km. from the U. S. Capitol, and I, together with both preceeding and contemporary colleagues, daily travel "downtown" to reach the Observatory from home.

On the other hand, before one rushes off to locate an astrometric observatory in some remote mountain fastness, or perhaps upon some island in a gentle clime, one ought to be aware that the mathematics governing, e.g., mountain lee waves, have an uncanny resemblance to those concerning urban heat islands.

Over one-half century ago, Emden (2) summed up one of his works by writing, "... a favorable location of the station, namely, at a site at which experience shows that no sloping of layers in the atmosphere sets in, even if only transient. Whether broad plains or mountain summits are to be preferred must be learned from experience".

Unfortunately, the fundamental astrometric quality of a site can be determined, usually, only a posteriori, after a considerable investment for installation and for perhaps five or ten years of observing. Thus the economics of the situation do not permit much in the way of "site testing", and so the fundamental astrometrist must investigate the site with which he may be blessed, or more likely, cursed, with any tools at hand, theoretical or observational.

THEORETICAL BACKGROUND

The basis for the present calculations is contained in an excellent paper by Olfe and Lee (3). The reader is referred to their work for the somewhat lengthy details.

In brief, their governing two-dimensional equation for the temperature perturbation field caused by the heat island's conduction and convection effects is:

$$\left[\frac{\partial^2}{\partial Z^2}\left(\frac{\partial}{\partial X}-\frac{\partial}{\partial Z^2}\right)+\tfrac{1}{4}\gamma\frac{\partial}{\partial X}\right]\psi = 0, \qquad (1)$$

where γ is a non-dimensional parameter depending upon various meteorlogical quantities, X and Z are non-dimensionalized horizontal and vertical coordinates, and ψ is the ratio of the perturbation at (X, Y) to the perturbation at the center of the heat island (0,0).

That is,

$$\psi = \frac{T'}{T'_o}, \qquad (2)$$

where primes indicate perturbation quantities.

Elementary solutions of the form,

$$RE\left[\exp(\sigma Z + ikX)\right], \qquad (3)$$

lead to the total solution,

$$\psi(X,Z) = \int_0^\infty g(k) RE\left\{\left[C_1 \exp(\sigma_1 Z) + C_2 \exp(\sigma_2 Z)\right] \exp(ikX)\right\} dk. \qquad (4)$$

The C's and σ's are complex, functions of κ and γ, and $g(\kappa)$ is the Fourier cosine transform of the assumed (symmetric) surface temperature distribution.

Following Scorer (4), it is possible to superpose two-dimensional solutions at varying angles, α, to the uniform flow direction, and thus generate a three-dimensional solution. In simple dimensional heat island cylindrical coordinates (where $\phi=0$ denotes the free flow direction) this has the form,

$$\psi(r,z,\phi) = \frac{1}{\pi}\int_{-\pi/2}^{+\pi/2}\int_0^\infty g(k) RE\left\{\left[C_1 \exp(\sigma_1 Z') + C_2 \exp(\sigma_2 Z')\right] \exp(ikX')\right\} dk\, d\alpha \qquad (5)$$

where $X' = \dfrac{r}{L_o} \cos(\phi + \alpha)$, (6)

$Z' = (\cos\alpha)^{\frac{1}{2}} Z$,

with L_o the half-width of the island.

THE PRESENT CASE

Referring to Figure 1, if $\overline{AB}$ is a tilted isopycnic induced by the perturbation, then for the density, ρ, one has,

$$\Delta\rho = -\left[\dfrac{\partial\rho'}{\partial q}\Delta q + \dfrac{\partial\rho'}{\partial z}\Delta z\right], \qquad (7)$$

where q is any horizontal coordinate, ρ_a is the unperturbed density along $\overline{AC}$, and $\Delta\rho$ is the normal change of density between A and B in the absence of any perturbation.

So,

$$\beta \simeq \dfrac{\Delta z}{\Delta q} = \dfrac{-\dfrac{\partial \rho'}{\partial q}}{\dfrac{\Delta\rho}{\Delta z} + \dfrac{\partial\rho'}{\partial z}}. \qquad (8)$$

Using the well known result that,

$$\rho(x) = \rho_o \exp\left(\dfrac{-gz}{RT}\right) = \rho_o \exp\left(\dfrac{3.4\times 10^{-2} z}{T}\right) \simeq \rho_o\left(1 - \dfrac{3.4\times 10^{-2} z}{T} + \ldots\right), \qquad (9)$$

with z in meters and T in $^o K$,

one has,

$$\Delta\rho = \rho_B - \rho_A = \rho_o\left(\dfrac{-3.4\times 10^{-2}\Delta z}{\overline{T}}\right) \qquad (10)$$

In order to work with a definite number, $\overline{T}$ will be taken as constant throughout the domain of the perturbation. This approximation is not critical in the present case since a $\pm 10^o$ swing of $\overline{T}$ can only change the results by less than $\pm 4\%$. Also, we are concerned with heights of less than one kilometer.

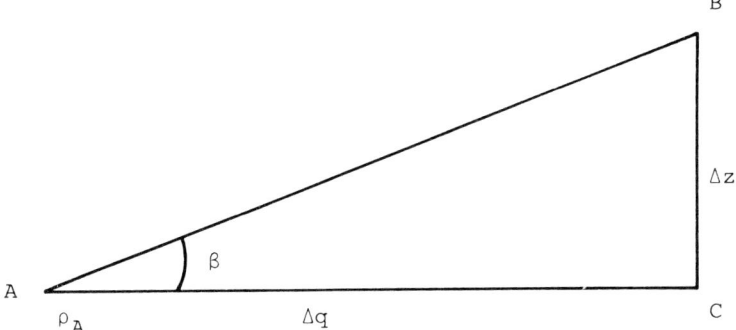

Figure 1

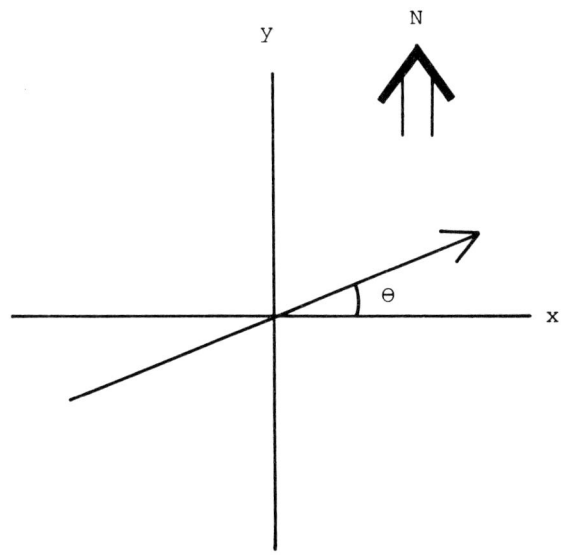

Figure 2

Taking $\overline{T} = 285°$, then,

$$\beta = \frac{-\partial\rho'/\partial q}{-1.2\times10^{-4}\rho_o + \partial\rho'/\partial z} . \qquad (11)$$

Since the Boussinesq approximation is valid in this case (5),

$$\frac{T'}{T_o} = -\frac{\rho'}{\rho_o} \qquad (12)$$

and

$$\psi = \frac{T'}{T'_o} = \frac{T'}{cT_o} = -\frac{\rho'}{c\rho_o} , \qquad (13)$$

where c ($\ll 1$) is used to represent the central perturbation as a fraction of the unperturbed central temperature or density.

Then, e.g.,

$$\frac{\partial\psi}{\partial q} = -\frac{1}{c\rho_o}\frac{\partial\rho'}{\partial q} \qquad (14)$$

and

$$\beta = -\frac{c\,\partial\psi/\partial q}{1.2\times10^{-4} + c\partial\psi/\partial z} \qquad (15)$$

Figure 2 illustrates the coordinate system used for the calculations. The positive y-axis is directed towards the north. The computations are carried out in a coordinate system which rotates in accordance with the azimuth of the wind, and hence which differs from the (x,y,z) system by a single rotation about z of amount θ. For example, θ = -45° indicates a NW wind.

The procedure consisted of choosing values for the parameters (which amounts to setting a value for γ), and then varying the location of the observatory, the wind direction and the zenith distance of the observations. Values of (x,y,z) were chosen and converted by Eq. (6) to X', Z' for use in Eq. (5).

The integration with respect to k is of the form,

$$\int_0^\infty e^{-x} f(x)\, dx, \qquad (16)$$

and the integration with respect to α can be brought to the form,

$$\int_{-1}^{+1} F(y)\, dy. \qquad (17)$$

The former is suitable for Laguerre integration and the latter for Gaussian integration. A cartesian product was formed so that,

$$\psi = \sum_j \sum_i L_i(x) G_j(y) f(x_i) F(y_j), \qquad (18)$$

where L and G are the appropriate weighting factors.

The partial derivatives could be expressed analytically by differentiation under the integral sign, however, since the values of ψ itself were of interest, the partials were evaluated by

$$\frac{\partial \psi}{\partial q} \approx \frac{\psi_{+\Delta q} - \psi_{-\Delta q}}{2\Delta q}, \qquad (19)$$

which for present purposes has no essential effect upon the numerical results. The quantity Δq was 25 meters for $q = x$ or y, and 10 meters for $q = z$.

The tilt angle was computed using Eq. (15), for vertical steps of 25 meters, up to a maximum height, z_{max}, of 0.8 or 1.0 km. That is;

$$\Delta z = 25m,$$

$$\Delta y = \tan \zeta \Delta z, \qquad (20)$$

$$\Delta x = 0,$$

where ζ is the zenith distance (positive to the north), and Δx is zero for meridian observations in the (x, y, z) system.

The tilt in right ascension is given by taking $q = x$, and the tilt in declination by taking $q = y$.

The integral of anomalous refraction,

$$R_a = \int_{n1}^{n2} \frac{\beta \, dn}{n\left[1 - \left(\frac{r_o n_o}{r n} \sin\zeta\right)^2\right]}, \quad (21)$$

can be considerably simplified in the present case by setting

$$\begin{array}{c} n = 1 \\ \text{and} \\ r_o n_o = r n. \end{array} \quad (22)$$

Since $\frac{dn}{dz} \simeq 3 \times 10^{-8}$ m^{-1},

$$R_a'' = \frac{1.1 \times 10^{-4}}{\cos^2 \zeta} \int_0^{z_{max}} \beta(z) \, dz \quad \text{sec. of arc.} \quad (23)$$

The angles given by Eq. (15) were very closely fitted by a polynomial in z and the integration carried out term by term to yield R".

PARAMETERS AND RESULTS

The following parameters were used in the calculations:

Stability coefficient	$\equiv s$	$= 1.0 \times 10^{-4}$	m^{-1}
Eddy diffusivity	$\equiv \kappa$	$= 10$	m^2/s
Wind velocity	$\equiv U_o$	$= 3$	m/s
Heat island half-width	$\equiv L_o$	$= 3$	km
Central perturbation	$\equiv c$	$= 2\%$	

Note that;

$$Z = \left(\frac{U_o}{\kappa L_o}\right)^{1/2} z$$

ENVIRONMENTAL SYSTEMATICS AND ASTRONOMICAL REFRACTION II

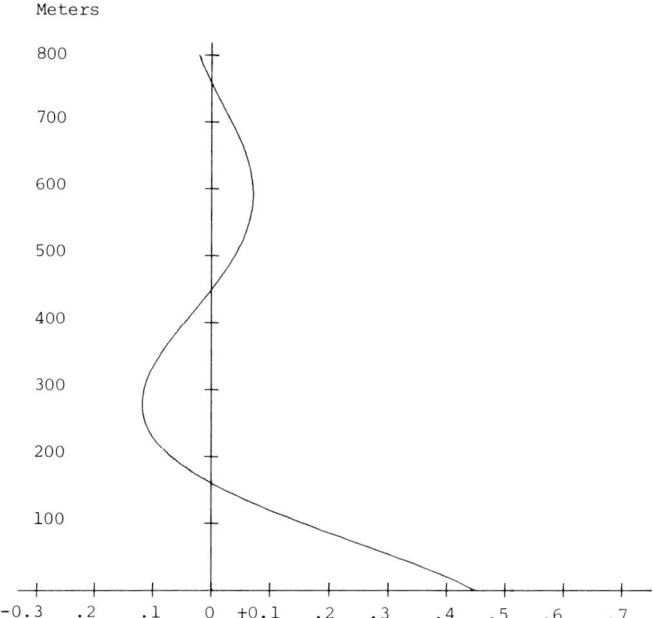

Figure 3   TILT ANGLE (DEGREES)

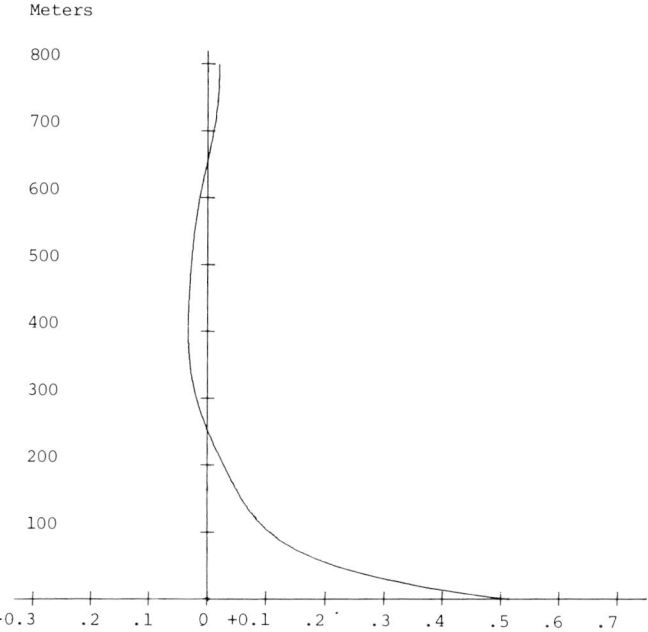

FIGURE 4   TILT ANGLE (DEGREES)

TABLE 1

	α	δ		α	δ
N	−5	+4		−2	+14
S	−5	+5		−7	+ 5

———————————————— x

	α	δ		α	δ
N	−18	− 2		+ 6	− 6
S	− 7	−20		+ 9	− 7

	α	δ		α	δ
N	−16	+ 1		+16	+ 4
S	−17	+ 1		+16	+28

———————————————— x

	α	δ		α	δ
N	−17	− 1		+16	−28
S	−16	−1		+16	− 4

	α	δ		α	δ
N	− 7	+20		+ 9	+ 7
S	−18	+ 2		+ 6	+ 6

———————————————— x

	α	δ		α	δ
N	− 5	− 5		− 7	− 5
S	− 5	− 4		− 2	−14

TABLE 2

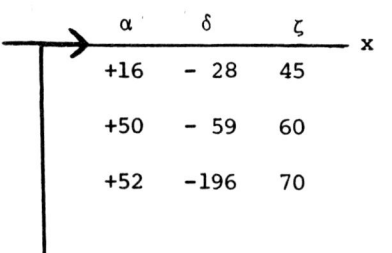

α	δ	ζ
+16	− 28	45
+50	− 59	60
+52	−196	70

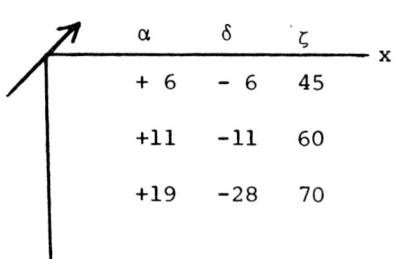

α	δ	ζ
+ 6	− 6	45
+11	−11	60
+19	−28	70

The quantity, γ, is given by

$$\gamma = 4 \frac{gs\kappa L_o}{(U_o \cos\alpha)^3} \tag{24}$$

where g is the acceleration of gravity.

Physically, this represents a relatively small, but intense, heat island with a stable temperature gradient upwind.

As in (I), the "mountain function" was used to represent the surface temperature distribution. In this case,

$$g(k) = 2/\pi \; \exp(-2k/\pi). \tag{25}$$

The integration of Eq. (5) gives a vertical perturbation which has the character of a heavily damped wave. This behavior is mirrored by the tilt angles given by Eq. (15). In the cases considered here, two dominant modes became apparent. These are illustrated in Fig. 3 and 4.

In Fig. 3 the area of the sign reversal of β (vis a vis the ground level sign) is much more pronounced than in Fig. 4. The former case adheres closely to the nature of the perturbation itself. The two cases are the result of the location of the observatory's meridian with respect to the center of the heat island, of the wind direction and of the zenith distance of the object observed. Since the anomalous refraction is proportional to the integral of these curves in the z direction, evidently the situation depicted in Fig. 3 gives, in general, a smaller effect than that in Fig. 4. It is also evident that in either case the lower levels (say z<250m) are the major contributors to the effect.

Calculations were initially made for fixed zenith distances, viz., $\zeta = \pm 45°$. The observatory was located in each quadrant of the (x,y) system in turn, for each of three wind directions. The results are shown schematically in Table 1. The unit is $0\overset{''}{.}001$. In all these cases the observatory's coordinates were $|x| = |y| = 4$ km., thus placing the observatory approximately 6 km. from the center of the heat island. The wind vector is also shown in Table 1.

The assumed symmetric heat island gives rise to the "anti-symmetry" in δ and the symmetry in α.

The quantities exhibited in Table 1 are not very large, varying between essentially zero and $\pm 0\overset{''}{.}03$. However if one recalls that transit circles engaged in fundamental work must determine a celestial pole, and hence must work at greater zenith distances, the import becomes more serious. Extending the zenith distance to $+60°$ and $+70°$ for the cases $\theta = 0°$ and $\theta = +45°$, for $x = +4$km and $y = -4$km, gives the results shown in Table 2. The $0\overset{''}{.}03$ alluded to earlier now increases as much as six-fold.

Increased zenith distances are not, however, the sole catalyst for obtaining large effects. The governing equation (1) represents the actions of two "competing" effects, conduction and a gravity wave. The latter is responsible for the negative temperature perturbation which leads to the tilt reversals. Very small changes in the perturbation profile lead to sensible changes in the tilt angles, and hence the integrated anomalous refraction.

The two most troublesome parameters are the stability coefficient and the eddy diffusivity. On the one hand, the gravity wavelength depends upon s, while the conduction profile depends upon κ. By way of experimentation, the observatory was placed on the y axis, 6 km. south of the origin. Looking downwind and upwind at $\zeta = +45°$, with $\kappa = 10$, gave (in declination), $0\overset{''}{.}001$ and $0\overset{''}{.}019$, respectively. Reducing κ to 1 gave $0\overset{''}{.}031$ and $0\overset{''}{.}073$ for the same cases. Since the deviations from pure conduction are proportional to γ, this indicates that reducing κ (and hence γ) reduces the negative temperature perturbation and hence increases the anomalous refraction. On the other hand, it is possible to vary both s and κ in such a manner as to keep γ constant and still change the anomalous refraction. This is so since the vertical coordinate is non-dimensionalized by the factor, $(U_0/\kappa L_0)$, thus changing κ changes the metric of the entire problem.

It would appear that theory can only indicate to us that we do indeed have a problem. Worse, it now appears that we should be thankful for convection effects for reducing the anomalous refraction by inducing low level temperature "cross-over"!

CONCLUSION

It appears to the author that it is absolutely necessary to devise some means of determining these pernicious systematic effects observationally. Experiments have been carried out at the Naval Observatory in which a Raman LIDAR was successfully used to probe the atmosphere for water vapor content. It now appears that this effort should be expanded to determine systematic isopycnic tilts, and perhaps, ultimately, to real time density profiling of the atmosphere. At the same time other efforts to solve these problems should proceed apace since any single solution is apt to introduce its own systematic errors.

The author would like to thank S. Long and B. Thacker for their assistance with the numerical integration of Eq. (23).

REFERENCES

(1) Hughes, J. A., 1974, Pub. Obs. Ast. Beograd, $\underline{18}$, 63.

(2) Emden, R., 1924, Astron. Nach., $\underline{45}$, 219.

(3) Olfe, D. B., and R. L. Lee, 1971, J. Astmos. Sci., $\underline{28}$, 1374.

(4) Scorer, R. S., 1956, Quart. J. Roy. Meteor. Soc., $\underline{82}$, 75.

(5) Spiegel, E. A., and G. Veronis, 1960, Ap. J., $\underline{131}$, 442.

AN ORIGIN OF THE VARIATIONS OF ASTRONOMICAL REFRACTION.

Haruo Yasuda and Rikinosuke Fukaya
Tokyo Astronomical Observatory, Mitaka, Tokyo, Japan.

Abstract. There exists an empirical relation between the anomalous refraction and the atmospheric density in the surface layer. From the relations the variations of scale height for each night can be determined by the temperature and pressure in the surface layer. A correction term to the refraction table is derived in an analytical expression.

1. EMPIRICAL RELATION BETWEEN THE ATMOSPHERIC DENSITY AND THE REFRACTION ANOMALY.

When the stars north and south of the zenith are observed with a meridian circle, there exists a relation

$$\phi_s - \phi_n = -\Delta r (\tan z_s + \tan z_n) \qquad (1)$$

between an observed latitude, ϕ, and a correction to the constant of refraction, Δr, under the assumption that the difference between ϕ_s and ϕ_n is due to refractions. Subscript, s and n, show the south and north of a zenith respectively and z is an absolute zenith distance. When the southern and northern stars have the same absolute zenith distances, the relation (1) is written in the form of

$$\phi_s - \phi_n = -2\Delta a_o \tan z \frac{\rho}{\rho_o} \qquad (2)$$

Here Δa_o is the correction to the astronomical refraction constant adopted in the Pulkovo Table, since the refractions applied to Tokyo observations are equivalent to those taken from the Pulkovo Table 4th Ed.. ρ_o is the air density of the standard air. From the observations of FK4 stars during the period, 1972 - 1976, Fukaya, one of the authors obtained a relation

$$a_o = \underset{\pm 0.005}{0\overset{\prime\prime}{.}049} \sin(T + \underset{\pm 3°}{277°}) + \underset{\pm 0.005}{0\overset{\prime\prime}{.}024} \sin(2T + \underset{\pm 13°}{202°}), \qquad (3)$$

where T is a fraction of a year in a degree of arc and its epoch is January 15. The value of ρ is calculated from the measured temperature and pressure in our observing pavilion. An annual term in (3) shows that the measured temperatures are too high in winter and too low in summer. It coincides with an annual term of the grounding inversion of temperature, $1.39°C \sin(T + 295°)$, at the Tateno Aerological Observatory. The observatory situates at the distance of about 65 km in the north-east direction from our observatory.

Using the observations of equatorial and circumpolar FK4 stars, Fukaya also obtained an empirical relation of

$$\delta(\Delta r) = 8\overset{"}{.}6 \Delta\rho \quad . \qquad (4)$$
$$\pm 1.8$$

See Figure 1. Here ρ is calculated in a unit of 10^{-4} gr/little from the temperature and pressure measured in an observing pavilion, and $\Delta\rho$ is designated as a residual of ρ for each observing tour from a smoothed average of densities for five tours around it. $\delta(\Delta r)$ is a residual of r in second of arc from its smoothed average for the same tour as used in the calculation of $\Delta\rho$. The standard deviations of $\Delta\rho$ and $\delta(\Delta r)$ are ± 0.0125 gr/little and $\pm 0\overset{"}{.}09/\tan^{-1} 1$. They show the differences between the stationary atmosphere adopted in the Refraction Table and the dynamical atmosphere at Mitaka.

Being calculated by the use of aerological data at the height of 1 km, there exists a relation of

$$\delta(\Delta r) = 7\overset{"}{.}7 \Delta\rho_1 , \qquad (5)$$
$$\pm 1.8$$

while, being calculated by the aerological data at the height of 5 km, there is no longer a proper correlation between them. See Figure 2.

The practical diminution law of the air density at Mitaka differs from that adopted in the Refraction Table. It is concluded that (1) Δr is quite due to the atmospheric density below the height of 5 km, and that (2) Δr may be calculated by the use of the density in the surface layer.

2. INTERPRETATION OF THE EMPIRICAL LAW.

To demonstrate qualitatively such anomalous refraction as $\delta(\Delta r)$, we shall express the refraction in such an analytical form as is used in the textbook of Newcomb (1906).

Teleki (1967) set up the following relations from the aerological data ;

$$\rho = \rho_o e^{-as} ,$$

AN ORIGIN OF THE VARIATIONS OF ASTRONOMICAL REFRACTION

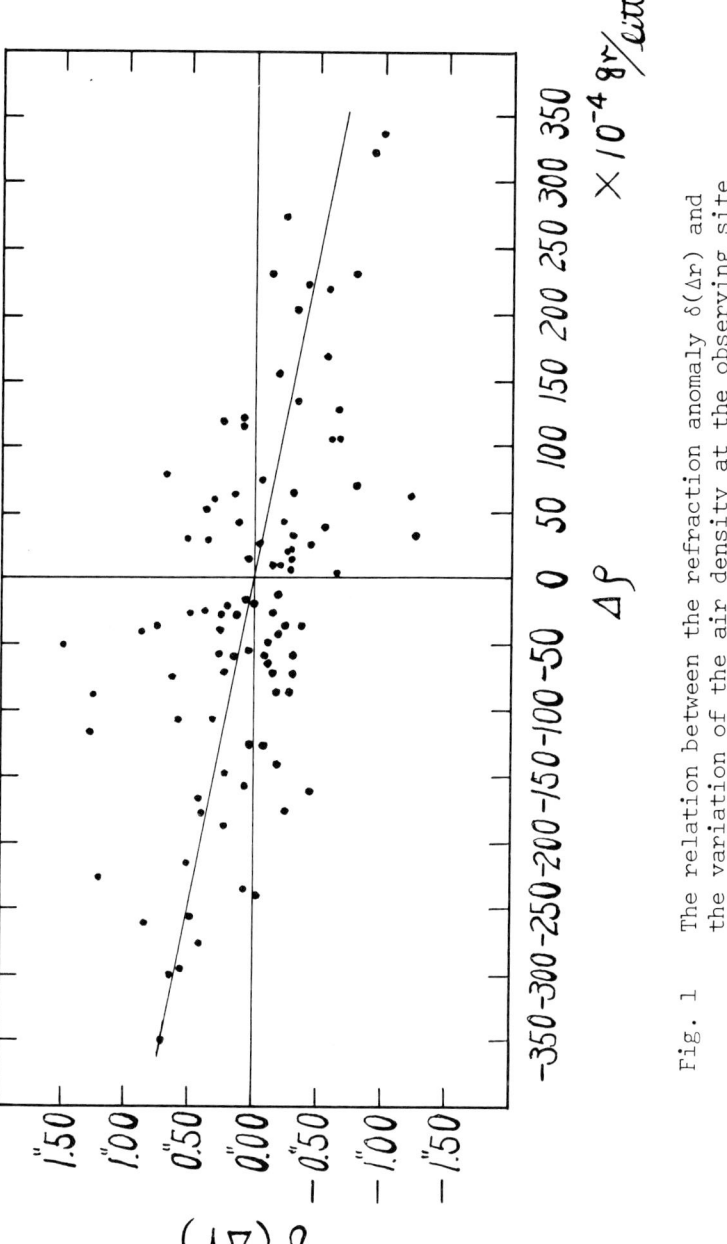

Fig. 1 The relation between the refraction anomaly $\delta(\Delta r)$ and the variation of the air density at the observing site $\Delta\rho$

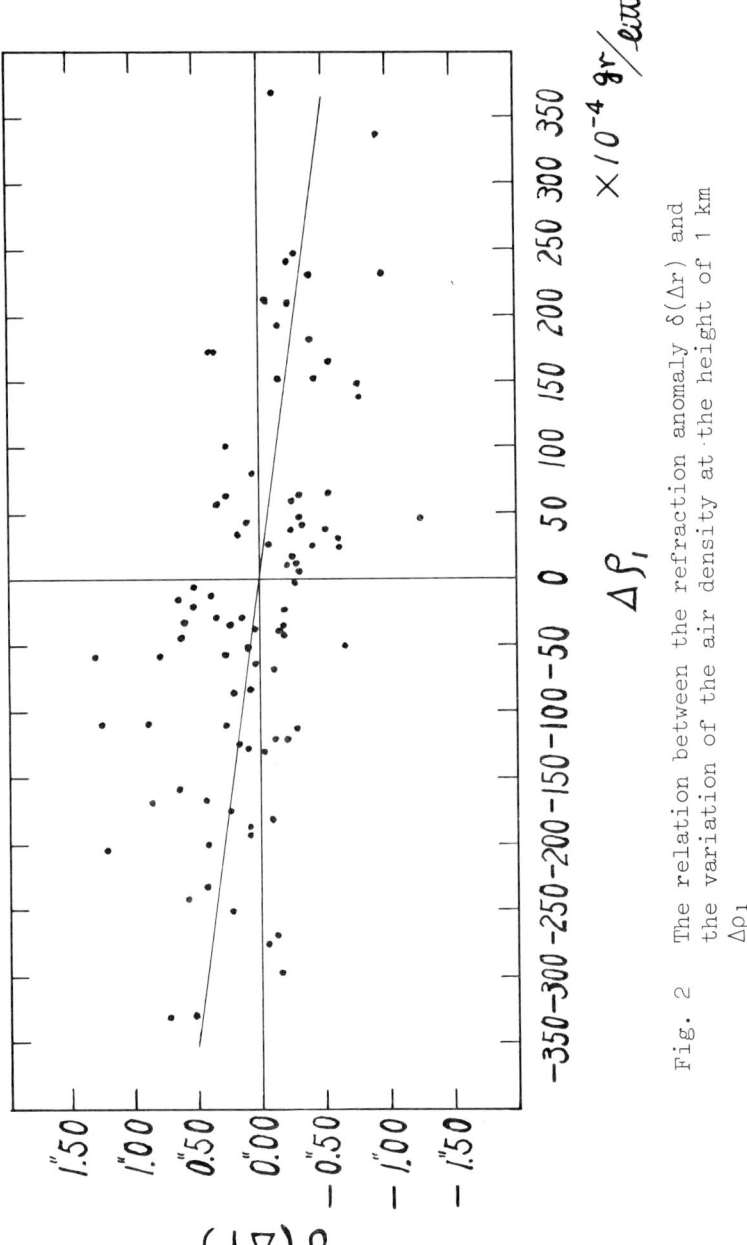

Fig. 2 The relation between the refraction anomaly $\delta(\Delta r)$ and the variation of the air density at the height of 1 km $\Delta \rho_1$

$$a = \left[\frac{\Gamma_h - \Gamma}{T_s}\right] r_o \quad , \qquad (6)$$

$$s = (r/r_o) - 1 = \frac{H}{r_o} \quad ,$$

where r and r_o are the radius of curvature of an equidensity layer and of the geoid at the station respectively, T_s the mean temperature of the layer, and H the height of the layer. Γ is the vertical gradient of an air layer.

The rigorous differential equation of astronomical refraction can be expressed by

$$dR = -\frac{d\mu}{\mu} \frac{\sin z}{\sqrt{(\mu r/\mu_o r_o)^2 - \sin^2 z}} \quad . \qquad (7)$$

From the Dale-Gladstone equation and the first equation of (6), the refractive index of the air, μ, is expressed by the formula of

$$\mu = 1 + c\rho_o e^{-as} \quad , \qquad (8)$$

where the coefficient c is a constant depending on the wave length and characterizing a given air medium. From (6) and (8), we have

$$\frac{\mu r}{\mu_o r_o} = (1 - \alpha\omega)(1 + s) = 1 + 2u \qquad (9)$$

and

$$d\mu = -c\rho_o\, d\omega \qquad (10)$$

where the constant α and the variable ω are

$$\alpha = \frac{c\rho_o}{1 + c\rho_o}$$

and $\quad \omega = 1 - \dfrac{\rho}{\rho_o}$

Substituting (9) and (10) in (7) and integrating from the point of observation to the outer limit of the atmosphere, we have

$$R = \alpha m \tan z \qquad (11)$$

where

$$m = \int_0^1 \frac{d\omega}{(1 - \alpha\omega)\sqrt{1 + 2u\sec^2 z}} \quad . \qquad (12)$$

This general formula for the astronomical refraction coincides with Newcomb's formula except of a term of $1 - \alpha\omega$.

From the equations (11) and (12), we have, keeping the accuracy of the refraction R to $1 \cdot 10^{-2}$ arc second,

$$R = \left(1 + \alpha - \frac{1}{a}\right) \tan z - \alpha\left(\frac{1}{a} - \frac{1}{2}\alpha\right) \tan^3 z \qquad (13)$$

in the case of zenith distances to 70°. α can be calculated by the temperature and pressure at the earth surface and contributes only to the mean value of the corrections to the Pulkovo Table at a given station. Taking only the differential correction of a, we have

$$\Delta R = \frac{\alpha}{a^2} \Delta a \, (\tan z + \tan^3 z) \, .$$

According to the difinitions of Δr and $\delta(\Delta r)$ in (1) and (4), we have

$$\Delta r = \frac{\alpha}{a^2} \Delta a \, . \qquad (14)$$

Substituting (4) in (14), we finally get

$$\Delta a = 1.08 \cdot 10^4 \Delta\rho \qquad (15)$$

where $\Delta\rho$ is expressed in (gr/cm^3). The numerical coefficient of (15) is calculated by putting

$$a = 850 \text{ and } \alpha = 0.28 \cdot 10^{-3}.$$

At a standard pressure 760 mm Hg, the change of temperature by 20°C is equivalent to that of $0.9 \cdot 10^{-4}$ gr/cm³ in air density and causes the change of a by 1.0. The results do not greatly differ from the seasonal variation of a in Teleki's paper (1967). The increment of Δa by 1.0 is equivalent to that of scale height by 9 gpm. The amplitude of the seasonal variation of a is expected to be about 1.0 at Tokyo. See Figure 3.

3. CONCLUSION.

The variation of the density distribution of atmospheric layers below the height of 3 or 5 km is likely the most effective triggers to the variation of refraction. From the empirical relation (4), the value of Δa at Tokyo can be calculated by the relation (15). In practice, after a correction term (14) has been applied to the refraction calculated by the Refraction Table, the correction to the refraction constant should be determined in conventional ways, since its origin is due to the difference between the mean atmospheric density at a given station and that adopted in the Refraction Table.

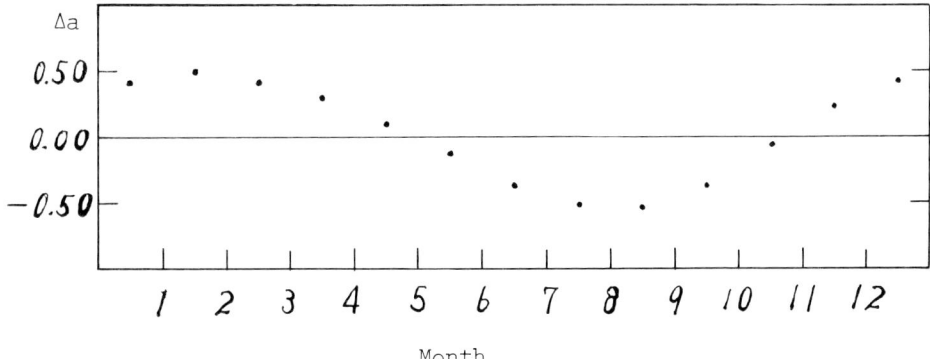

Fig. 3 The seasonal variation of a at Mitaka.
The value of a is assumed to be 850.

REFERENCES.

Newcomb, E. 1907, A Compendium of Spherical Astronomy (Dover Publ.
             Inc. NewYork) p. 173.
Teleki, G. 1967, Publ. Astron. Obs. Beograd, No. 13.

DISCUSSION

E. Tengström: remarked that the secular change of the atmosphere is
 also one of the sources which can influence the refractional calculation during the long catalogue work.

ON ASTRONOMICAL REFRACTION FOR A THREE-DIMENSIONAL MODEL ATMOSPHERE.

V.I.SERGIENKO
Institute for Metrology, Irkutsk, U.S.S.R.

The existing refraction theories for a one-dimensional model atmosphere do not account for anomalous refraction. The latter can be calculated using a three-dimensional model atmosphere.

The author has deduced formulae for calculation of refraction from the zenith distance R_Z and azimuth R_A for a three-dimensional model atmosphere. For the considered model atmosphere the variation of the refractive index n in the vertical plane is described by the exponential function (cases of some other simple functions are also considered). In the horizontal plane the variation of n against x and y (rectangular coordinate system) is accounted by horizontal gradients $\mathcal{E}$ and ω respectively. The $\mathcal{E}$ and ω can be calculated from the data of aerological probing. While gradients of meteorological element variation in the vertical plane is by three-four orders larger than that in the horizontal plane (Leichtman, D.L., 1970)., $\mathcal{E}$ and ω are small compared with the vertical gradient of n. Then the path of a beam of light in the three-dimensional model atmosphere will be expressed by

$$\begin{aligned}\frac{d\alpha}{ds} &= \omega(1-\alpha^2) - \mathcal{E}\alpha\beta - \bar{B}\alpha\sqrt{1-\alpha^2-\beta^2} \\ \frac{d\beta}{ds} &= \mathcal{E}(1-\beta^2) - \omega\alpha\beta - \bar{B}\beta\sqrt{1-\alpha^2-\beta^2}\end{aligned}\Bigg\} \quad (1)$$

where α and β are cosines of the tangent to the trajectory of the beam,

$$\left.\begin{aligned}\omega &= \frac{1}{n}\frac{\partial n}{\partial y} \\ \mathcal{E} &= \frac{1}{n}\frac{\partial n}{\partial x}\end{aligned}\right\} \quad \text{small parameters,}$$

$\bar{B} = \ell n B$, B is taken from $n = AB^H$; H is the height above sea level.

System of differential equations (1) is a non-linear dynamic system of the second order with two small parameters. The equation system is solved by the analytical method with application of Bernoulli's method (Danko P.E., Popov A.T.,1974). or methods of perturbation theory.

For the solution of the equations by the former method an assumption of was made on the invariability of n within each atmospheric layer. The practical application of this method is expedient if there are sufficient meteorological data describing in detail the structure of the atmosphere. The utilization of perturbation theory methods permitted to solve system (1) in general form. Series expansion of the solutions was done following the powers of the two small parameters and separately up to the terms of the first and second order. Boundary conditions for the adopted model atmosphere are values of n on the Earth's surface and at a 75 km height. Initial conditions are values of direction cosines of a light beam in the medium with $n = 1 \pm 10-$

Refraction was calculated using the following definitions;

<u>Definition I</u>. Refraction R_Z for the zenith distance Z is assumed to be an angle between two tangents of the trajectory of a light beam at the end points of the beam path in the atmosphere.

<u>Definition II</u>. Refraction R_A for azimuth A is an angle between two binormals to the spatial curve of a beam of light at the end points of the beam path in the atmosphere.

R_Z calculated from the proposed formulae for a case when there are no horizontal gradients of the refractive index n coincides to $\pm$ 0″.02 with the refraction calculated using Gylden's theory which was basic for the Pulkovo refraction tables.

For the three-dimensional model atmosphere the length of the trajectori to the beam has to be calculated to an accuracy of ± 0.10m cosec Z at zenith distances Z 80° in order to determine refraction with an accuracy of $\pm 0″.01$.

The practical application of the deduced formulae is possible only the case when there are data of free atmospheric probing. In the ground and boundary atmospheric layers these may be obtained using the collected meteorological information for the moments of observations of each star (Kudeeva V.S., Pavlov B.A., Sergienko V.I.).

The proposed formulae of the refraction calculation give a chance to account for the asymmetry of the atmosphere, calculate refractions for different azimuths and reach a higher accuracy of the results in dependence on the atmosphere.

References

1. Danko P.E., Popov A.T. Vysschaya Matematika v Uprazhneniyakh i Zadachakh, part II, M., Izd-vo "Vysschaya Schkola, 1974, pp.160-161.

2. Leichtman D.L. Phisika Pogranichnogo Sloya Atmosphery. L., Hidrometeoizdat, 1970, pp.41-43.

3. Sergienko V.I., Pavlov B.A., Kudeeva V.S., Trudy VNIIFTRI, vypusk 31 (61) pp.38-43.

DISCUSSION

G. Teleki: noticed the importance of such investigations but he expects the application of these theories in practice. The basic problem is how to get a real information on the variations (with hight) of inclinations of the equal density layers.

LES EFFETS DE LA REFRACTION ATMOSPHERIQUE SUR LES COORDONNEES
TANGENTIELLES EN ASTROMETRIE [1]

J. Dommanget

Observatoire Royal de Belgique
Uccle, Bruxelles.

ABSTRACT

   The astrometric plate reduction needs the knowledge of formulae expressing the transformation of the configuration of stellar positions on the celestial sphere into the corresponding observed figure of the photographic images on the plate. This transformation includes phenomena such as the atmospheric refraction, the optical distortion, the plate position on the instrument, etc. of which the relative importance may vary from one case to another and may lead thus to various formulae.

   Among these phenomena, the atmospheric refraction seems to be one of the most important.

   In this paper, we give expressions for the refracted tangential coordinates (generalisation of the standard coordinates) under the form :

$$X_r = X + a\, X_a + R_x \quad ; \quad Y_r = Y + a\, Y_a + R_y$$

where X, Y are the unrefracted coordinates, a, the coefficient of the term of the first order of the classical expression for the atmospheric refraction, and R_x, R_y, the rests of the developments of X_r and Y_r respectively. The coefficients X_a and Y_a have simple expressions which are easily computed when one knows the celestial coordinates of both the tangential point and the star concerned.

   One must first mention that any point of the sky may be chosen as celestial pole and that all formulae remain the same at the condition that pseudo-equatorial coordinates related to that pole are introduced.

   Of course, the choice of the zenith as celestial pole introduces many simplifications, not only in the expressions X'_a, Y'_a, R'_x and R'_y of the corresponding expressions : X_a, Y_a, R_x and R_y, but also in these expressions themselves, because they are more easily written as functions of the zenithal coefficients.

The aim of this study appears when giving finally for each of the above mentioned expressions, their maximum values, for different cases corresponding to given values of the field of the instrument (which may reach 90°) and various values of the zenithal distance of the center of the plate. It is important to show that the rests R_x, R_y, R'_x and R'_y which, in fact, are functions of the coefficients of the expression for the refraction, <u>are generally not sensible to important errors on these coefficients and may thus be computed as constants when one knows approximated values of them</u>. Thus, in the final expressions of the refracted tangential coordinates, only the coefficient a of the first term of the expression for the refraction must be introduced as an unknown. This is of great importance when establishing the formulae to be used in plate reduction. Also important is the knowledge of the order of magnitude of all the terms concerned for a given width of the field and a known zenithal distance. This may help in withdrawing some terms which necessarily appear to be considered as negligible in connection with the expected accuracy of the plate reduction.

I. INTRODUCTION

La réduction des clichés astrographiques implique l'usage de formules de transformation permettant le passage des positions des objets célestes, aux positions de leurs images sur le chiché et vice-versa. Le choix de ces formules peut varier d'un cas à l'autre suivant l'importance des phénomènes concourant à la transformation de la configuration céleste dans celle apparaissant sur le cliché. La connaissance des ordres de grandeur de ces phénomènes et des divers termes de leurs expressions est donc primordiale dans la recherche des formules de réduction à considérer. Parmi ces phénomènes, la réfraction atmosphérique joue un rôle particulièrement important.

Bien que de nombreuses études aient été consacrées aux effets de la réfraction atmosphérique sur la position des astres et sur leurs positions relatives dans tous systèmes de coordonnées tangentielles [2], nous nous proposons de reprendre ici ce problème dans le but de donner pour les coordonnées tangentielles réfractées, des expressions simples, formées de termes dont les ordres de grandeur sont aisément calculables à priori pour des champs de diverses étendues pouvant atteindre jusqu'à 45° autour de leurs centres.

Ce faisant, nous serons amené à discuter les expressions de l'effet de la réfraction dans le plan tangent à la sphère céleste pour diverses sortes de coordonnées tangentielles. Nous examinerons également la sensibilité des termes des développements considérés, à des variations petites des coefficients de la formule donnant la réfraction atmosphérique.

II. EXPRESSIONS DES COORDONNEES TANGENTIELLES [2] :

Nous préciserons avant tout la définition et les expressions des coordonnées tangentielles telles que nous les considérons ici.

Soit P, un point <u>quelconque</u> de la sphère céleste, défini comme pôle d'un système de coordonnées pseudo-équatoriales α, δ (fig. 1);

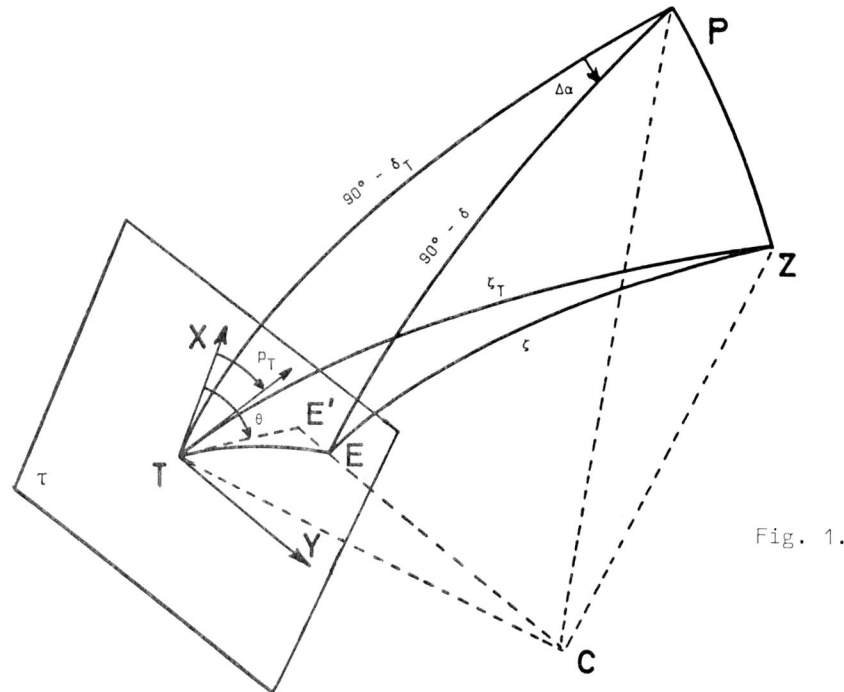

Fig. 1.

soient aussi T, un point de coordonnées (α_T, δ_T) où l'on considère un plan tangent à la sphère céleste et E, la position d'une étoile de coordonnées (α, δ). Dans ce plan, on définit les coordonnées tangentielles de E comme étant les coordonnées rectangulaires X et Y de la projection de E dans ce plan, depuis le centre de la sphère, par rapport à des axes qui sont respectivement : la trace du plan du grand cercle passant par P et T et celle du plan qui lui est perpendiculaire au point T. Les sens positifs correspondent respectivement, pour l'axe X, à la direction du pôle P et, pour l'axe Y, à celle conférant au système T.XY, le sens direct pour un observateur situé au centre C de la sphère céleste. Ce dernier correspond également au sens croissant de la coordonnée pseudo-équatoriale α.

Si θ est l'angle dièdre des plans de grand cercle passant d'une part par T et E et d'autre part, par T et P, et compté positivement de TX vers TY, on a :

$$\begin{cases} X = \text{tg TE} \cos\theta = \dfrac{\sin\text{TE}\cos\theta}{\cos\text{TE}} & (1) \\ \\ Y = \text{tg TE} \sin\theta = \dfrac{\sin\text{TE}\sin\theta}{\cos\text{TE}} & (2) \end{cases}$$

avec ($\Delta\alpha = \alpha - \alpha_T$) :

$$\cos\text{TE} = \sin\delta \sin\delta_T + \cos\delta \cos\delta_T \cos\Delta\alpha \qquad (3)$$

et

$$\begin{cases} \sin\text{TE} \sin\theta = \cos\delta \sin\Delta\alpha & (4) \\ \\ \sin\text{TE} \cos\theta = \sin\delta \cos\delta_T - \cos\delta \sin\delta_T \cos\Delta\alpha & (5) \end{cases}$$

Notons que l'arc TE est essentiellement positif.

Les expressions (1) à (5) sont classiques. Elles sont utilisées pour le calcul des coordonnées tangentielles X et Y.

Pour préciser l'aspect d'un nombre représentant l'une des coordonnées X ou Y, il faut remarquer que si toutes les étoiles sont situées à moins de 45° du point T (tg TE ≤ 1), les coordonnées tangentielles X et Y sont inférieures à 1 et si, de plus, les positions sont données avec une précision de 0",01, elles possèdent au maximum sept à huit décimales.

III. FORMULES DE LA REFRACTION

Si on se limite à des distances zénithales ne dépassant pas 75°, l'angle de réfraction r s'exprime par la relation :

$$r = a \, \text{tg} \, \zeta + b \, \text{tg}^3 \, \zeta \qquad (6)$$

avec une précision théorique généralement de l'ordre de 0",01 à 0",02, pour autant que l'on adopte pour a et b, des valeurs correspondant à la pression p (en mm) et à la température t (en degrés Celsius) de l'air au sol, au moment de l'observation et données par les relations :

$$\begin{cases} a = + 0",024 + 0",079017 \, p - 0",08260 \, pt' & (7) \\ b = + 0",0040 - 0",0001101 \, p + 0",000028 \, pt' & (8) \end{cases}$$

où $t' = \dfrac{t}{273° + t}$, ainsi que nous l'avons montré ailleurs [3]. Quelques valeurs de a et de b sont rassemblées dans le tableau I.

TABLEAU I

Coefficients a et b de l'expression (6) de la
réfraction atmosphérique.

p \ t		-30°	0°	+30°
790mm	a	+70",503	+62",447	+55",987
	b	- 0,0857	- 0,0830	- 0,0808
640mm	a	+57,121	+50,595	+45,361
	b	- 0,0687	- 0,0665	- 0,0647
490mm	a	+43,739	+38,742	—
	b	- 0,0516	- 0,0499	—

Notons que les pressions à considérer sont celles trouvées en apportant aux valeurs observées, d'une part, les corrections de gravité dues à l'altitude et à la latitude du lieu et d'autre part, celles relatives à la température du baromètre, comme indiqué dans la Connaissance des Temps.

Les valeurs extrêmes de a et de b susceptibles d'être rencontrées sont donc respectivement, en valeurs absolues :

$$\begin{cases} |a| = 71" = 3{,}4\ 10^{-4}\ \text{rad} \\ |b| = 0",086 = 4{,}2\ 10^{-7}\ \text{rad} \end{cases} \text{et} \begin{cases} |a| = 39" = 1{,}9\ 10^{-4}\ \text{rad} \\ |b| = 0",050 = 2{,}4\ 10^{-7}\ \text{rad} \end{cases}$$

tandis que les valeurs correspondant à la réfraction normale (p = 760 mm et t = 0°) sont de l'ordre de :

$$\begin{cases} |a| = 60" = 2{,}9\ 10^{-4}\ \text{rad} \\ |b| = 0",080 = 3{,}9\ 10^{-7}\ \text{rad} \end{cases}$$

En tenant compte de l'expression (6), on peut développer sin r en série, fonction de a et de b, exprimés en radians. En négligeant ensuite les termes plus petits que 1.10^{-8} dans les cas extrêmes où $\zeta = 75°$, c'est-à-dire où tg ζ = 3,7, on a :

$$\sin r = a\ \text{tg}\ \zeta + b\ \text{tg}^3 \zeta \qquad (9)$$

De même, pour cos r, on trouve :

$$\cos r = 1 - \frac{a^2}{2}\ \text{tg}^2 \zeta \qquad (10)$$

avec une précision équivalant à celle des relations (6) et (9).

IV. EXPRESSIONS DE QUELQUES LIGNES TRIGONOMETRIQUES RELATIVES AUX POSITIONS REFRACTEES DES ASTRES :

IV.1. Remarque préliminaire :

Dans ce qui suit, nous serons amené à établir plusieurs fois le développement en série de l'une des fonctions $\sin g_r$ et $\cos g_r$ d'une grandeur angulaire g affectée de la réfraction (g_r), alors que le développement de l'autre est connu. Nous croyons utile de rappeler que si l'on a :

$$\sin g_r = \sin g + a S + R_S$$

où a est le coefficient principal de la réfraction et R_S, le reste du développement, on a nécessairement :

$$\cos g_r = \cos g + a C + R_C$$

avec :

$$C \cos g = - S \sin g \tag{11}$$

et vice-versa. La relation (11) s'établit facilement en égalant à zéro, le coefficient du terme en a de l'expression : $\sin^2 g_r + \cos^2 g_r - 1$, identiquement nulle.

IV.2. Formules générales :

IV.2.1. Cas d'un arc sur la sphère :
Soient Z le zénith et A, B, les positions de deux astres sur le ciel (fig. 2). Désignons par A_r et B_r, les positions réfractées de ces astres.
Dans le triangle sphérique $A_r B_r Z$, on a :

$$\cos A_r B_r = \cos(\zeta_A - r_A) \cos(\zeta_B - r_B) + \sin(\zeta_A - r_A) \sin(\zeta_B - r_B) \cos \Delta A$$

Or, en tenant compte des expressions (9) et (10), on trouve d'une part :

$$\cos(\zeta - r) = \cos \zeta \cos r + \sin \zeta \sin r$$
$$= \cos \zeta + a \sin \zeta \, \text{tg} \, \zeta + (b \sin \zeta \, \text{tg} \, \zeta - \frac{a^2 \cos \zeta}{2}) \, \text{tg}^2 \zeta + \ldots$$

que l'on peut poser :

$$= \cos \zeta + a \, C \quad\quad + R_C$$

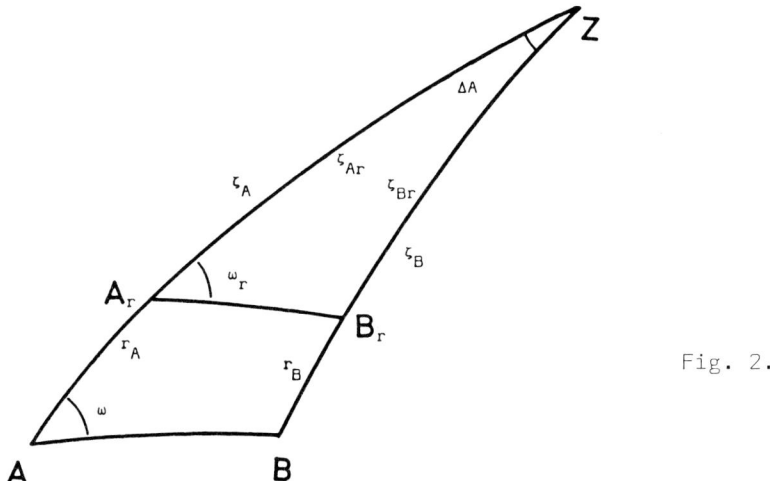

Fig. 2.

et d'autre part :

$$\sin(\zeta - r) = \sin \zeta \cos r - \cos \zeta \sin r$$
$$= \sin \zeta - a \sin \zeta - (\frac{a^2}{2} + b) \sin \zeta \, \text{tg}^2 \zeta + \ldots$$

que l'on peut poser :

$$= \sin \zeta + a\, S + R_S$$

Il vient dès lors :

$$\cos A_r B_r = (\cos \zeta_A + a\, C_A + R_{CA})(\cos \zeta_B + a\, C_B + R_{CB}) +$$
$$+ (\sin \zeta_A + a\, S_A + R_{SA})(\sin \zeta_B + a\, S_B + R_{SB}) \cos \Delta A$$

que l'on écrira provisoirement sous la forme :

$$\cos A_r B_r = \cos AB + a\, C + R_{c,AB}$$

en posant :

$$C = \mathcal{C}_A \cos \zeta_B + \mathcal{C}_B \cos \zeta_A + (\mathcal{S}_A \sin \zeta_B + \mathcal{S}_B \sin \zeta_A) \cos \Delta A$$
$$= \frac{\sin^2(\zeta_A - \zeta_B)}{\cos \zeta_A \cos \zeta_B} + 2 \sin \zeta_A \sin \zeta_B (1 - \cos \Delta A) \qquad (12)$$

et

$$R_{c,AB} = a^2 (C_A C_B + S_A S_B \cos \Delta A) +$$

$$+ (R_{CA} \cos \zeta_B + R_{CB} \cos \zeta_A) + (R_{SA} \sin \zeta_B + R_{SB} \sin \zeta_A) \cos \Delta A +$$

$$+ \ldots$$

Remarquons en passant que, sous les conditions ζ_A, $\zeta_B \leq 75°$, on a

$$\left. \begin{array}{ll} a\, C_A & \text{et} \quad a\, C_B \\ \\ a\, S_A & \text{et} \quad a\, S_B \\ \\ R_{CA} & \text{et} \quad R_{CB} \\ \\ R_{SA} & \text{et} \quad R_{SB} \end{array} \right\} \text{d'un ordre de grandeur inférieur ou égal respectivement à} \left\{ \begin{array}{l} 1.10^{-3} \\ \\ 3.10^{-4} \\ \\ 2.10^{-5} \\ \\ 3.10^{-6} \end{array} \right.$$

Dès lors, le terme $a\, C$ du développement de $\cos A_r B_r$ atteint au plus un ordre de grandeur de 2.10^{-3} tandis que le reste $R_{c,AB}$ - dont l'expression est compliquée, puisque fonction de a^2, de b et de puissances supérieures - a une valeur qui ne peut dépasser en aucune façon l'ordre de grandeur du maximum de son terme principal, soit : 4.10^{-5}, toujours sous les conditions ζ_A, $\zeta_B \leq 75°$.

En tenant compte ensuite de la relation

$$\sin \zeta_A \sin \zeta_B \cos \Delta A = \cos AB - \cos \zeta_A \cos \zeta_B ,$$

(12) peut s'écrire :

$$C = \frac{\cos^2 \zeta_B + \cos^2 \zeta_A - 2 \cos \zeta_A \cos \zeta_B \cos AB}{\cos \zeta_A \cos \zeta_B}$$

et comme :

$$\cos \zeta_B = \cos \zeta_A \cos AB + \sin \zeta_A \sin AB \cos \omega,$$

il vient finalement après quelques simplifications faciles :

$$C = \frac{\sin^2 AB\, (1 - \sin^2 \zeta_A \sin^2 \omega)}{\cos \zeta_A \cos \zeta_B}$$

L'expression de $\cos A_r B_r$ se présente alors comme suit :

$$\cos A_r B_r = \cos AB + a \frac{\sin^2 AB}{\cos AB} U_{AB} + R_{c,AB} \qquad (13)$$

en posant

$$U_{AB} = \frac{\cos AB (1 - \sin^2 \zeta_A \sin^2 \omega)}{\cos \zeta_A \cos \zeta_B} \qquad (14)$$

le choix de la forme (13) s'imposant, comme nous aurons l'occasion de le constater, pour des raisons d'homogénéité et de simplification notable des relations que nous établirons par la suite. Remarquons que U_{AB} a le signe de $\cos AB$ et dès lors, que le coefficient de a dans (13), est toujours positif (AB < 90°).

Par application de la relation (11), on trouve ensuite :

$$\sin A_r B_r = \sin AB - a \sin AB \cdot U_{AB} + R_{s,AB} \qquad (15)$$

où $R_{s,AB}$ est le reste du développement en série du sinus et comporte, comme $R_{c,AB}$, des termes en a^2, en b et de puissances supérieures.

Par ailleurs, on établit encore facilement la relation suivante :

$$\operatorname{tg} A_r B_r = \operatorname{tg} AB (1 - a U_{AB} + \ldots)(1 - a U_{AB} \operatorname{tg}^2 AB + \ldots)$$

$$= \operatorname{tg} AB - a \frac{\operatorname{tg} AB}{\cos^2 AB} U_{AB} + R_{t,AB} \qquad (16)$$

sous la condition que AB soit peu voisin du point de discontinuité AB = 90°, de la fonction tg AB.

IV.2.2. <u>Cas d'un angle dièdre</u> : Considérons l'angle ω dont un des côtés passe par Z (fig. 2). On a :

$$\frac{\sin \omega_r}{\sin \zeta_{Br}} = \frac{\sin \Delta A}{\sin A_r B_r}$$

et

$$\frac{\sin \omega}{\sin \zeta_B} = \frac{\sin \Delta A}{\sin AB}$$

On en déduit :

$$\sin \omega_r = \sin \omega \frac{\sin \zeta_{Br}}{\sin \zeta_B} \cdot \frac{\sin AB}{\sin A_r B_r}$$

Or, par application de (15) à l'arc BZ et en tenant compte de (14), on a :

$$\sin \zeta_{Br} = \sin \zeta_B (1 - a + \ldots)$$

En se servant ensuite de cette expression, ainsi que de l'expression (15) de sin $A_r B_r$, il vient :

$$\sin \omega_r = \sin \omega - a \sin \omega (1 - U_{AB}) + R_{s,\omega} \quad (17)$$

Par ailleurs, de la relation (11), on déduit l'expression :

$$\cos \omega_r = \cos \omega + a \frac{\sin^2 \omega}{\cos \omega} (1 - U_{AB}) + R_{c,\omega} \quad (18)$$

valable, quel que soit ω, puisque pour $\omega = 90°$, on trouve : $1 - U_{AB} = 0$ et que l'expression $(1 - U_{AB})/\cos \omega$ tend vers : $\sin AB \sin \zeta_A / \cos \zeta_B$.

IV.3. Application à quelques lignes trigonométriques intéressantes pour la présente étude :

Par application de certaines des formules établies ci-dessus, on trouve :

a) pour l'arc TE (supposé $\leq 45°$) :

d'après la formule (16) :

$$\text{tg } T_r E_r = \text{tg TE} - a \frac{\text{tg TE}}{\cos^2 TE} V + R_{t,TE} \quad (19)$$

en posant

$$V = U_{TE} = \frac{\cos TE (1 - \sin^2 \zeta_T \sin^2 \theta')}{\cos \zeta_T \cos \zeta} \quad (20)$$

b) pour l'angle $\theta' = ZTE$:

d'après les formules (17) et (18) :

$$\sin \theta'_r = \sin \theta' - a \sin \theta' (1 - V) + R_{s,\theta'} \quad (21)$$

et

$$\cos \theta'_r = \cos \theta' + a \frac{\sin^2 \theta'}{\cos \theta'} (1 - V) + R_{c,\theta'} \quad (22)$$

c) pour l'angle parallactique p_T :

d'après les formules (17) et (18) :

$$\sin p_{Tr} = \sin p_T - a \sin p_T (1 - W) + R_{s,p_T} \quad (23)$$

et
$$\cos P_{Tr} = \cos p_T + a \frac{\sin^2 p_T}{\cos p_T} (1 - W) + R_{c,p_T} \quad (24)$$

avec
$$W = U_{TP} = \frac{\sin \delta_T (1 - \sin^2 \zeta_T \sin^2 p_T)}{\cos \zeta_T \cos PZ} \quad (25)$$

d) pour l'angle de position $\theta = p_T + \theta'$:

$$\sin \theta_r = \sin p_{Tr} \cos \theta'_r + \cos p_{Tr} \sin \theta'_r$$

et en tenant compte des relations (21) à (24) :

$$\sin \theta_r = \sin \theta - a \cos \theta \left\{ \tg \theta' (1 - V) + \tg p_T (1 - W) \right\} + R_{s,\theta} \quad (26)$$

De même, on trouve :

$$\cos \theta_r = \cos \theta + a \sin \theta \left\{ \tg \theta' (1 - V) + \tg p_T (1 - W) \right\} + R_{c,\theta} \quad (27)$$

L'expression de V est toujours positive (puisque TE est supposé au plus égal à 45°), tandis que W a le signe de δ_T.

V. EXPRESSIONS GENERALES DES COORDONNEES TANGENTIELLES REFRACTEES :

A partir des formules que nous venons d'établir, il est maintenant aisé d'exprimer les coordonnées tangentielles réfractées en fonction de ces mêmes coordonnées non réfractées et des coefficients de la réfraction.

On trouve :

$$X_r = \tg T_r E_r \cos \theta_r = X + a X_a + R_X \quad (28)$$

avec :

$$X_a = -\frac{\tg TE}{\cos^2 TE} \cos \theta \cdot V + \tg TE \sin \theta \left\{ \tg \theta' (1-V) + \tg p_T (1-W) \right\} \quad (29)$$

et
$$R_X = \text{fonction de } a^2, a^3, \ldots b; b^2, \ldots a^i b^j, \ldots ,$$

ainsi que :

$$Y_r = \tg T_r E_r \sin \theta_r = Y + a Y_a + R_y \quad (30)$$

avec :

$$Y_a = -\frac{\text{tg TE}}{\cos^2 \text{TE}} \sin \theta \cdot V - \text{tg TE} \cos \theta \left[\text{tg } \theta'(1-V) + \text{tg } p_T(1-W)\right] \quad (31)$$

et

$$R_y = \text{fonction de } a^2, a^3, \ldots b, b^2, \ldots a^i b^j, \ldots$$

VI. COORDONNEES TANGENTIELLES POLAIRES (Standard) ET ZENITHALES :

Telles que nous les avons traitées jusqu'ici, les coordonnées tangentielles X et Y ont un caractère général parce qu'elles sont rapportées à un point P quelconque du ciel auquel on rattache un système de coordonnées (α, δ) pseudo-équatoriales.

Mais il est, bien entendu, deux positions du point P particulièrement intéressantes. Ce sont le pôle céleste et le zénith du lieu d'observation.

Si l'on impose au point P d'être le pôle céleste, on aura, dans l'hémisphère boréal : PZ = 90° - ϕ et les coordonnées tangentielles correspondantes pourraient être appelées des <u>coordonnées tangentielles polaires</u>, lesquelles se confondent avec les <u>coordonnées standard</u> lorsque l'on se donne un équinoxe standard.

Dans l'hémisphère austral, on a : PZ = 90° + ϕ, car le pôle à choisir ne peut plus être le pôle céleste nord, mais bien le pôle céleste sud, le seul visible dans le ciel et auquel il est permis d'appliquer la réfraction atmosphérique (voir VIII).

Pour faciliter la présentation de cette étude, nous poursuivrons d'abord les développements dans le cas d'une latitude boréale, l'autre cas s'en déduisant aisément comme indiqué en X.

Si le point P est confondu avec le zénith Z, les formules établies ci-dessus se simplifient considérablement et permettent même des simplifications des expressions apparaissant dans le cas général comme nous allons le voir. Nous proposons d'appeler ce système, le système de <u>coordonnées tangentielles zénithales</u>.

Dans ce cas, PZ = 0°, p_T = 0° (d'où : $\theta = \theta'$) et δ = 90° - ζ. On trouve alors :

1) coordonnées tangentielles zénithales :

$$X' = \text{tg TE} \cos \theta' \quad (32)$$

$$Y' = \text{tg TE} \sin \theta' \quad (33)$$

2) coordonnées tangentielles zénithales réfractées :

$$X'_r = X' + a X'_a + R'_x \qquad (34)$$

avec

$$X'_a = \frac{\text{tg TE}}{\cos^2 \text{TE} \cos \theta'} \left\{ \cos^2 \text{TE} \sin^2 \theta' - V (1 - \sin^2 \theta' \sin^2 \text{TE}) \right\}$$

et en remplaçant V par son expression (20) :

$$X'_a = \frac{\text{tg TE}}{\cos^2 \text{TE} \cos \theta'} \left\{ \cos^2 \text{TE} \sin^2 \theta' + \right. \\ \left. - \frac{\cos \text{TE} (1-\sin^2 \zeta_T \sin^2 \theta')(1-\sin^2 \text{TE} \sin^2 \theta')}{\cos \zeta_T \cos \zeta} \right\}$$

Par réduction au même dénominateur et par simplification du numérateur ainsi formé, en considérant l'expression (fig. 1) :

$$\cos \zeta = \cos \zeta_T \cos \text{TE} + \sin \zeta_T \sin \text{TE} \cos \theta' \qquad (35)$$

une première fois, dans le sens d'une substitution de $\cos \zeta$ par son expression et une seconde fois, dans celui - inverse - d'une simplification d'écriture, il vient :

$$X'_a = \frac{\sin \text{TE}}{\cos \zeta_T \cos \zeta \cos^2 \text{TE}} (\cos \zeta \sin \zeta_T \sin \text{TE} \sin^2 \theta' - \cos \theta') \qquad (36)$$

En éliminant $\sin \theta'$ et $\cos \theta'$ par l'emploi de (32) et de (33), cette dernière expression s'écrit encore :

$$X'_a = Y'^2 \text{tg } \zeta_T - \frac{X'}{\cos \zeta_T \cos \zeta \cos \text{TE}} \qquad (37)$$

De même pour l'ordonnée Y', on obtient :

$$Y'_r = Y' + a Y'_a + R'_y \qquad (38)$$

avec :

$$Y'_a = - \frac{\text{tg TE} \sin \theta'}{\cos^2 \text{TE}} (\cos^2 \text{TE} + V \sin^2 \text{TE})$$

ou, en remplaçant V par son expression (20), en réduisant au même dénominateur et en se servant de la relation (35) comme pour X'_a :

$$Y'_a = - \frac{\sin \text{TE} \sin \theta'}{\cos \zeta_T \cos \zeta \cos^2 \text{TE}} (\cos^2 \zeta_T + \cos \zeta \sin \zeta_T \sin \text{TE} \cos \theta') \qquad (39)$$

En substituant à sin θ' et à cos θ', leurs expressions tirées de (32) et de (33), Y'_a peut encore s'écrire :

$$Y'_a = -Y' \frac{\cos \zeta_T}{\cos \zeta \cos TE} - X' Y' \tg \zeta_T \qquad (40)$$

Remarquons que l'on peut encore remplacer dans (37) et (40), $\cos \zeta$ par :

$$\cos \zeta = \cos TE (\cos \zeta_T + X' \sin \zeta_T)$$

et l'on trouve respectivement :

$$X'_a = Y'^2 \tg \zeta_T - \frac{X'}{\cos^2 TE \cos^2 \zeta_T (1 + X' \tg \zeta_T)} \qquad (41)$$

et

$$Y'_a = -\frac{Y'}{\cos^2 TE (1 + X' \tg \zeta_T)} - X' Y' \tg \zeta_T \qquad (42)$$

VII. EXPRESSIONS CANONIQUES DES COEFFICIENTS X_a ET Y_a :

La considération des relations (32) à (40) relatives aux coordonnées tangentielles zénithales, permet de conférer aux coefficients X_a et Y_a, des formes canoniques avantageuses.

On les obtient facilement, soit par voie géométrique, soit par voie algébrique.

1) Par voie géométrique, on remarquera tout d'abord (fig. 3) que l'on a, pour la coordonnée X [4] :

$$X_r = X'_r \cos p_{Tr} - Y'_r \sin p_{Tr}$$

Ensuite, par les relations (34) et (37) d'une part, et (24) et (23) d'autre part, il vient pour le coefficient X_a :

$$X_a = (X'_a \cos p_T - Y'_a \sin p_T) + \tg p_T (1 - W)(X' \sin p_T + Y' \cos p_T)$$

En remplaçant d'abord W par son expression (25), puis en considérant pour cos PZ, son expression (fig. 1) :

$$\cos PZ = \cos \zeta_T \sin \delta_T + \sin \zeta_T \cos \delta_T \cos p_T \qquad (43)$$

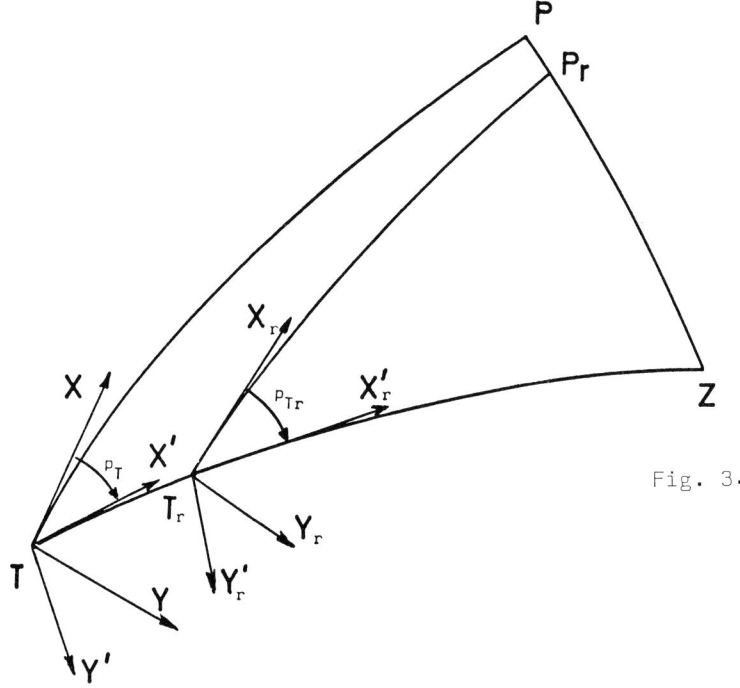

Fig. 3.

et en posant enfin :

$$K = -\cos \zeta_T \cos \delta_T + \sin \zeta_T \sin \delta_T \cos p_T \qquad (44)$$

il vient :

$$X_a = (X'_a \cos p_T - Y'_a \sin p_T) - \sin p_T \, \mathrm{tg}\, \zeta_T \frac{K}{\cos PZ}(X' \sin p_T + Y' \cos p_T) \qquad (45)$$

De même, on trouve pour Y_a :

$$Y_a = (X'_a \sin p_T + Y'_a \cos p_T) + \sin p_T \, \mathrm{tg}\, \zeta_T \frac{K}{\cos PZ}(X' \cos p_T - Y' \sin p_T) \qquad (46)$$

2) Ces relations s'établissent algébriquement, en remplaçant dans les expressions (29) et (31), θ par θ' + p_T.

VIII. CHOIX DES COORDONNEES TANGENTIELLES CONVENANT LE MIEUX A LA REDUCTION DES CLICHES - ORDRES DE GRANDEUR DE X'_a, Y'_a, R'_x ET R'_y ET DE X_a, Y_a, R_x ET R_y :

VIII.1. Considérations générales :

Le choix des formules de réduction des clichés astrographiques dépend des termes que leurs ordres de grandeur obligent à retenir dans les développements en série exprimant les effets des phénomènes à prendre en considération. Il est donc indispensable dans chaque cas, de prévoir ces ordres de grandeur, lesquels varient d'un cliché à l'autre en fonction de l'instrument utilisé (champ, distorsion), de la zone céleste photographiée (distance zénithale) et du lieu d'observation (position du pôle). En particulier, en ce qui concerne la réfraction atmosphérique, il y a donc lieu de connaître les ordres de grandeur des coefficients X_a et Y_a et des restes R_x et R_y, des expressions (28) et (30).

Or, à l'examen des expressions canoniques (45) et (46) de X_a et de Y_a, il apparaît que ces coefficients peuvent devenir relativement grands dans certains cas particuliers. En effet, pour un point T donné et une valeur maximale imposée de TE, correspondant à l'importance du champ instrumental, X_a et Y_a varient non seulement en fonction de θ, mais aussi en fonction de la position du pôle (p_T et δ_T) laquelle peut conduire à des valeurs de cos PZ voisines de zéro.

A ce sujet, une remarque importante paraît devoir être faite ici : elle concerne la définition même du système des coordonnées tangentielles <u>réfractées</u>. En effet, ces coordonnées dépendent essentiellement de la <u>définition</u> de l'orientation de l'axe X_r, laquelle peut se faire relativement, soit au pôle non réfracté, soit au pôle réfracté.

Dans le premier cas, le pôle réfracté est un élément commun à tous les systèmes de coordonnées tangentielles réfractées, mais il y apparaît comme un point de discontinuité particulièrement gênant lors de réductions de clichés pris dans ses environs, car les développements en séries considérés dans le problème, y perdent toute leur signification.

Dans le second cas, - <u>uniquement considéré ici</u> - cet inconvénient est éliminé, mais il apparaît alors que, dans les régions équatoriales où le pôle céleste, sud ou nord, peut être situé à moins de 15° de l'horizon, PZ est supérieur à 75° et la relation (6) donnant la réfraction, n'est plus applicable au pôle, dans les conditions où nous avons développé la présente étude. Mathématiquement d'ailleurs, on obtient alors des valeurs tendant vers l'∞, pour les seconds termes de (45) et de (46) - termes que l'on pourrait appeler <u>les termes polaires</u> - et dès lors aussi, pour X_a et Y_a.

Aussi, devant l'impossibilité de généralisation totale dans un cas comme dans l'autre, c'est la considération du système de <u>coordonnées</u>

tangentielles zénithales - où le pôle, confondu avec le zénith, ne subit plus aucun effet dû à la réfraction - qui s'impose en premier lieu. Elle exigera chaque fois, bien entendu, une transformation de coordonnées (X,Y) en (X',Y') avant de procéder à la réduction d'un cliché, puis une transformation inverse pour revenir au système des coordonnées équatoriales de départ.

Ce sont alors les expressions (34) et (38) qu'il y a lieu de retenir ainsi que les ordres de grandeur des plus grandes valeurs de X'_a et de Y'_a et aussi de R'_x et de R'_y, qu'il y a lieu de connaître.

VIII.2. Maximums de X'_a et de Y'_a :

L'expression du maximum de X'_a, pour toutes valeurs données de ζ_t et de TE, s'établit facilement. En effet, d'après (36), on a :

$$\frac{\partial X'_a}{\partial \theta'} = \frac{\sin TE}{\cos \zeta_T \cos^2 TE} \cdot \frac{\partial}{\partial \theta'} \left\{ \sin \zeta_T \sin TE \sin^2 \theta' - \frac{\cos \theta'}{\cos \zeta} \right\}$$

et, en tenant compte de (35) :

$$= \frac{\sin TE \sin \theta'}{\cos \zeta_T \cos^2 TE} \cdot \frac{2 \sin TE \sin \zeta_T \cos \theta' \cos^2 \zeta + \cos \zeta_T \cos TE}{\cos^2 \zeta} \qquad (47)$$

Or, le numérateur du second facteur de (47) est positif dans tout l'intervalle de variation de ζ, de $\zeta_T -$ TE à $\zeta_T +$ TE, bornes comprises, et ne s'y annule donc pas.

En effet :

a) pour $\zeta = \zeta_T -$ TE ($\theta' = 0°$) :

ce numérateur vaut :

$$2 \sin TE \sin \zeta_T \cos^2 (\zeta_T - TE) + \cos \zeta_T \cos TE$$

et est toujours positif, puisque tous les éléments qui le composent le sont (TE essentiellement positif), et :

pour $\zeta = \zeta_T +$ TE ($\theta' = \pm 180°$) :

il vaut :

$$- 2 \sin TE \sin \zeta_T \cos^2 (\zeta_T + TE) + \cos \zeta_T \cos TE$$

et est également toujours positif. En effet, on a d'une part :

$$2 \sin TE \cos TE \leq 1 \qquad \text{(si TE} \leq 45°)$$

d'où :
$$2 \sin TE \sin \zeta_T \cos TE \cos \zeta_T \leq 1$$

et d'autre part :
$$\cos(\zeta_T + TE) < \cos \zeta_T \cos TE$$

(puisque $\sin \zeta_T \sin TE > 0$). Ces deux dernières inégalités permettent ensuite d'écrire la suivante :
$$2 \sin TE \sin \zeta_T \cos(\zeta_T + TE) < 1$$

d'où, par multiplication membre à membre avec la seconde, l'on tire :
$$2 \sin TE \sin \zeta_T \cos^2(\zeta_T + TE) < \cos \zeta_T \cos TE$$

b) le numérateur dont il est question, est régulièrement décroissant de $\zeta = \zeta_T - TE$ à $\zeta = \zeta_T + TE$, comme le sont les fonctions $\cos \theta'$ et $\cos^2 \zeta$.

L'expression (47) s'annule donc seulement pour :

$$\begin{cases} \theta' = 0° \\ \text{et pour} \\ \theta' = \pm 180° \end{cases} \text{ et donc pour : } \begin{cases} \zeta = \zeta_T - TE \\ \text{et pour} \\ \zeta = \zeta_T + TE \end{cases} \quad (48)$$

c'est-à-dire lorsque le point E se trouve sur le grand cercle vertical passant par T.

On trouve alors à partir de (36), pour extremums de X'_a, l'expression :
$$X'_a = \frac{\mp \sin TE}{\cos \zeta_T \cos(\zeta_T \mp TE) \cos^2 TE}$$

De ces deux extremums, c'est le maximum - algébriquement parlant - qui l'emporte en valeur absolue et l'on a finalement :
$$|X'_a|_{max} = \frac{\sin TE}{\cos \zeta_T \cos(\zeta_T + TE) \cos^2 TE} \quad (49)$$

Les valeurs de cette expression sont données dans le tableau II pour diverses valeurs de ζ_T et de TE (avec $\zeta_T + TE \leq 75°$).

Quant au maximum de $|Y'_a|$, il est solution d'une équation assez compliquée du quatrième degré en $\cos \theta'$, solution dont nous n'avons pu

TABLEAU II

Maximums de $|X'_a|$

ζ_T / TE	0°	10°	20°	30°	40°	50°	60°	70°	74°
1°	0,017	0,018	0,020	0,024	0,030	0,043	0,072	0,16	0,24
5°	0,088	0,092	0,10	0,12	0,16	0,24	0,42	0,99	
10°	0,18	0,19	0,22	0,27	0,36	0,56	1,0		
15°	0,29	0,31	0,36	0,45	0,63	1,0	2,1		
20°	0,41	0,45	0,54	0,70	1,0	1,8			
25°	0,57	0,64	0,77	1,0	1,6	3,1			
30°	0,77	0,88	1,1	1,5	2,5				
35°	1,0	1,2	1,6	2,3	4,3				
40°	1,4	1,7	2,3	3,7					
45°	2,0	2,5	3,6	6,3					

$\zeta_T + TE > 75°$

TABLEAU III

Maximums de $|Y'_a|$

ζ_T / TE	0°	10°	20°	30°	40°	50°	60°	70°	74°
1°	0,017	0,017	0,017	0,017	0,017	0,017	0,017	0,017	0,017
5°	0,088	0,088	0,088	0,088	0,088	0,088	0,088	0,088	
10°	0,18	0,18	0,18	0,18	0,18	0,18	0,18		
15°	0,29	0,29	0,29	0,29	0,29	0,29	0,29		
20°	0,41	0,41	0,41	0,41	0,41	0,41			
25°	0,57	0,57	0,57	0,57	0,57	0,58			
30°	0,77	0,77	0,77	0,77	0,78				
35°	1,0	1,0	1,0	1,1	1,1				
40°	1,4	1,4	1,4	1,5					
45°	2,0	2,0	2,0	2,2					

$\zeta_T + TE > 75°$

établir l'expression. Les maximums donnés dans le tableau III ont été obtenus en calculant à l'aide d'un ordinateur, les valeurs de Y'_a pour toutes valeurs de θ' de 10° en 10°, comprises entre 0° et 180° (Y'_a change simplement de signe pour θ' compris entre 180° et 360°).

On constate que :

$$|Y'_a|_{max} \leq |X'_a|_{max} \qquad (50)$$

pour toutes valeurs tabulées de TE et de ζ_T.

VIII.3. Maximums de R'_x et de R'_y :

Les restes R'_x et R'_y ont des expressions compliquées et celles de leurs extrémums ne le sont pas moins. Aussi, la seule façon pratique d'en connaître les ordres de grandeur, est de procéder à l'aide d'un ordinateur, au calcul de diverses valeurs de ces restes, comme indiqué ci-dessus pour Y'_a, pour des valeurs convenablement échelonnées de ζ_T, de TE et de θ'. Ainsi, en faisant varier θ' de 10° en 10°, nous avons pu établir les tableaux IV et V pour des valeurs de ζ_T et de TE identiques à celles considérées dans les tableaux II et III.

Mais comme on ne possède pas les expressions de ces restes, on les calcule par les relations (34) et (38), mises sous les formes :

$$R'_x = X'_r - X' - a X'_a \qquad (51)$$

$$R'_y = Y'_r - Y' - a Y'_a \qquad (52)$$

où X'_r et Y'_r sont calculées comme X' et Y', par les relations (1) à (5), mais après correction des coordonnées de T et de E, par application de la formule de réfraction (6) et en adoptant pour a et b, les valeurs correspondant à la réfraction normale (p = 760 mm; t = 0°). Pour des valeurs différentes, celles de R'_x et de R'_y changent peu et comme ce sont ici des ordres de grandeur que l'on cherche, les tableaux IV et V reflètent correctement la situation. Nous discuterons plus loin la sensibilité des restes R'_x et R'_y, aux variations admissibles de a et de b et sa signification quant à l'intérêt des formules proposées.

On constate aussi que l'on a :

$$|R'_y|_{max} \leq |R'_x|_{max} \qquad (53)$$

pour toutes valeurs tabulées de TE et de ζ_T.

VIII.4. Remarques :

Lorsque le pôle céleste n'est pas trop éloigné du zénith, c'est-à-dire sous des latitudes relativement élevées, les ordres de grandeur des

TABLEAU IV

Maximums de $|R'_x|.10^7$, calculés pour une réfraction normale correspondant à : a = + 0,00029 rad. et b = - 3,9 10^{-7} rad.

TE \ ζ_T	0°	10°	20°	30°	40°	50°	60°	70°	74°
1°	0,0	0,0	0,0	0,1	0,3	0,7	2,6	14,6	37,3
5°	0,0	0,1	0,2	0,6	1,6	4,8	18,3	122,	
10°	0,0	0,2	0,6	1,6	4,4	14,1	62,6		
15°	0,1	0,4	1,3	3,4	9,8	34,3	196,		
20°	0,3	0,9	2,5	6,8	20,6	84,9			
25°	0,6	1,7	4,7	13,2	44,8	241,			
30°	1,2	3,3	9,0	26,8	107,				
35°	2,4	6,4	17,6	58,3	305,				
40°	4,8	12,4	36,4	142,					
45°	9,5	25,2	81,8	420,					

$\zeta_T + TE > 75°$

TABLEAU V

Maximums de $|R'_y|.10^7$, calculés pour une réfraction normale correspondant à : a = + 0,00029 rad. et b = - 3,9 10^{-7} rad.

TE \ ζ_T	0°	10°	20°	30°	40°	50°	60°	70°	74°
1°	0,0	0,0	0,0	0,0	0,0	0,1	0,2	0,5	0,8
5°	0,0	0,0	0,0	0,1	0,2	0,5	1,0	2,9	
10°	0,0	0,1	0,1	0,3	0,6	1,2	3,0		
15°	0,1	0,2	0,3	0,7	1,3	3,0	9,2		
20°	0,3	0,4	0,7	1,4	2,8	7,2			
25°	0,6	0,8	1,4	2,7	6,2	20,3			
30°	1,2	1,6	2,8	5,6	14,5				
35°	2,4	3,2	5,5	12,0	39,4				
40°	4,8	6,2	11,3	28,2					
45°	9,5	12,5	24,9	78,0					

$\zeta_T + TE > 75°$

coefficients X_a et Y_a ne diffèrent pas beaucoup de ceux de X'_a et de Y'_a. Il y a alors intérêt - après s'être assuré de ce fait - à conserver le système de coordonnées tangentielles polaires (X,Y) pour la réduction des clichés : cela évite la transformation des (X,Y) en (X',Y') et vice-versa.

Toutefois, devant la complexité des formules à considérer pour établir les maximums des fonctions X_a et Y_a - ainsi que ceux de R_x et de R_y -, il apparaît avantageux de déterminer une fois pour toutes ces maximums, ici encore, à l'aide d'un ordinateur, pour le lieu d'observation considéré et pour le champ de l'instrument utilisé. On procédera pour R_x et R_y, comme indiqué plus haut pour R'_x et R'_y. On n'omettra cependant pas de faire varier dans ce cas, l'angle θ de 0° à 360°, car la symétrie axée sur TX' ne se retrouve pas dans le système T.XY.

Ainsi, pour la latitude de l'Observatoire Royal de Belgique, les calculs ont été faits pour deux cas distincts :
TE = 5°, voisin de celui représenté par l'Astrographe Double dont le champ est de 8° (TE = 4°);
TE = 15°, voisin de celui représenté par la Camera Balistique IGN dont le champ est de 35° (TE = 17°,5).

Les résultats en sont donnés dans le tableau VI. Leur comparaison

TABLEAU VI

Maximums de X_a, Y_a, R_x et R_y pour la latitude d'Uccle et pour deux valeurs de TE

ζ_T	TE = 5°				TE = 15°			
	X_a	Y_a	$R_x \cdot 10^{-7}$	$R_y \cdot 10^{-7}$	X_a	Y_a	$R_x \cdot 10^{-7}$	$R_y \cdot 10^{-7}$
0°	0,088	0,088	0,0	0,0	0,29	0,29	0,1	0,1
10°	0,092	0,093	0,1	0,1	0,31	0,31	0,4	0,5
20°	0,10	0,10	0,3	0,3	0,36	0,37	1,3	1,4
30°	0,12	0,13	0,6	0,6	0,45	0,46	3,4	3,5
40°	0,16	0,16	1,6	1,6	0,63	0,62	9,8	9,6
50°	0,24	0,20	4,8	3,8	1,0	0,86	34,	28
60°	0,42	0,31	18,	13,	2,1	1,6	196,	142
70°	0,99	0,67	123,	81,	—	—	—	—

avec ceux donnés pour TE = 5° et TE = 15° dans les tableaux II à V, est instructive. Ainsi, on notera par exemple, pour TE = 15° et ζ_T = 20°, que les maximums correspondants sont de :

$X'_a = 0,36 \qquad R'_x = 1,3.10^{-7} \qquad Y'_a = 0,29 \qquad R'_y = 0,3.10^{-7}$

$X_a = 0,36 \qquad R_x = 1,3.10^{-7} \qquad Y_a = 0,37 \qquad R_y = 1,4.10^{-7}$

c'est-à-dire que les coefficients X_a, Y_a et les restes R_x et R_y relatifs au système de coordonnées tangentielles polaires, ont des maximums du même ordre que leurs correspondants X'_a, Y'_a, R'_x et R'_y dans le système de coordonnées tangentielles zénithales et dès lors, que l'étude de clichés pris à la latitude d'Uccle, peut être menée avec autant de facilité dans l'un comme dans l'autre système, du point de vue des effets de la réfraction atmosphérique.

IX. CONCLUSION :

Nous avons ainsi finalement établi des expressions <u>rigoureuses</u> des coordonnées tangentielles réfractées polaires X_r et Y_r, et zénithales X'_r et Y'_r, aucun terme de leurs développements n'ayant été négligé. Nous avons précisé par ailleurs, les ordres de grandeur de chacun de ces termes, pour tous les cas pouvant se présenter en pratique, tant en position du centre du cliché sur le ciel ($\zeta_T \leq 75°$) qu'en grandeur du champ instrumental (TE $\leq 45°$).

Il convient maintenant de préciser les avantages de l'emploi de ces expressions dans l'établissement des formules de réduction des clichés destinés à l'astrométrie.

La loi (6) n'est jamais qu'une loi moyenne à laquelle des écarts sensibles locaux s'observent couramment. Aussi est-il préférable de définir expérimentalement - autant que faire se peut - les paramètres a et b permettant de représenter au mieux les effets de la réfraction atmosphérique, pour chaque observation et pour la région du ciel couverte par le cliché photographique et donc, de définir ces paramètres à partir des constantes mêmes du cliché, plutôt que d'introduire leurs valeurs calculées à priori à l'aide des formules (7) et (8). Nous allons voir que la forme des expressions (34) et (38) s'y prête parfaitement et, par la même occasion, dans quelle mesure il convient d'appliquer ce principe.

Pour préciser les idées, notons d'abord que les anomalies de réfraction ne peuvent généralement dépasser ni même sans doute atteindre celles correspondant, par l'emploi des relations (7) et (8), à des variations de la température de ± 10° et de la pression, de ± 10 mm. Dans ces conditions extrêmes, les erreurs sur a et sur b sont, en moyenne, de l'ordre de :

	avec $\Delta t = \pm 10°$	avec $\Delta p = \pm 10$ mm
pour a :	de ± 4 %	de ± 1 à 2 %
pour b :	de ± 1 %	de ± 1 à 2 %

TABLEAU VII

Variations de $R'_x \cdot 10^7$ et de $R'_y \cdot 10^7$ dues à des variations de a et de b de 10 %

ζ_T		0°		10°		20°		30°		40°		50°		60°		70°		74°	
TE		R'_x	R'_y	R'_x	R'_y	R'_x	R'_y	R'_x	R'_y	R'_x	R'_y	R'_x	R'_y	R'_x	R'_y	R'_x	R'_y	R'_x	R'_y
1°	da	0,0	0,0	0,0	0,0	0,0	0,0	0,0	0,0	0,0	0,0	0,0	0,0	0,0	0,0	0,0	0,0	0,0	0,0
	db	0,0	0,0	0,0	0,0	0,0	0,0	0,0	0,0	0,0	0,0	0,0	0,0	0,3	0,0	1,5	0,0	3,7	0,1
5°	da	0,0	0,0	0,0	0,0	0,0	0,0	0,0	0,0	0,0	0,0	0,0	0,0	0,0	0,0	0,0	0,1		
	db	0,0	0,0	0,0	0,0	0,0	0,0	0,1	0,0	0,2	0,0	0,5	0,1	1,8	0,1	12,3	0,3		
10°	da	0,0	0,0	0,0	0,0	0,0	0,0	0,0	0,0	0,0	0,0	0,0	0,0	0,0	0,1				
	db	0,0	0,0	0,0	0,0	0,0	0,0	0,2	0,0	0,4	0,1	1,4	0,1	6,2	0,3				
15°	da	0,0	0,0	0,0	0,0	0,0	0,0	0,0	0,0	0,0	0,0	0,0	0,0	0,2	0,1				
	db	0,0	0,0	0,1	0,0	0,1	0,0	0,3	0,1	1,0	0,1	3,4	0,3	19,6	1,0				
20°	da	0,0	0,0	0,0	0,0	0,0	0,0	0,0	0,0	0,1	0,0	0,2	0,1						
	db	0,0	0,0	0,1	0,0	0,2	0,1	0,7	0,1	2,0	0,3	8,4	0,8						
25°	da	0,0	0,0	0,0	0,0	0,0	0,0	0,1	0,0	0,2	0,0	0,6	0,1						
	db	0,1	0,1	0,2	0,1	0,4	0,1	1,3	0,3	4,4	0,6	23,9	2,1						
30°	da	0,0	0,0	0,0	0,0	0,1	0,0	0,2	0,0	0,5	0,0								
	db	0,1	0,1	0,3	0,1	0,8	0,3	2,6	0,6	10,5	1,4								
35°	da	0,1	0,1	0,1	0,1	0,2	0,1	0,4	0,1	1,5	0,1								
	db	0,2	0,2	0,6	0,3	1,7	0,5	5,6	1,2	29,8	3,9								
40°	da	0,2	0,2	0,3	0,2	0,5	0,2	1,2	0,2										
	db	0,4	0,4	1,1	0,5	3,4	1,1	13,6	2,7										
45°	da	0,4	0,4	0,6	0,4	1,1	0,4	3,4	0,6										
	db	0,8	0,8	2,3	1,1	7,6	2,3	40,3	7,5										

$\zeta_T + TE \leq 75°$

ce qui indique, d'après la table VII donnant les variations de R'_x et de R'_y calculées pour des variations de a et de b de 10 %, que ces restes sont insensibles à 1.10^{-7} (= $0''02$) près, à des écarts : $\Delta t = \pm 10°$ et $\Delta p = \pm 10$ mm, sauf peut-être pour des valeurs extrêmes de ζ_T et de TE, auxquels cas d'ailleurs, la précision des réductions est nécessairement moins bonne et atteint difficilement 1.10^{-6}, voire 1.10^{-5}.

On en conclut donc que les restes R'_x et R'_y peuvent être calculés à priori dans chaque cas, avec toute la précision désirée, à l'aide des formules (51) et (52), en utilisant les meilleures valeurs des coefficients a et b définis sur base des informations disponibles concernant la température et la pression atmosphérique au sol au moment de l'observation.

Par contre, en ce qui concerne les termes aX'_a et aY'_a figurant dans les expressions (34) et (38) de X'_r et de Y'_r, il ne peuvent être connus à priori avec la même précision que R'_x et R'_y, puisqu'une erreur de 1 % sur a entraîne - d'après les valeurs de X'_a et de Y'_a figurant dans les tableaux II et III - des erreurs pouvant atteindre plusieurs secondes de degré dans certaines régions de clichés couvrant des champs d'étendue moyenne et pris à des distances zénithales pas tellement grandes.

Aussi, dans les expressions (34) et (38) de X'_r et de Y'_r, on aura intérêt à laisser au coefficient a le caractère d'un paramètre à définir parmi les constantes du cliché, mais à se servir des valeurs approchées de a et de b, pour calculer par (52) et (53), les restes R'_x et R'_y avec toute la précision désirable.

Après réduction du cliché, on vérifiera si la concordance est suffisante entre la valeur approchée de a adoptée pour le calcul des restes R'_x et R'_y et celle trouvée par les constantes du cliché. Une discordance trop grande entraînant des écarts sensibles sur les valeurs calculées de R'_x et de R'_y pourrait alors - exceptionnellement sans doute - inciter à procéder à une nouvelle réduction générale du cliché à partir de la valeur améliorée de a.

Il en va de même pour les restes R_x et R_y lorsque la réduction est effectuée directement dans le système T.XY.

X. CAS DES OBSERVATIONS EFFECTUEES DANS L'HEMISPHERE AUSTRAL :

Les relations établies ici pour les coordonnées tangentielles polaires l'ont été en faisant référence au pôle céleste nord.

Il est évident que dans l'hémisphère austral où seul le pôle céleste sud est visible dans le ciel, c'est à celui-ci qu'il faut rattacher le système de coordonnées tangentielles polaires, puisqu'ainsi que nous l'avons montré au paragraphe VII., c'est au pôle réfracté que l'on rattache le système réfracté correspondant et que dès lors, des deux pôles célestes, seul le pôle sud, de par sa position au dessus de l'horizon, peut être pris en considération.

Toutefois, toutes les expressions établies ici restent valables dans ce cas, si l'on remplace α par $-\alpha$, δ par $-\delta$ et ϕ par $-\phi$, et l'on se souvient que dans ces conditions, la coordonnée X est dirigée vers le pôle céleste sud, et la coordonnée Y, dans le sens du mouvement diurne, c'est-à-dire, dans le sens opposé à celui des ascensions droites croissantes.

Cette étude a nécessité de nombreux calculs et de délicates programmations pour l'ordinateur IBM 1620 dont nous disposions à l'époque de la réalisation de ce travail. M. O. NYS, Calculateur à l'Observatoire, s'est acquitté du travail avec maîtrise : nous lui sommes reconnaissant pour sa collaboration sûre et efficace.

---

[1] Ce travail a été présenté respectivement :
   1) devant la section d'Astronomie et de Mécanique Céleste du Congrès de l'Association Française pour l'Avancement des Sciences, tenu à Besançon en juillet 1969 ("Sciences", tome I, n°4, 1970, p. 205);
   2) lors du Symposium n° 61 de l'Union Astronomique Internationale, intitulé "New Problems in Astrometry", tenu à Perth (Australie) en août 1973 (D. Reidel Publishing Cy.).

[2] Les coordonnées tangentielles dont il est question ici ne peuvent être confondues avec celles considérées en géométrie analytique. Nous avons préféré la présente terminologie à celle de "coordonnées standard" car notre travail porte sur des coordonnées rectangulaires à caractère tout-à-fait général situées dans le plan tangent à la sphère céleste en un point choisi pour origine du système de coordonnées. Ce n'est que dans certaines conditions particulières, qu'elles se confondent avec les coordonnées standard telles qu'elles ont été définies par l'Union Astronomique Internationale (Transactions, VI, 1938, p. 347).

[3] J. DOMMANGET et O. NYS.- Expressions de la réfraction atmosphérique dans le cas d'une distance zénithale inférieure à 75°.- Bulletin astronomique de l'Observatoire Royal de Belgique, VII, fasc. 2, 1970, p. 89 (N.B. : Une erreur typographique systématique s'est glissée dans cette publication : la lettre ξ doit être remplacée partout par la lettre ζ, utilisée couramment pour désigner la distance zénithale, comme nous le faisons ici).

[4] Il est à remarquer que les angles parallactiques p_T et p_{Tr} sont les angles de position du zénith dans les systèmes respectifs $T.XY$ et $T_r.X_rY_r$. Dans l'hémisphère nord où l'on choisira pour système de coordonnées tangentielles polaires, celui rattaché au pôle céleste nord, ces angles sont donc compris entre 0° et 180°, pour des positions de T et de T_r situées à l'Ouest, et entre 180° et 360°, pour des positions de T et de T_r situées à l'Est. Dans l'hémisphère austral, où l'on choisira le pôle céleste sud comme pôle du système de coordonnées tangentielles, c'est l'inverse qui se passe.

DISCUSSION

G. Teleki: Do you have the possibility to correct your calculations of coordinates for possible influence of anomalous refraction if you observe at great zenith distances (at which the differential measurements can also be influenced by the anomalous refraction)?

J. Dommanget: answered that he uses the normal refraction formulae which are adjusted for the observed astrographic field by the plate reduction. This may give information on the local departure from a mean representation of the refraction effect on a more extended field in zenith distance. Anyway, usually no observation at very great zenith distance.

D.G. Currie: asked for the definition of the normal and anomalous refraction.

G. Teleki: repeated the definition of these values mentioned in his paper "Investigation of Astronomical Refraction - Today and Tomorrow", which he presented during the first session.

ATMOSPHERIC TURBULENCE EFFECTS ON ASTROMETRIC
DETERMINATIONS

V.I.Ivanov
Siberian Institute of Terrestrial Magnetism, Ionosphere and Radio Wave Propagation, Irkutsk, USSR

1. INTRODUCTION

It is well known that the turbulent state of the atmosphere gives rise to fluctuations of the coefficient of light refraction. In astrometric observations these fluctuations produce the phenomenon of image motion which serves as the source of accidental errors of observations. The need for the study of the effect of atmospheric turbulence is documented in (G.Teleki, 1967) and others.
The crucial point of the problem in question is the dependence of the accidental error on the length of observation. In searching for this dependence different methods have been applied: empirical data of various authors (K.Lambeck, 1968), autocorrelation functions of image motion (I.G.Kolchinsky, 1973), and power spectra (E.Høg, 1968). In the last paper for the external and internal accidental mean errors the following expressions have been obtained:

$$m_{ext.}^2 = m_\tau^2 + (m_t^2 - m_{\Delta t}^2)/n \qquad (1)$$

$$m_{int.}^2 = m_t^2 - (n \cdot m_\tau^2 - m_{\Delta t}^2)/(n-1) \qquad (2)$$

where t is the length of one coordinate count, $t = \tau/n$ is the time between counts, τ is the general length of observations, n is the number of counts.
All errors in (1) and (2) are found from data on power spectrum of image motion W(f), for example:

$$m_\tau^2 = \int_0^\infty W_\tau(f) \cdot df = \int_0^\infty W(f) \cdot \left(\frac{\sin \pi f \tau}{\pi f \tau}\right)^2 \cdot df \qquad (3)$$

The method of spectral analysis of the motion process has also been used in our investigations and in order to study the general properties of spectra data on normalized spectral density have been obtained. Then equation (3) will be of the form:

$$\frac{m_\tau^2}{\sigma_\infty^2} = \int_0^\infty \frac{W(f)}{\sigma_\infty^2} \cdot \left(\frac{\sin \pi f \tau}{\pi f \tau}\right)^2 \cdot df \qquad (3a)$$

where σ_∞^2 means the dispersion of motion in the frequency interval from 0 to ∞.
From formulas (1)-(3a) it follows that for the estimation of accidental errors of observations one must know the amplitude (or dispersion) of image motion σ. The peculiar feature of this paper is that the calculation of σ and consequently of the value m_τ is carried out for specific conditions of observations and instruments.

2. PROCEDURE OF CALCULATING THE AMPLITUDE OF IMAGE MOTION

The statistical theory of wave propagation in a turbulent medium (V.I.Tatarsky, 1967) provides for σ the expression that is valid in the case of a steady-state random process. In astrometric determinations due to differing times of observation τ the stationarity is not realized, therefore the dependence σ on τ should be taken into account. Denoting it through $I(\tau)$ for σ^2 we shall have the expression:

$$\sigma^2 = 2.84 \cdot D^{-1/3} \cdot \sec Z \cdot I(\tau) \int_0^\infty C_n^2(h) \cdot dh \qquad (4)$$

where D is the objective diameter, z is the zenithal distance, C_n is the structural constant of pulsations of the refraction coefficient, h is the height above Earth surface. In order to employ practically formula (4) one should verify theoretical dependences σ on D and Z, find the function $I(\tau)$ and define the influence of meteorological conditions characterized by the integral of C_n^2. For this purpose, observations have been carried out of motion of stars and the artificial light source (mire) in the ground layer of the atmosphere with the AZT-14 telescope (D=48 cm, F=7.7 m), as well as that of the Sun and Moon with the ACU-5 telescope (D=44 cm, F=17 m). Observations were accomplished during a year at the Sayan Observatory of SibIZMIR, Siberian Department, USSR Academy of Sciences (H=2000 m). Photographic and photoelectric methods of recording the image motion were used.
Verification of the dependences of σ on D and Z has shown that on the average they satisfy well the theory and data of other authors (I.G.Kolchinsky, 1967). For the apertures 3-48 cm, $\sigma_D \sim D^{-1/6}$. The power index at sec Z obtained from analysis of short and long realizations of motions of stars became equal to 0.49 ± 0.10. The amplitude of motion in the zenith for our point equals $\sigma_0 = \pm 0".33$.
To find the function $I(\tau)$ we made use of our spectra of image motions of point and extended light sources and $I(\tau)$

is expressed in the form:

$$I(\tau) = \int_{1/\tau}^{f_o} \frac{W(f)}{\sigma^2} \, df \qquad (5)$$

The spectral density $W(f)/\sigma^2$ is defined over a broad frequency range from 10^{-5} to 400 Hz. Table 1 lists the values of $I(\tau)$.

Table 1

f, Hz	τ, s	$I(\tau)$
0.00001	100000	4.0850
0.0001	10000	2.2890
0.001	1000	1.8917
0.01	100	1.6975
0.1	10	1.3342
1	1	0.7485
10	0.1	0.3545
100	0.01	0.0167
400	0.0025	0.0000

If the frequency f_u is the upper and f_1 the lower boundary of the range, then for the reduction of the measured σ for a given frequency interval it would be sufficient to determine the subtractions $\Delta I(\tau) = I(\tau_u) - I(\tau_1)$. They allow σ to be reduced to the frequency interval for which $\Delta I = 1$.
In order to determine the influence of meteorological conditions one should know the profile $C_n^2(h)$. Since it is unknown we have used the characteristics being local or common criteria of the turbulent state of the atmosphere. For them we have taken Richardson's gradient number Ri (S.S.Zilitinkevich, 1970), Monin's stratification parameter μ (S.S.Zilitinkevich, 1970) and the thickness Δh of the turbulent layer. For comparison with them we have searched for reduced dispersions:

$$\sigma_{01}^2 = \frac{\sigma^2 \cdot D^{-1/3}}{\sec Z \cdot \Delta I(\tau)} \qquad (6)$$

Then, by approximating, e.g., the dependence of σ_{01}^2 on Ri, we have

$$\sigma_{01}^2 = 0.045 + 0.0018/Ri \qquad (7)$$

Similar expressions have been derived also for the characteristics of μ and Δh.
Taking into account the above data, the procedure of calculating the amplitude of motion is as follows. Using the differences of temperature and wind speed at the two levels 1 and 4 m — $\Delta T(1-4)$ and $\Delta V(1-4)$ characteristics of turbulent behaviour of Ri, μ or Δh are determined. Then from

formulas (7) and (6), the amplitude of motion for the set aperture of the telescope, zenithal distance, frequency range and conditions of observation, are found. The procedure of the calculation of σ has been tested using observations of stars. The correlation coefficient of the measured and calculated values of σ made up $r = 0.84\pm0.03$.

3. DEPENDENCE OF THE RANDOM ERROR ON THE LENGTH OF OBSERVATION

Applying numerical methods and data on normalized spectrum from formula (3a) were found the values of m_τ/σ_∞ which are listed in Table 2 for the broad frequency range. In order to obtain directly the value m_τ one should know σ_∞. It is obtained from the formula:

$$\sigma_\infty^2 = \Delta I_\infty(\tau) \cdot \frac{\sigma_0^2}{\Delta I_0(\tau)} \qquad (8)$$

where $\Delta I_0(\tau)$ and $\Delta I_\infty(\tau)$ are subtractions of $I(\tau)$ functions corresponding to the amplitudes σ_0 and σ_∞.
It has been found that for the calculation of σ_∞ one should **practically take only the frequency range from 10^{-3} to 400 Hz**. At more high frequencies the amplitude of motion is very small, lower frequencies belong to the region of refraction motions which in the main are taken into account in astrometric determinations from meteorological data.

Table 2

τ, sec	m_τ/σ_∞	m_τ'' at $\sigma_\infty = 0''.45$ ($\sigma_0 = 0''.33$)
0.003	1.366	0''.615
0.01	1.342	0.604
0.1	1.173	0.528
1	0.957	0.431
10	0.633	0.285
100	0.349	0.157
1000	0.196	0.088
10000	0.110	0.050
100000	0.064	0.029

Since $\sigma_0 = 0''.33$, and from Table 1 $\Delta I_0(\tau) \approx 1.0$, $\Delta I_\infty(\tau) = 1.89$, we obtain $\sigma_\infty = 0''.45$. Table 2 by using this quantity lists the **values of m_τ as a function of observation time.**
Thus having the amplitude of motion of σ_0 measured or calculated by the above procedure, one can obtain the value m_τ and from formulas (1) and (2) the random errors of astrometric determinations applied to either program of observation.

The scheme just considered has been applied for several works. Thus for the Pulkovo photographic vertical circle (PVC) the estimate of determination of the zenithal distance was $\pm 0".161$, while its real value from 1975-76 observations is $\pm 0".18$ (for zenith).

4. CONCLUSIONS

Let us emphasize the main results of essential importance for astrometry.

a) From the dependence of the amplitude of motion on the behaviour of the aperture of the telescope it follows that **D=20 cm should be considered as the optimum objective diameter** of astrometric instruments. With further increase in D, the value σ decreases little. This conclusion confirms the correctness of choice of the value D for present instruments (e.g. PVC).

b) The optimum length of astrometric determinations (see **Table 2) on the average, is the 100 s interval (of about** 2 min). Continuous averaging of readings should be considered as the best variant of measurements. In discrete readings, it is more advantageous to cover a greater time interval of observations rather than to maximize the number of readings.

c) It has been found that the normalized spectral density of motions is some universal function of frequency since it practically is independent of D,Z and the conditions of observations. This implies that the character of the dependence of the random error of astrometric determinations on the time of observations will be a function but the value of error itself will be a function of astroclimatic conditions. One should search for places with minimum amplitude of image motion.

d) The developed procedure for calculating the amplitude of motion allows one to do estimation of the random error for specific instruments and conditions of observations. It has been found that the error due to atmospheric turbulence makes up about 37% contribution to the dispersion of the random error of a single measurement of coordinates with PVC and almost fully accounts for the random error of determination of clock correction for Pulkovo transit instruments.

The above results should be utilized in evaluating the length and the program of astrometric observations, in developing new instruments and methods of coordinate determinations and their automation.

References

Høg,E.:1968,Zeitschrift für Astroph., 69,pp.313-325.
Kolchinsky,I.G.:1967,Optical instability of the Earth atmo-

sphere from observations of stars. Kiev,"Naukova dumka".
Kolchinsky,I.G.:1973,Astrometriya i astrofizika, 20, pp.19-39.
Lambeck,K.:1968,SAO special report, 269,pp.1-27.
Tatarsky,V.I.:1967,Propagation of waves in a turbulent atmosphere. M.,"Nauka".
Teleki,G.:1967,Proceedings of the 17 Astron. Conf. USSR. L.,"Nauka", pp.101-106.
Zilitinkevich,S.S.:1970,Atmosphere boundary layer dynamics. L.,Gidrometeoizdat.

ON THE CALCULATION OF REFRACTION IN MODEL ATMOSPHERES

J. Saastamoinen
National Research Council of Canada

ABSTRACT

Approximations have been removed from the derivation of the coefficients in the binomial expansion for astronomical refraction, $\Delta z_1 = \tan z_1 (Y_0 - \frac{1}{2} Y_1 \sec^2 z_1 + \frac{3}{8} Y_2 \sec^4 z_1 - \ldots)$, allowing the calculation of any number of terms to any precision desired. The range of the refraction formula has been extended to greater zenith distances ($<90°$) by inserting a damping factor into the binomial formula, truncating the expansion at a proper point, and rearranging the terms. Another, computer-manipulated series has been developed for zenith distances at or near the horizon. Further applications include the calculation of photogrammetric and parallactic refractions, as well as range corrections in satellite geodesy.

NOTATION

The notation has been taken from an earlier work (Saastamoinen, 1972-73) on the same subject matter, although important changes and additions have been made to facilitate the derivation of general formulas not presented before. The following symbols appear in the text without explanation:

β	vertical gradient of temperature
e	base of natural logarithms
g	intensity of gravity
i, k	$0, 1, 2, \ldots$
j	$1, 2, 3, \ldots$
n	refractive index
r	radius vector
R	gas constant of dry air
T	absolute temperature
z	zenith distance
$n_1, r_1,$ etc.	values at the point of observation

r_a, r_b, etc.   values at the lower and upper limits, respectively, of an atmospheric layer with β const.

The same symbol may occasionally be used in different meanings (e also for partial pressure of water vapor) if confusion is unlikely and the notation is familiar from geodetic and meteorological literature .

1. ASTRONOMICAL REFRACTION

1.1 Introduction.

For the evaluation of astronomical refraction

$$\Delta z_1 = \int_0^{\log_e n_1} \tan z \, d\log_e n \tag{1}$$

in a spherically symmetric model atmosphere, the law of refraction

$$y \sin z = \sin z_1 \tag{2}$$

where

$$y = nr/(n_1 r_1)$$

gives

$$\tan z = \tan z_1 [1 + (y^2-1)\sec^2 z_1]^{-\frac{1}{2}}$$

$$= \tan z_1 \left[1 + (-)^k \sum_{k=1}^{\infty} \frac{1 \cdot 3 \cdot 5 \ldots (2k-1)}{2 \cdot 4 \cdot 6 \ldots (2k)} (y^2-1)^k \sec^{2k} z_1 \right]. \tag{3}$$

Binomial series (3) is convergent if $(y^2-1)\sec^2 z_1 < 1$.

Providing that the condition for convergence is satisfied, the contribution to the astronomical refraction of a layer of air between radii vectors r_a and r_b is consequently

$$\Delta z_1(a;b) = \tan z_1 (Y_0 - \frac{1}{2} Y_1 \sec^2 z_1 + \frac{3}{8} Y_2 \sec^4 z_1 - \frac{5}{16} Y_3 \sec^6 z_1 + \ldots) \tag{4}$$

where the coefficients

$$Y_k = \int_{\log_e n_b}^{\log_e n_a} (y^2 - 1)^k d\log_e n \tag{5}$$

are functions of the model atmosphere employed. Total astronomical refraction (1) is the sum of the contributions from all the layers between the point of observation and the top of the atmosphere.

Power series (4), or the variant

$$\Delta z_1(a;b) = \tan z_1(Y_0' - \frac{1}{2} Y_1' \tan^2 z_1 + \frac{3}{8} Y_2' \tan^4 z_1 - \frac{5}{16} Y_3' \tan^6 z_1 + \ldots) \quad (4')$$

for which binomial series

$$\tan z = \frac{1}{y} \tan z_1 \left[1 + \frac{1}{y^2}(y^2-1)\tan^2 z_1 \right]^{-\frac{1}{2}}$$

$$= \tan z_1 \left[\frac{1}{y} + (-)^k \sum_{k=1}^{\infty} \frac{1 \cdot 3 \cdot 5 \ldots (2k-1)}{2 \cdot 4 \cdot 6 \ldots (2k)} \frac{1}{y^{2k+1}} (y^2-1)^k \tan^{2k} z_1 \right] \quad (3')$$

gives the coefficients

$$Y_k' = \int_{\log_e n_b}^{\log_e n_a} (1/y)^{2k+1} (y^2-1)^k d\log_e n, \quad (5')$$

is the common foundation of most formulas proposed for the calculation of astronomical refraction (Teleki, 1974). Essential differences are found, however, in the way of calculation of the coefficients.

1.2 Calculation of coefficients Y_k in terms of atmospheric integrals.

Multiplication of the identity,

$$y^2 - 1 = \frac{2}{n_1^2 r_1}(r - r_1) - \frac{2}{n_1}(n_1 - n) + \frac{1}{n_1^2 r_1^2}(r - r_1)^2$$

$$+ \frac{4}{n_1^2 r_1}(n-1)(r-r_1) + \frac{1}{n_1^2}(n_1-n)^2 + \frac{2}{n_1^2 r_1^2}(n-1)(r-r_1)^2$$

$$+ \frac{2}{n_1^2 r_1}(n-1)^2(r-r_1) + \frac{1}{n_1^2 r_1^2}(n-1)^2(r-r_1)^2,$$

by

$$d\log_e n = dn/n = dn - (n-1)dn + (n-1)^2 dn - (n-1)^3 dn + \ldots$$

gives

$$(y^2-1)d\log_e n = \frac{2}{n_1^2 r_1}(r-r_1)dn - \frac{2}{n_1}(n_1-n)d\log_e n$$

$$+ \frac{1}{n_1^2 r_1^2}(r-r_1)^2 dn + \frac{2}{n_1^2 r_1}(n-1)(r-r_1)dn + \frac{1}{n_1^2}(n_1-n)^2 d\log_e n$$

$$+ \frac{1}{n_1^2 r_1^2}(n-1)(r-r_1)^2 dn. \tag{6}$$

It is easy to verify that equation (6) is also a mathematical identity.

We can now write, in terms of atmospheric integrals given by the definitions

$$P(i,j) = \frac{1}{r_1^j} \int_{n_b}^{n_a} (n-1)^i (r-r_1)^j dn \tag{7}$$

$$P'(i,j) = P(i,j) + \frac{j}{i+1} P(j-1, i+1) \tag{7'}$$

and

$$Q(i) = \int_{\log_e n_b}^{\log_e n_a} (n_1 - n)^i d\log_e n, \tag{8}$$

the formulas

$$Y_0 = Q(0) \tag{9}$$

$$n_1^2 Y_1 = 2 P(0,1) - 2n_1 Q(1) + P'(0,2) + Q(2) + P(1,2) \tag{10}$$

for the calculation of the first two coefficients in series expansion (4).

Similarly, by actual multiplication, is obtained the identity

$$(y^2-1)^2 d\log_e n = \frac{4}{n_1^4 r_1}(1 - n_1^2)(r - r_1)dn + \frac{2}{n_1^4 r_1^2}(3 - n_1^2)(r - r_1)^2 dn$$

$$+ \frac{4}{n_1^4 r_1}(3 - n_1^2)(n-1)(r-r_1)dn + \frac{4}{n_1^2}(n_1 - n)^2 d\log_e n + \frac{4}{n_1^4 r_1^3}(r - r_1)^3 dn$$

$$+ \frac{2}{n_1^4 r_1^2}(9 - n_1^2)(n-1)(r-r_1)^2 dn + \frac{12}{n_1^4 r_1}(n-1)^2(r-r_1)dn$$

$$- \frac{4}{n_1^3}(n_1 - n)^3 d\log_e n + \frac{1}{n_1^4 r_1^4}(r - r_1)^4 dn + \frac{12}{n_1^4 r_1^3}(n-1)(r-r_1)^3 dn$$

$$+ \frac{18}{n_1^4 r_1^2}(n-1)^2(r-r_1)^2 dn + \frac{4}{n_1^4 r_1}(n-1)^3(r-r_1)dn + \frac{1}{n_1^4}(n_1 - n)^4 d\log_e n$$

$$+ \frac{3}{n_1^4 r_1^4}(n-1)(r-r_1)^4 dn + \frac{12}{n_1^4 r_1^3}(n-1)^2(r-r_1)^3 dn$$

$$+ \frac{6}{n_1^4 r_1^2}(n-1)^3(r-r_1)^2 dn + \frac{3}{n_1^4 r_1^4}(n-1)^2(r-r_1)^4 dn$$

$$+ \frac{4}{n_1^4 r_1^3}(n-1)^3(r-r_1)^3 dn + \frac{1}{n_1^4 r_1^4}(n-1)^3(r-r_1)^4 dn \tag{11}$$

which gives the formula

$$n_1^4 Y_2 = 4(1-n_1^2)P(0,1) + 2(3-n_1^2)P'(0,2) + 4n_1^2 Q(2)$$
$$+ 4P'(0,3) + 2(9-n_1^2)P(1,2) - 4n_1 Q(3) + P'(0,4)$$
$$+ 12P'(1,3) + Q(4) + 3P'(1,4) + 12P(2,3)$$
$$+ 3P'(2,4) + P(3,4) \qquad (12)$$

for the third coefficient in series expansion (4).

It can be shown that, for any positive integer k, coefficient $n_1^{2k} Y_k$ formed in this way will consist of a sum of $1 + 2k(1+k)$ terms, as follows:

i. The terms containing integrals $Q(i)$ are

$$(-)^k \left[(2n_1)^k Q(k) - k(2n_1)^{k-1} Q(k+1) + \frac{k(k-1)}{2!}(2n_1)^{k-2} Q(k+2) \right.$$
$$\left. - \frac{k(k-1)(k-2)}{3!}(2n_1)^{k-3} Q(k+3) + \ldots (-)^k Q(2k) \right].$$

The total number of these terms is $k+1$.

ii. Each coefficient $n_1^{2k} Y_k$ contains the odd integrals

$$k(k-1)\ldots(k-j+2)(k-j+1) \left[\frac{2^j}{(2j-1)!}(2k-1)(2k-3)\ldots(2k-2j+5)(2k-2j+3) \right.$$
$$- \frac{2^{j-1}}{1!(2j-3)!}(2k-3)(2k-5)\ldots(2k-2j+5)(2k-2j+3) n_1^2$$
$$+ \frac{2^{j-2}}{2!(2j-5)!}(2k-5)(2k-7)\ldots(2k-2j+5)(2k-2j+3) n_1^4$$
$$- \frac{2^{j-3}}{3!(2j-7)!}(2k-7)(2k-9)\ldots(2k-2j+5)(2k-2j+3) n_1^6$$
$$\left. + \ldots (-)^{j-1} \frac{2^1}{(j-1)!1!} n_1^{2j-2} \right] (1-n_1^2)^{k-2j+1} P(0,2j-1)$$

and the even integrals

$$k(k-1)\ldots(k-j+2)(k-j+1) \left[\frac{2^j}{(2j)!}(2k-1)(2k-3)\ldots(2k-2j+3)(2k-2j+1) \right.$$
$$- \frac{2^{j-1}}{1!(2j-2)!}(2k-3)(2k-5)\ldots(2k-2j+3)(2k-2j+1) n_1^2$$
$$+ \frac{2^{j-2}}{2!(2j-4)!}(2k-5)(2k-7)\ldots(2k-2j+3)(2k-2j+1) n_1^4$$
$$- \frac{2^{j-3}}{3!(2j-6)!}(2k-7)(2k-9)\ldots(2k-2j+3)(2k-2j+1) n_1^6$$
$$\left. + \ldots (-)^j \frac{2^0}{j!0!} n_1^{2j} \right] (1-n_1^2)^{k-2j} P(0,2j).$$

There will be k terms of each kind, or a total of 2k terms.

If the multipliers of integrals $P(0,2j)$ and $P(0,2j-1)$ are written out as polynomials, in ascending powers of n_1^2, it will be found that their corresponding terms, taken in the order from left to right, have the ratios $(1/2j)(2k-2j+1)$, $(1/2j)(2k-2j-1)$, ..., $(3/2j)$, and $(1/2j)$, respectively. This will be shown by the symbolic notation

$$P(0,2j) = \frac{1}{2j}(2k-2j+1, 2k-2j-1, \ldots, 3, 1)P(0,2j-1) \qquad (k \geq j)$$

the numbers in parentheses indicating term-by-term multiplication of the polynomial coefficient of $P(0,2j-1)$ in order to form that of $P(0,2j)$. As also holds

$$P(0,2j+1) = (\frac{2}{2j+1})(k-j, k-j-1, \ldots, 1, 0)P(0,2j) \qquad (k > j)$$

all the terms given by the direct formulas can be calculated in succession from the first one, $2k(1-n_1^2)^{k-1}P(0,1)$, the latter developed into a polynomial by the aid of the binomial theorem.

iii. As y is a symmetric function of n and r, numerous relationships exist that aid the calculation of further terms. Using the symbolic notation, previously introduced, we have

for $k \geq j > i$, a total of $\frac{1}{2}k(k+1)$ terms

$$P(2i+1, 2j) = (\frac{1}{2i+1})(2k-2i-1, 2k-2i-3, \ldots, 2j-2i+1, 2j-2i-1)P(2i, 2j)$$

for $k > j > i$, a total of $\frac{1}{2}k(k-1)$ terms

$$P(2i+1, 2j+1) = (\frac{1}{2i+1})(2k-2i-1, 2k-2i-3, \ldots, 2j-2i+1)P(2i, 2j+1)$$

for $k \geq j > i > 0$, a total of $\frac{1}{2}k(k-1)$ terms

$$P(2i, 2j) = \frac{1}{i}(k-i, k-i-1, \ldots, j-i+1, j-i)P(2i-1, 2j)$$

and for $k \geq j+1 > i > 0$, a total of $\frac{1}{2}k(k-1)$ terms

$$P(2i, 2j+1) = \frac{1}{i}(k-i, k-i-1, \ldots, j-i+2, j-i+1)P(2i-1, 2j+1)$$

which altogether add $k(2k-1)$ terms to those previously determined.

iv. All the remaining terms are combinative; they will be included simply by priming the integrals $P(i,j)$, excepting those of the form $P(i,i+1)$, which do not possess a counterpart.

We shall not dwell with all the arguments needed in a rigorous proof of these rules by double induction; let it suffice here to show their general validity in a single instance, say, for integral $P(2,3)$.

Successive application of the symbolic equations, starting from the last one, gives

$$P(2,3) = \frac{1}{3}[(2k-1)^2(k-1)^2,(2k-3)^2(k-2)^2,\ldots,9,0]P(0,1).$$

We know already, from equation (12), that this equality is true if $k=2$. It remains to be shown that if the equality is true for $k=p$, it is also true for $k=p+1$.

To form the polynomial multiplier of $P(2,3)$ in the coefficient $n_1^{2p+2}Y_{p+1}$ it is convenient to use a scheme

	$n_1^2(y^2-1) =$
$P(2,2) = \frac{1}{2}[(2p-1)^2(p-1),(2p-3)^2(p-2),\ldots,9,0]P(0,1)$	$(2/r_1)(r-r_1)$
$P(2,1) = [(2p-1)(p-1),(2p-3)(p-2),\ldots,3,0]P(0,1)$	$+(1/r_1^2)(r-r_1)^2$
$P(1,2) = \frac{1}{2}[(2p-1)^2,(2p-3)^2,\ldots,9,1]P(0,1)$	$+(4/r_1)(n-1)(r-r_1)$
$P(1,1) = (2p-1,2p-3,\ldots,3,1)P(0,1)$	$+(2/r_1^2)(n-1)(r-r_1)^2$
$P(0,2) = \frac{1}{2}(2p-1,2p-3,\ldots,3,1)P(0,1)$	$+(2/r_1)(n-1)^2(r-r_1)$
$P(0,1) = (1,1,\ldots,1,1)P(0,1)$	$+(1/r_1^2)(n-1)^2(r-r_1)^2$
$P(0,3) = \frac{1}{3}[(2p-1)(p-1),(2p-3)(p-2),\ldots,3,0]P(0,1)$	$+(n-1)^2$
$P(1,3) = \frac{1}{3}[(2p-1)^2(p-1),(2p-3)^2(p-2),\ldots,9,0]P(0,1)$	$+2(n-1)$
$P(2,3) = \frac{1}{3}[(2p-1)^2(p-1)^2,(2p-3)^2(p-2)^2,\ldots,100,9,0]P(0,1)$	$+(1-n_1^2)$

which displays the identity for y^2-1, slightly modified from the form given previously, together with a set of symbolic equations derived by the rules we assume valid. It is now easy to comprehend that the required polynomial is obtained by multiplying each term of the identity into the equation on the left, and adding up the results. These calculations, shown in a separate table, establish the polynomial in the form

$$P(2,3) = \frac{2}{3} p(p+1)[p(2p+1)^2,(p-1)(2p-1)^2,\ldots,50,9](1-n_1^2)^{p-1}$$

which is found equivalent to

$$P(2,3) = \frac{1}{3}[p^2(2p+1)^2,(p-1)^2(2p-1)^2,\ldots,9,0](2p+2)(1-n_1^2)^p$$

if both formulas are expanded using the binomial theorem. Consequently, the original equality is true for $k=p+1$.

1.3 Calculation of atmospheric integrals $P(i,j)$ and $Q(i)$.

The concept of the atmospheric integrals was introduced in an earlier work (Saastamoinen, 1972-73), where formulas for some of the integrals $P(i,j)$ have been derived on the basis of equations

Table 1-1. Calculation of the multiplier of integral $P(2,3)$ in $n_1^{2p+2} Y_{p+1}$

$(2p-1)^2(p-1)$,	$(2p-3)^2(p-2)$,	............,9,		0
$(2p-1)(p-1)$,	$(2p-3)(p-2)$,	............,3,		0
$2(2p-1)^2$,	$2(2p-3)^2$,	............,18,		2
$2(2p-1)$,	$2(2p-3)$,	............,6,		2
$2p-1$,	$2p-3$,	............,3,		1
1,	1,	............,1,		1
$\frac{1}{3}(2p-1)(p-1)$,	$\frac{1}{3}(2p-3)(p-2)$,	............,1,		0
$\frac{2}{3}(2p-1)^2(p-1)$,	$\frac{2}{3}(2p-3)^2(p-2)$,	............,6,		0
$\frac{1}{3}(2p-1)^2(p-1)^2$,	$\frac{1}{3}(2p-3)^2(p-2)^2$,	............,3,		0
0,	$\frac{1}{3}(2p-1)^2(p-1)^2(\frac{1}{p-1})$,	......,$\frac{100}{3}(\frac{p-2}{2!})$,		$3(\frac{p-1}{1})$

Sum: $\{\frac{1}{3}p(p+1)(2p+1)^2,\ \frac{1}{3}(p-1)(p+1)(2p-1)^2, \ldots\ldots,\ \frac{50}{3}(p+1),\ 3(p+1)\}\ P(0,1)$

$$n - 1 = (n_a - 1)\left(\frac{T}{T_a}\right)^{-\frac{g}{R\beta} - 1} \qquad (\beta \neq 0) \qquad (13a)$$

$$n - 1 = (n_a - 1)e^{-\frac{g}{RT}(r - r_a)} \qquad (\beta = 0) \qquad (13b)$$

in an atmosphere consisting of two layers, the troposphere and the stratosphere. We shall no longer restrict the number of layers that may be taken into the atmospheric model, otherwise retaining the various assumptions implicit in equations (13a) and (13b).

In the derivation of a general formula for integrals $P(i,j)$ it is best to take either

$$H = RT/g$$

or

$$h = r - r_1$$

as the independent variable in terms of which all the other quantities are expressed. Setting for brevity

$$m' = -\frac{g}{R\beta} - 1$$

we have then, for $\beta \neq 0$,

$$n-1 = (n_a - 1)H_a^{-m'}H^{m'} = (n_a - 1)(m'+1)^{-m'}H_a^{-m'}[h_a + (m'+1)H_a - h]^{m'}$$

$$dn = (n_a - 1)m'H_a^{-m'}H^{m'-1}dH = -(n_a - 1)m'(m'+1)^{-m'}H_a^{-m'}[h_a + (m'+1)H_a - h]^{m'-1}dh$$

$$r - r_1 = h_a + (m'+1)(H_a - H) = h_b - (m'+1)(H - H_b) = h$$

and

$$r_1{}^j P(i-1,j) = (n_a - 1)^i m' H_a^{-im'} \int_{H_b}^{H_a} [h_a + (m'+1)H_a - (m'+1)H]^j H^{im'-1} dH \quad (i \neq 0)$$

$$= (n_a - 1)^i m' (m'+1)^{-im'} H_a^{-im'} \int_{h_a}^{h_b} [h_a + (m'+1)H_a - h]^{im'-1} h^j dh$$

The solution of either integral is given by the formula

$$r_1{}^j P(i-1,j) = \int_{n_b}^{n_a} (n-1)^{i-1}(r-r_1)^j dn$$

$$\quad (i \neq 0)$$

$$= \frac{j!}{0!i^j}(im')_j (H_a{}^j A_i - H_b{}^j B_i) + \frac{j!}{1!i^{j-1}}(im')_{j-1}(H_a{}^{j-1}h_a A_i - H_b{}^{j-1}h_b B_i)$$

$$+ \frac{j!}{2!i^{j-2}}(im')_{j-2}(H_a{}^{j-2}h_a{}^2 A_i - H_b{}^{j-2}h_b{}^2 B_i)$$

$$+ \ldots + \frac{j}{i}(im')_1 (H_a h_a{}^{j-1} A_i - H_b h_b{}^{j-1} B_i) + h_a{}^j A_i - h_b{}^j B_i \quad (14)$$

where

$$A_i = \frac{1}{i}(n_a - 1)^i \qquad B_i = \frac{1}{i}(n_b - 1)^i$$

and

$$\frac{im'+i}{im'+1} = (im')_1, \quad \left(\frac{im'+i}{im'+2}\right)(im')_1 = (im')_2, \quad \ldots, \quad \left(\frac{im'+i}{im'+j}\right)(im')_{j-1} = (im')_j.$$

Because

$$\lim_{\beta \to 0} (im')_j = 1,$$

it is evident that equation (14) is also valid if $\beta = 0$, in which case the factors containing m' are simply left out.

For integrals $Q(i)$, the immediate solution is

$$Q(i) = \int_{n_b}^{n_a} \frac{1}{n}(n_1 - n)^i dn$$

$$= n_1^i \log_e(n_a/n_b) - i n_1^{i-1}(n_a - n_b) + \frac{1}{2}\frac{i(i-1)}{2!} n_1^{i-2}(n_a^2 - n_b^2)$$

$$- \frac{1}{3}\frac{i(i-1)(i-2)}{3!} n_1^{i-3}(n_a^3 - n_b^3) + \ldots (-)^i \frac{1}{i}(n_a^i - n_b^i).$$

This expression may be transformed, setting

$$A = n_a - n_b \qquad C = \frac{1}{2}(n_a + n_b)$$

and

$$\log_e(n_a/n_b) = (A/C) + \frac{1}{2^2 \cdot 3}(A/C)^3 + \frac{1}{2^4 \cdot 5}(A/C)^5 + \ldots,$$

into the following formula suitable for numerical evaluation,

$$Q(i) = (n_1 - C)^i (A/C)$$

$$+ \frac{1}{2^2 \cdot 3}\left[n_1^2 + (i-2)n_1 C + \frac{1}{2!}(i-1)(i-2)C^2\right](n_1-C)^{i-2}(A/C)^3$$

$$+ \frac{1}{2^4 \cdot 5}\left[n_1^4 + (i-4)n_1^3 C + \frac{1}{2!}(i-3)(i-4)n_1^2 C^2 + \frac{1}{3!}(i-2)(i-3)(i-4)n_1 C^3 \right.$$

$$\left. + \frac{1}{4!}(i-1)(i-2)(i-3)(i-4)C^4\right](n_1 - C)^{i-4}(A/C)^5$$

$$+ \frac{1}{2^6 \cdot 7}\left[n_1^6 + (i-6)n_1^5 C + \ldots + \frac{1}{6!}(i-1)(i-2)\ldots(i-6)C^6\right](n_1-C)^{i-6}(A/C)^7$$

$$+ \ldots, \tag{15}$$

where i may take any value ($i = 0, 1, 2, \ldots$).

1.4 Extension of the range of the refraction formula.

We shall now go back to the binomial expansion for $\tan z$, and replace equation (3) by

$$\tan z = f \tan z_1 [1 + f^2(y^2-1)\sec^2 z_1 - (1-f^2)]^{-\frac{1}{2}}$$

$$= f \tan z_1 \left\{1 + (-)^k \sum_{k=1}^{\infty} \frac{1 \cdot 3 \cdot 5 \ldots (2k-1)}{2 \cdot 4 \cdot 6 \ldots (2k)}[f^2(y^2-1)\sec^2 z_1 - (1-f^2)]^k\right\}$$

$$\tag{16}$$

where f stands for a positive number less than 1. This series is convergent if $|f^2(y^2-1)\sec^2 z - (1-f^2)| < 1$; consequently, it can be made convergent for any zenith distance $z_1 < 90°$ by taking f sufficiently small.

The idea behind this arrangement is to truncate series (16), after k+1 terms, whereupon it is permissible to restore the original order of terms in ascending powers of y^2-1. The contribution to the astronomical refraction of a layer of air between radii vectors r_a and r_b then becomes

$$\Delta z_1(a;b) = \tan z_1 \left[F_0 Y_0 - \frac{1}{2} F_1 Y_1 \sec^2 z_1 + \frac{3}{8} F_2 Y_2 \sec^4 z_1 - \frac{5}{16} F_3 Y_3 \sec^6 z_1 \right.$$
$$\left. + \ldots (-)^k \frac{1.3\ldots(2k-1)}{2.4\ldots(2k)} F_k Y_k \sec^{2k} z_1 \right] + R_{k+1} \quad (17)$$

where

$$F_0 = f \left\{ 1 + \frac{1}{2}(1-f^2) + \ldots + \frac{1.3\ldots(2k-1)}{2.4\ldots(2k)}(1-f^2)^k \right\},$$

$$F_1 = f^3 \left\{ 1 + \frac{3}{2}(1-f^2) + \ldots + \frac{3.5\ldots(2k-1)}{2.4\ldots(2k-2)}(1-f^2)^{k-1} \right\},$$

$$F_2 = f^5 \left\{ 1 + \frac{5}{2}(1-f^2) + \ldots + \frac{5.7\ldots(2k-1)}{2.4\ldots(2k-4)}(1-f^2)^{k-2} \right\},$$

$$\ldots\ldots\ldots\ldots, \quad (18)$$

$$F_{k-1} = f^{2k-1} \left\{ 1 + \frac{2k-1}{2}(1-f^2) \right\},$$

$$F_k = f^{2k+1}.$$

Formula (17) is valid if

$$f^2 < \frac{2}{(y_b^2-1)\sec^2 z_1 + 1}$$

but in order to keep remainder R_{k+1} small, without the necessity of including an excessive number of terms, f should be chosen so that

$$\left| f^2 - \frac{1}{(y^2-1)\sec^2 z_1 + 1} \right| = \min. \quad (19)$$

considering all the values of y and z_1 involved.

With given numerical values of f and k, equations (18) provide a set of <u>damping factors</u>, $F_0, F_1, \ldots, F_k$, by which the coefficients of the first k+1 terms of refraction formula (4) are multiplied.

Upon truncation of series (16), the remaining terms of tan z are

$$(-)^{k+1} \frac{1.3.5\ldots(2k+1)}{2.4.6\ldots(2k+2)} f \tan z_1 \left[1 - \left(\frac{2k+3}{2k+4}\right) V + \frac{(2k+3)(2k+5)}{(2k+4)(2k+6)} V^2 \right.$$
$$\left. - \frac{(2k+3)(2k+5)(2k+7)}{(2k+4)(2k+6)(2k+8)} V^3 + \ldots \right] V^{k+1}$$

where

$$V = f^2(y^2-1)\sec^2 z_1 - (1-f^2).$$

This gives the approximate formula

$$R_{k+1} \sim (-)^{k+1} \frac{1.3.5...(2k+1)}{2.4.6...(2k+2)} f \tan z_1 \int_{n_b}^{n_a} \frac{v^{k+1}}{1 + \frac{2k+3}{2k+4} v} dn \qquad (20)$$

with

$$dn = -(\frac{m'}{m'+1})(\frac{n-1}{H})dr$$

for the evaluation of the magnitude of remainder R_{k+1} by numerical integration. This evaluation is necessary because the (k+1)th term of formula (17), unlike that of equation (4), does not give an indication of the accuracy achieved.

If $\tan z_1$ is numerically large, we may choose to substitute $f \cos z_1$ for f in equation (16), and consider the series

$$\tan z = f \sin z_1 [1 + f^2(y^2 - 1) - (1 - f^2\cos^2 z_1)]^{-\frac{1}{2}} \qquad (16')$$
$$= f \sin z_1 \{1 + (-)^k \sum_{k=1}^{\infty} \frac{1.3.5...(2k-1)}{2.4.6...(2k)} [f^2(y^2-1) - (1-f^2\cos^2 z_1)]^k\}$$

where f now stands for a positive number greater than 1. The contribution to the astronomical refraction of a layer of air between radii vectors r_a and r_b then becomes

$$\Delta z_1(a;b) = \sin z_1 \Big[F_0 Y_0 - \frac{1}{2} F_1 Y_1 + \frac{3}{8} F_2 Y_2 - \frac{5}{16} F_3 Y_3$$
$$+ ...(-)^k \frac{1.3.5...(2k-1)}{2.4.6...(2k)} F_k Y_k \Big] + R_{k+1} \qquad (17')$$

where the damping factors

$$F_0 = f\{1 + \frac{1}{2}(1-f^2\cos^2 z_1) + ... + \frac{1.3...(2k-1)}{2.4...(2k)}(1-f^2\cos^2 z_1)^k\}$$

$$F_1 = f^3\{1 + \frac{3}{2}(1-f^2\cos^2 z_1) + ... + \frac{3.5...(2k-1)}{2.4...(2k-2)}(1-f^2\cos^2 z_1)^{k-1}\}$$

$$F_2 = f^5\{1 + \frac{5}{2}(1-f^2\cos^2 z_1) + ... + \frac{5.7...(2k-1)}{2.4...(2k-4)}(1-f^2\cos^2 z_1)^{k-2}\}$$

$$\qquad (18')$$
....

$$F_{k-1} = f^{2k-1}\{1 + \frac{2k-1}{2}(1-f^2\cos^2 z_1)\}$$

$$F_k = f^{2k+1}$$

may be written out as polynomials of $\cos^2 z_1$. Series (16') is convergent if

$$f^2 < \frac{2}{y_b^2 - \sin^2 z_1}$$

but again, f should be chosen so that

$$\left|f^2 - \frac{1}{y^2 - \sin^2 z_1}\right| = \min. \tag{19'}$$

for all the values of y and z_1 involved. Remainder R_{k+1} is evaluated by numerical integration using the formulas

$$W = f^2(y^2 - 1) - (1 - f^2\cos^2 z_1)$$

$$dn = -\left(\frac{m'}{m'+1}\right)\left(\frac{n-1}{H}\right)dr$$

and

$$R_{k+1} \sim (-)^{k+1} \frac{1\cdot 3\cdot 5\cdots(2k+1)}{2\cdot 4\cdot 6\cdots(2k+2)} f \sin z_1 \int_{n_b}^{n_a} \frac{W^{k+1}}{1 + \frac{2k+3}{2k+4}W} dn. \tag{20'}$$

If $z_1 = 90°$, equation (17) can not be applied unless the point of observation is moved along the light ray to the base (r_a) of a higher layer, where

$$\tan z_a = \{[n_a r_a/(n_1 r_1 \sin z_1)]^2 - 1\}^{-\frac{1}{2}}$$

has a suitable value. The refraction component for the nearly horizontal section of the light ray may be calculated using the series expansions in 1.5.

1.5 Calculation of astronomical refraction at or near the horizon.

We shall now consider the integral

$$\Delta z_1(a;b) = \sin z_a \int_{\log_e n_b}^{\log_e n_a} (\eta^2 - \sin^2 z_a)^{-\frac{1}{2}} d\log_e n \tag{21}$$

where

$$\eta = nr/(n_a r_a)$$

in the calculation of astronomical refraction for any zenith distance $0° \leq z_a \leq 90°$, especially for $z_a = 90°$. In the solution given below, the integrand will be expressed in the form of a convergent series, and integrated term by term.

Let

$$x = (r - r_a)/H_a$$

be chosen as the independent variable. We have then, from equation (13a),

$$(n - 1)/(n_a - 1) = (1 - \frac{x}{m'+1})^{m'}$$
$$= 1 - m_1 x + m_1 m_2 x^2 - m_1 m_2 m_3 x^3 + \ldots$$

and

$$n/n_a = 1 - a_0 m_1 x + a_0 m_1 m_2 x^2 - a_0 m_1 m_2 m_3 x^3 + \ldots \qquad (22)$$

with

$$a_0 = (n_a - 1)/n_a; \quad m_j = \frac{1}{j} - \frac{1}{m'+1} \quad (\text{or } m_j = \frac{1}{j}, \text{ if } \beta = 0).$$

The derivative of series (22)

$$\frac{dn}{dx}/n_a = -a_0 m_1 + 2 a_0 m_1 m_2 x - 3 a_0 m_1 m_2 m_3 x^2 + \ldots$$

divided by the series itself gives

$$d\log_e n = -a_0 m_1 [1 - (2m_2 - a_0 m_1)x + (3m_2 m_3 - 3a_0 m_1 m_2 + a_0^2 m_1^2)x^2$$
$$- (4m_2 m_3 m_4 - 4a_0 m_1 m_2 m_3 - 2a_0 m_1 m_2^2 + 4a_0^2 m_1^2 m_2 - a_0^3 m_1^3)x^3 + \ldots]dx. \qquad (23)$$

Multiplication of series (22) by the binom

$$r/r_a = 1 + (H_a/r_a)x = 1 + b_0 x$$

further gives

$$\eta = 1 + (b_0 - a_0 m_1)x + a_0 m_1 (m_2 - b_0)x^2 - a_0 m_1 m_2 (m_3 - b_0)x^3 + \ldots$$

whence

$$\eta^2 - \sin^2 z_a = \cos^2 z_a + 2(b_0 - a_0 m_1)x + 2(a_0 m_1 m_2 - 2a_0 m_1 b_0$$
$$+ \frac{1}{2} a_0^2 m_1^2 + \frac{1}{2} b_0^2)x^2 - 2a_0 m_1 (m_2 m_3 - 2b_0 m_2 + a_0 m_1 m_2 - a_0 m_1 b_0 + b_0^2)x^3 + \ldots$$
$$= \cos^2 z_a + c_1 x + c_2 x^2 - c_3 x^3 + \ldots \qquad (24)$$

The remaining part of the calculation consists of the extraction of the inverse square root of series (24), multiplication by differential (23), and integration of the product according to equation (21).

If $z_a = 90°$, we have $\sin^2 z_a = 1$, $\cos^2 z_a = 0$

$$\eta^2 - 1 = c_1 x(1 + \frac{c_2}{c_1} x - \frac{c_3}{c_1} x^2 + \ldots) = c_1 x(1 + X)$$

$$(\eta^2 - 1)^{-\frac{1}{2}} = \frac{1}{\sqrt{c_1}} x^{-\frac{1}{2}} (1 - \frac{1}{2} X + \frac{3}{8} X^2 - \frac{5}{16} X^3 + \ldots) \qquad (X < 1)$$

The integrand is obtained in the form

$$(\eta^2 - 1)^{-\frac{1}{2}} d\log_e n = - \frac{a_0 m_1}{\sqrt{c_1}} x^{-\frac{1}{2}} (1 - d_1 x + d_2 x^2 - \ldots) dx$$

with

$$d_1 = 2m_2 - a_0 m_1 + \frac{1}{2}(c_2/c_1)$$

$$d_2 = 3m_2(m_3 - a_0 m_1) + \frac{1}{2c_1}(2c_2 m_2 + c_3 - a_0 c_2 m_1) + \frac{3}{8}(c_2/c_1)^2 + a_0^2 m_1^2$$

....

which gives

$$\Delta z_1(a;b) = \frac{2a_0 m_1}{\sqrt{c_1}} \sqrt{x_b} (1 - \frac{1}{3} d_1 x_b + \frac{1}{5} d_2 x_b^2 - \ldots) \quad (25)$$

as the final result.

If $\cos^2 z_a$ is numerically small, the extraction of the inverse square root succeeds similarly if we first find (by iteration) a number, x_0, such that

$$\cos^2 z_a - c_1 x_0 + c_2 x_0^2 + c_3 x_0^3 + \ldots = 0$$

and substitute a new variable $w = x + x_0$ that eliminates the constant term in equation (24). This procedure, which is equivalent to extending the light ray to its lowest point where the tangent line is horizontal, also finds application in the calculation of refraction if $z_a > 90°$.

(to be continued)

REFERENCES

Saastamoinen, J.: 1972-73, "Contributions to the theory of atmospheric refraction", Bulletin Géodésique, Nos. 105-107

Teleki, G.: (edited by) "The present state and future of the astronomical refraction investigations". Proceedings of the Study Group on Astronomical Refraction of the International Astronomical Union Commission 8, Belgrade, 1974.

APPENDIX A

Numerical applications

A few illustrative samples are given on the calculation of astronomical refraction in a spherically symmetric model (Atmospheric Model No. 2, (Saastamoinen, 1972-73)) specified as follows:

$r_1 = 6380$ km $\quad$ $T_1 = 285.08$ K $\quad$ $n_1 = 1.000280868$

Troposphere $0 - 10.4$ km: $\quad$ $\beta = -6.45$ K.km^{-1} $\quad$ $R/g = (2.8704/98)$ km.K^{-1}

Stratosphere $10.4 - \infty$ km: $\quad$ $\beta = 0$ $\quad\quad\quad\quad\quad$ $R/g = (2.8704/98)$ km.K^{-1}

The diminution of the refractive index with height is given by equations (13a) and (13b); for the purpose of calculation, the specified values are assumed to be exact.

1. Coefficients Y_k.

Layer:	$0 - 10.4$ km	$10.4 - 24$ km	$24 - \infty$ km	Binomial multiplier
Y_0	39″614630	16″134229	2″176190	1
$10^2 Y_1$	4.768940	6.821834	1.956960	$-1/2$
$10^4 Y_2$	0.834121	3.067383	1.847449	$3/8$
$10^6 Y_3$	0.170238	1.466399	1.854364	$-5/16$
$10^8 Y_4$	0.037758	0.742198	2.007800	$35/128$
$10^{10} Y_5$	0.008818	0.394937	2.379461	$-63/256$
$10^{12} Y_6$	0.002132	0.219156	3.124625	$231/1024$
$10^{14} Y_7$	0.000529	0.125831	4.581027	$-429/2048$
$10^{16} Y_8$	0.000133	0.074246	7.511767	$6435/32768$
$10^{18} Y_9$	0.000034	0.044772	13.731743	$-12155/65536$

2. Examples of formulas for astronomical refraction.

$s = 10^{-2} \sec^2 z_1$

Zenith distances $0° \leq z_1 \leq 80°$. $\hfill (f = 1)$

$\Delta z_1 = \tan z_1 (57″92505 - 6″77387 s + 2″15586 s^2 - 1″09094 s^3$
$\quad\quad\quad + 0″76228 s^4 - 0″68493 s^5 + 0″75479 s^6 - 0″98607 s^7$
$\quad\quad\quad + 1″48977 s^8 - 2″55514 s^9)$. $\hfill \max R_{10} \sim 0″0003$

Zenith distances $0° \le z_1 \le 82°$. $\hfill (f = 0.9)$

$$\Delta z_1 = \tan z_1 (57\overset{''}{.}92505 - 6\overset{''}{.}77386s + 2\overset{''}{.}15580s^2 - 1\overset{''}{.}09052s^3$$
$$+ 0\overset{''}{.}75966s^4 - 0\overset{''}{.}67073s^5 + 0\overset{''}{.}68851s^6 - 0\overset{''}{.}72595s^7$$
$$+ 0\overset{''}{.}64970s^8 - 0\overset{''}{.}34516s^9).$$

$\hfill \max R_{10} \sim 0\overset{''}{.}0001$

Zenith distances $82° \le z_1 \le 84°$. $\hfill (f = 0.9, 0 - 10.4 \text{ km};$
$\hfill f = 0.75, 10.4 - \infty \text{ km})$

$$\Delta z_1 = \tan z_1 (57\overset{''}{.}92398 - 6\overset{''}{.}76680s + 2\overset{''}{.}13081s^2 - 1\overset{''}{.}02570s^3$$
$$+ 0\overset{''}{.}61974s^4 - 0\overset{''}{.}40724s^5 + 0\overset{''}{.}25657s^6 - 0\overset{''}{.}13688s^7$$
$$+ 0\overset{''}{.}05285s^8 - 0\overset{''}{.}01080s^9).$$

$\hfill \max R_{10} \sim 0\overset{''}{.}0001$

Zenith distances $84° \le z_1 \le 86°$. $\hfill (f = 0.9, 0 - 10.4 \text{ km};$
$\hfill f = 0.75, 10.4 - 24 \text{ km}; f = 0.56, 24 - \infty \text{ km})$

$$\Delta z_1 = \tan z_1 (57\overset{''}{.}909480 - 6\overset{''}{.}696487s + 1\overset{''}{.}962573s^2 - 0\overset{''}{.}758515s^3$$
$$+ 0\overset{''}{.}3019568s^4 - 0\overset{''}{.}1082585s^5 + 0\overset{''}{.}03197350s^6 - 0\overset{''}{.}007131299s^7$$
$$+ 0\overset{''}{.}001056695s^8 - 0\overset{''}{.}000077801s^9).$$

$\hfill \max R_{10} \sim 0\overset{''}{.}001$

3. Contributions of the atmospheric integrals to the coefficients

Integral	k = 0	k = 1	k = 2	k = 3	k = 4
Q(0)	57".92504 93				
P(0,1)		7".57784 80	-0".63824 36	0".04479 67	-0".00293 46
Q(1)		-0.81327 43			
P'(0,2)		0.00921 64	2.76257 89	-0.38785 11	0.03811 30
Q(2)		0.00007 61	0.02283 64		
P'(0,3)			0.00823 12	1.36993 73	-0.26934 02
P(1,2)		0.00000 04	0.00045 03	0.05619 56	-0.01105 00
Q(3)			-0.00000 48	-0.00080 15	
P'(0,4)			0.00000 84	0.00835 36	0.97252 94
P'(1,3)			0.00000 06	0.00028 54	0.02490 88
Q(4)				0.00000 03	0.00003 15
P'(0,5)				0.00002 10	0.00979 73
P'(1,4)				0.00000 06	0.00016 61
P(2,3)					0.00001 09
Q(5)					
P'(0.6)					0.00004 42
P'(1,5)					0.00000 06
P'(2,4)					
Q(6)					
P(0.7)					0.00000 01
P'(1,6)					
P'(2,5)					
P(3,4)					
P(0,8)					
P(1,7)					
P(2,6)					
P'(3,5)					
P(0,9)					
P(1,8)					
P(2,7)					
P(3,6)					
P(4,5)					
P(0,10)					
P(1,9)					
P(2,8)					
P(0,11)					
P(1,10)					
P(0,12)					
P(0,13)					
Totals	57".92504 93	6".77386 66	2".15585 74	1".09093 79	0".76227 71

in the formula for astronomical refraction.

k = 5	k = 6	k = 7	k = 8	k = 9	Integral
					Q(0)
0''.00018 54	-0''.00001 14	0''.00000 07			P(0,1)
					Q(1)
-0.00321 01	0.00024 78	-0.00001 81	0''.00000 13	-0''.00000 01	P'(0,2)
					Q(2)
0.03403 25	-0.00350 36	0.00031 97	-0.00002 69	0.00000 21	P'(0,3)
0.00139 63	-0.00014 38	0.00001 31	-0.00000 11	0.00000 01	P(1,2)
					Q(3)
-0.24594 10	0.03798 70	-0.00462 21	0.00048 67	-0.00004 65	P'(0,4)
-0.00630 09	0.00097 33	-0.00011 84	0.00001 25	-0.00000 12	P'(1,3)
					Q(4)
0.87915 29	-0.27188 64	0.04963 91	-0.00696 97	0.00083 17	P'(0,5)
0.01146 23	-0.00354 63	0.00064 76	-0.00009 09	0.00001 09	P'(1,4)
0.00069 95	-0.00021 64	0.00003 95	-0.00000 55	0.00000 07	P(2,3)
-0.00000 13					Q(5)
0.01324 24	0.96729 38	-0.35378 42	0.07454 60	-0.01186 38	P'(0,6)
0.00011 41	0.00657 48	-0.00240 61	0.00050 71	-0.00008 07	P'(1,5)
0.00000 64	0.00033 15	-0.00012 13	0.00002 56	-0.00000 41	P'(2,4)
	0.00000 01				Q(6)
0.00009 29	0.02039 33	1.25605 98	-0.53052 56	0.12672 79	P(0,7)
0.00000 06	0.00008 78	0.00436 42	-0.00184 46	0.00044 07	P'(1,6)
	0.00000 34	0.00015 02	-0.00006 35	0.00001 52	P'(2,5)
	0.00000 02	0.00001 06	-0.00000 45	0.00000 11	P(3,4)
0.00000 04	0.00020 43	0.03533 45	1.88095 44	-0.90128 51	P(0,8)
	0.00000 06	0.00007 53	0.00329 85	-0.00158 19	P(1,7)
		0.00000 20	0.00008 03	-0.00003 85	P(2,6)
		0.00000 01	0.00000 52	-0.00000 25	P'(3,5)
	0.00000 12	0.00047 80	0.06810 53	3.19123 05	P(0,9)
		0.00000 06	0.00007 17	0.00280 77	P(1,8)
			0.00000 14	0.00004 74	P(2,7)
			0.00000 01	0.00000 23	P(3,6)
				0.00000 01	P(4,5)
		0.00000 40	0.00119 47	0.14461 76	P(0,10)
			0.00000 08	0.00007 51	P(1,9)
				0.00000 10	P(2,8)
			0.00001 32	0.00319 04	P(0,11)
				0.00000 10	P(1,10)
			0.00000 01	0.00004 47	P(0,12)
				0.00000 04	P(0,13)
0''.68493 24	0''.75479 12	0''.98606 88	1''.48977 26	2''.55514 42	Totals

4. Sample calculation of R_{k+1}.

$$(k = 9; z_1 = 85°)$$

h, km	$-10^6 \frac{dn}{dr}$	$10^2(y^2 - 1)$	V	$-0\overset{''}{.}4154(\frac{dn}{dr})f\ V^{10}/(1 + \frac{21}{22} V)$	
0	27.28	0.0000	-0.1900	$0\overset{''}{.}764 \times 10^{-6}$	
2.6	22.34	0.0686	-0.1168	0.004	
5.2	18.07	0.1397	-0.0410	0.000	f = 0.9
7.8	14.39	0.2128	0.0369	0.000	
10.4	11.28	0.2878	0.1168	0.002	
10.4	13.90	0.2878	-0.2244	1.786	
13.8	8.163	0.3871	-0.1508	0.018	
17.2	4.793	0.4896	-0.0749	0.000	f = 0.75
20.6	2.814	0.5939	0.0023	0.000	
24	1.652	0.6993	0.0804	0.000	
24	1.6523	0.6993	-0.3977	61.31	
36	0.2523	1.075	-0.242	0.05	
48	0.0385	1.453	-0.086	0.00	f = 0.56
60	0.0059	1.833	0.070	0.00	
72	0.0009	2.212	0.227	0.00	

Integrals; $\left. \begin{array}{l} 0 - 10.4 \text{ km:} \quad 0\overset{''}{.}5 \\ 10.4 - 24 \text{ km:} \quad 2'' \\ 24 - 72 \text{ km:} \quad 195'' \end{array} \right\} \times 10^{-6}$

$R_{10} \sim 0\overset{''}{.}0002$

5. Astronomical refraction in the troposphere at $z_1 = 90°$.

$$x = 0.11976\ 14110\ h \qquad (h \text{ in km})$$

$$\Delta z_1(r_1;r) = 2020\overset{''}{.}53687\ \sqrt{x}\ (1 - 0.21827\ 814\ x + 0.03226\ 0119\ x^2$$
$$- 0.00286\ 4275\ x^3 + 0.00021\ 7379\ x^4 - 0.00002\ 3471\ x^5 + 0.00000\ 2731\ x^6$$
$$- 0.00000\ 0321\ x^7 + 0.00000\ 0040\ x^8 - 0.00000\ 0005\ x^9).$$

(Total $1743\overset{''}{.}330$).

APPENDIX B

Coefficients Y_k

1. Polynomial multipliers of integrals $P(i,j)$ in $n_1^{2k}Y_k$.

$$k(k-1)\ldots(k-p+1)(k-p)(P_n(i,j)(1-n_1^2)^{k-i-j}P(i,j)$$

where

$$p = \tfrac{1}{2}(j-2) \text{ if } j \text{ is even, or}$$
$$p = \tfrac{1}{2}(j-1) \text{ if } j \text{ is odd.}$$

$P_n(0,1) = 2$

$P_n(0,2) = (2k-1) - n_1^2$

$\tfrac{3}{2} P_n(0,3) = (2k-1) - 3n_1^2$

$P_n(1,2) = (2k-1)^2 - (8k-6)n_1^2 + n_1^4$

$6 P_n(0,4) = (2k-1)(2k-3) - 6(2k-3)n_1^2 + 3n_1^4$

$\tfrac{3}{2} P_n(1,3) = (2k-1)^2 - 2(8k-7)n_1^2 + 9n_1^4$

$15 P_n(0,5) = (2k-1)(2k-3) - 5(4k-6)n_1^2 + 15n_1^4$

$6 P_n(1,4) = (2k-1)^2(2k-3) - (2k-3)(26k-25)n_1^2 + 3(22k-35)n_1^4 - 9n_1^6$

$\tfrac{3}{2} P_n(2,3) = (k-1)(2k-1)^2 - (28k^2-58k+31)n_1^2 + (41k-55)n_1^4 - 9n_1^6$

$6 P_n(0,6) = \tfrac{1}{15}(2k-1)(2k-3)(2k-5) - (2k-3)(2k-5)n_1^2 + 3(2k-5)n_1^4 - n_1^6$

$15 P_n(1,5) = (2k-1)^2(2k-3) - (2k-3)(38k-39)n_1^2 + 5(34k-57)n_1^4 - 75n_1^6$

$2 P_r(2,4) = \tfrac{1}{3}(k-1)(2k-1)^2(2k-3) - (2k-3)(14k^2-31k+18)n_1^2$
$\qquad + (74k^2-229k+180)n_1^4 - (47k-82)n_1^6 + 3n_1^8$

$3 P_n(0,7) = \tfrac{1}{105}(2k-1)(2k-3)(2k-5) - \tfrac{1}{5}(2k-3)(2k-5)n_1^2 + (2k-5)n_1^4 - n_1^6$

$3 P_n(1,6) = \tfrac{1}{30}(2k-1)^2(2k-3)(2k-5) - \tfrac{2}{15}(2k-3)(2k-5)(13k-14)n_1^2$
$\qquad + (6k-15)(4k-7)n_1^4 - (22k-56)n_1^6 + 25n_1^8$

$5 P_n(2,5) = \tfrac{1}{3}(k-1)(2k-1)^2(2k-3) - \tfrac{1}{3}(2k-3)(58k^2-135k+83)n_1^2$
$\qquad + (158k^2-521k+441)n_1^4 - 5(39k-77)n_1^6 + 50n_1^8$

$$3\,P_n(3,4) = \tfrac{1}{6}(k-1)(2k-1)^2(2k-3)^2 - \tfrac{1}{2}(2k-3)^2(20k^2-48k+31)n_1^2 + (172k^3$$
$$-852k^2+1433k-819)n_1^4 - (232k^2-866k+819)n_1^6 + \tfrac{3}{2}(49k-93)n_1^8$$
$$-\tfrac{3}{2}n_1^{10}$$

$$6\,P_n(0,8) = \tfrac{1}{420}(2k-1)(2k-3)(2k-5)(2k-7) - \tfrac{1}{15}(2k-3)(2k-5)(2k-7)n_1^2$$
$$+ \tfrac{1}{2}(2k-5)(2k-7)n_1^4 - (2k-7)n_1^6 + \tfrac{1}{4}n_1^8$$

$$105\,P_n(1,7) = \tfrac{1}{3}(2k-1)^2(2k-3)(2k-5) - \tfrac{4}{3}(2k-3)(2k-5)(17k-19)n_1^2$$
$$+ 14(2k-5)(16k-29)n_1^4 - 20(35k-91)n_1^6 + 245 n_1^8$$

$$90\,P_n(2,6) = (k-1)(2k-1)^2(2k-3)(2k-5) - 2(2k-3)(2k-5)(38k^2-92k+59)n_1^2$$
$$+ 10(2k-5)(88k^2-305k+273)n_1^4 - 60(59k^2-276k+322)n_1^6$$
$$+ 15(137k-361)n_1^8 - 150 n_1^{10}$$

$$15\,P_n(3,5) = \tfrac{1}{3}(k-1)(2k-1)^2(2k-3)^2 - \tfrac{2}{3}(2k-3)^2(40k^2-102k+71)n_1^2$$
$$+ 2(332k^3-1740k^2+3121k-1917)n_1^4 - 4(374k^2-1543k+1638)n_1^6$$
$$+ 25(43k-99)n_1^8 - 150 n_1^{10}$$

....

DISCUSSION

J. Saastamoinen: replied to a question from Garfinkel, that in the refraction calculations we need atmosphere models with at least 5-6 layers. Two layers models are not enough.

CALCULATING ASTRONOMICAL REFRACTION BY MEANS OF CONTINUED FRACTIONS

S. Mikkola
Dept. of Astronomy, Univ. of Turku, Finland

0. Abstract: A continued fraction was derived for the summation of the asymptotic expansion of astronomical refraction. Using simple approximations for the last denominator of the fraction, accurate formulae, useful down to the horizon, were obtained. The method is not restricted to any model of the atmosphere and can thus be used in calculations based on actual aerological measurements.

1. Introduction

The usual way to evaluate the integral of astronomical refraction is to expand it into a series in powers of the tangent or secant of the zenith distance (Newcomb 1906, Oterma 1960, Joshi and Mueller 1974 and enclosed references). This series may, mathematically, have a non-zero radius of convergence, or else it may be a totally divergent asymptotic series, depending upon the model used for the upper atmosphere. However, for the first coefficients of the series there are not, in practice, significant differences in the numerical values, so that it always behaves like an asymptotic expansion. Thus it can be used only up to a certain zenith distance.

According to the theory of continued fractions the formal power series of a continued fraction is usually an asymptotic expansion, and vice versa. In this paper we shall investigate the use of a continued fraction to sum the expansion of astronomical refraction.

2. Development of the refraction integral

2.1. Notations

μ is the index of refraction of the air
r is the distance from the centre of the earth
z is the apparent zenith distance
μ_0, r_o, z_o are the above quantities at the place of observation

$\psi = (r\mu/r_0\mu_0)^2 - 1$

$S = \sin z_0$

$C = \cos z_0$

Δz is the astronomical refraction

2.2. Expansion into a continued fraction

The usual asymptotic expansion of the integral of the refraction is

$$\Delta z = S \sum_{n=0} (-1)^n \alpha_n C^{-2n-1} \qquad (2.1)$$

where the coefficients α_n are the moment integrals

$$\alpha_0 = \int_1^{\mu_0} \frac{d\mu}{\mu} = \log \mu_0$$

$$\alpha_n = \frac{1 \times 3 \times \cdots \times (2n-1)}{2 \times 4 \times \cdots \times (2n)} \int_1^{\mu_0} \psi^n \frac{d\mu}{\mu} \qquad (2.2)$$

Now, an asymptotic series of this type has what is called an S-fraction expansion (Wall 1948, p. 200), i.e. a continued fraction expansion of the form

$$\Delta z = \cfrac{\alpha_0 S}{C + \cfrac{b_1}{C + \cfrac{b_2}{C + \cfrac{b_3}{C + \cdots}}}} \qquad (2.3)$$

A convenient algorithm for computing the partial numerators b_k from the coefficient α_n is the quotient-difference algorithm (Henrici, 1967). The formulae needed in this case are given in section 3.1. It is difficult to investigate the mathematical convergence of this fraction, but as was shown in an earlier paper (Mikkola 1978) this fraction greatly resembles that of the error function and thus is very probably convergent for $C \neq 0$. However, to get a formula that can also be used at the horizon we first write fraction (2.3) in the recursive form

$$\Delta z = \alpha_0 S / g_1(C)$$

$$g_k(C) = C + b_k / g_{k+1}(C) \qquad (2.4)$$

and construct for $g_k(C)$ an asymptotic approximation useful for large k (say for k=n). If we put

$$Q_n = g_n(C) / g_{n+1}(C) \qquad (2.4)$$

then

$$g_n(C) = C + b_n Q_n / g_n(C) \tag{2.5}$$

from which follows the formula

$$g_n(C) = \frac{1}{2}(C + \sqrt{(C^2 + 4b_n Q_n)}) \tag{2.6}$$

As shown by Mikkola, 1978, the terms b_n are approximately of the form $b_n = n\beta$, where β is a number of the order $\beta \sim 10^{-3}$. Now it is not difficult to see that for $C >> 0$, $Q_n = 1 + O(\beta)$, and for $C = 0$, $Q_n = 1 + O(1/n)$ so that a good first approximation is obtained by replacing Q_n by its horizon value, giving

$$g_n(C) \approx \frac{1}{2}(C + \sqrt{(C^2 + q_n)}) \tag{2.7}$$

Here q_n is chosen in order to obtain the correct value for the horizontal refraction. Numerical experiments show that for large n (say n > 6) this approximation is, in practice, sufficient. However, to obtain better approximations we may write

$$g_n(C) = C + \frac{b_n}{\frac{1}{2}(C + \sqrt{(C^2 + q_{n+1})})} \tag{2.8}$$

and chose the parameters b_n and q_{n+1} to give suitable values for both the refraction and its derivative at the horizon. On the other hand (2.8) can be written in the form

$$g_n(C) = C(1 - \frac{2b_n}{q_{n+1}}) + \frac{2b_n}{q_{n+1}} \sqrt{(C^2 + q_{n+1})} \tag{2.9}$$

thus, due to the above fitting conditions we in fact have formula

$$g_n(C) = g_n'(0)C + \sqrt{(C^2(1 - g_n'(0))^2 + g_n^2(0))} \tag{2.10}$$

Here the prime indicates a derivative with respect to C. If approximation of this type is also used for g_{n+1}, then, due to the similarity of the formulae, their errors are quite similar and a good value for the ratio g_n/g_{n+1} is obtained. Thus we have for the quantity Q_n in formula (2.6) the very good approximation

$$Q_n \approx \frac{g_n'(0)C + \sqrt{(C^2(1 - g_n'(0))^2 + g_n^2(0))}}{g_{n+1}'(0)C + \sqrt{(C^2(1 - g_{n+1}'(0))^2 + g_{n+1}^2(0))}} \tag{2.11}$$

From formulae (2.4) it is easy to obtain recursion formulae for the quantities $g_n(0)$ and $g_n'(0)$ (given in section 3.1). However, to start the recursions the horizon refraction

$$\Delta z_\perp = \int_1^{\mu_0} \psi^{-1/2} \frac{d\mu}{\mu} \tag{2.11}$$

and the derivative

$$\Delta z_\perp' = \left\{\frac{d}{dC}\int_1^{\mu_0}(C^2+\psi)^{-1/2}\frac{d\mu}{\mu}\right\}_{C=0} \qquad (2.11)$$

are needed. Making the formal substitution $C^2 + \psi = \Theta^2$ we obtain

$$\Delta z_\perp' = 2\left\{\frac{d}{dC}\int_\infty^{C}\frac{d\log\mu(\psi=\Theta^2-C^2)}{d\psi}d\Theta\right\}_{C=0} = 2\left(\frac{d\log\mu}{d\psi}\right)$$

or
(2.12)

$$\Delta z_\perp' = \frac{\mu_0'/\mu_0}{1+\mu_0'/\mu_0}$$

Here μ_0' is the derivative of the refractive index with respect to the height (at ground and r_o = unit of distance).

3. Results and discussion

3.1. Collection of formulae

The quotient-difference algorithm for computing the partial numerators b_k of the continued fraction:

Using the expansion coefficients α_ℓ defined in (2.2) we start with

$$\left.\begin{array}{l} B_{1,\ell} = \alpha_\ell/\alpha_{\ell-1} \\ B_{2,\ell} = B_{1,\ell} - B_{1,\ell-1} \end{array}\right\} \quad \ell = 1,2,\ldots,n \qquad (3.1)$$

and continued by means of

$$B_{k,\ell} = \begin{cases} \dfrac{B_{k-1,\ell}}{B_{k-1,\ell-1}} B_{k-2,\ell-1}, & \text{if } k \text{ is odd} \\ B_{k-1,\ell} - B_{k-1,\ell-1} + B_{k-2,\ell-1}, & \text{if } k \text{ is even} \end{cases} \qquad (3.2)$$

$$k = 3,4,\ldots,n$$

which gives b_k values of:

$$b_k = B_{k,k} \qquad (3.3)$$

The recursion formulae for the horizontal values of g_n and g_n'. We start with

$$\Delta z_\perp = \int_1^{\mu_0} \psi^{-1/2}\frac{d\mu}{\mu}$$

$$\Delta z_\perp' = \frac{\mu_0'/\mu_0}{1+\mu_0'/\mu_0} \qquad (3.4)$$

and use the recursions

$$g_1(0) = \alpha_0/\Delta z_\perp \; ; \quad g_1'(0) = -\alpha_0^{-1} g_1^2(0) \Delta z_\perp'$$

$$g_{k+1}(0) = b_k/g_k(0) \; ; \quad g_{k+1}'(0) = \frac{g_{k+1}(0)}{g_k(0)} (1-g_k'(0)) \quad (3.5)$$

$$k = 1,2,3,\ldots,n$$

The approximation for g_n is

$$Q_n = \frac{g_n'(0)C + \sqrt{C^2(1-g_n'(0))^2 + g_n^2(0)}}{g_{n+1}'(0)C + \sqrt{C^2(1-g_{n+1}'(0))^2 + g_{n+1}^2(0)}} \quad (3.6)$$

$$g_n(C) = \frac{1}{2}(C + \sqrt{C^2 + 4b_n Q_n})$$

Now the refraction can be evaluated by means of the formula

$$\Delta z = \cfrac{\alpha_0 S}{C + \cfrac{b_1}{C + \cfrac{b_2}{C + \cfrac{\ddots}{\ddots \cfrac{}{C + \cfrac{b_{n-1}}{g_n(C)}}}}}} \quad (3.7)$$

3.2. Results of some numerical tests

To test the reliability of the approximations for g_n the results obtained using the continued fraction formula were compared with a direct numerical integration using a polytropic model atmosphere. Table 1 gives the errors for different z and n values when approximations of the type (2.7) were used for g_n in (3.7). Table 2 shows the errors when formulae (3.6) are used. As we can see, formulae (3.6) are surprisingly accurate as they yields an negligible error even for n = 1.

More details about the method using a polytropic model atmosphere are given in an earlier paper of the author (Mikkola 1978).

4. Acknowledgements

The author is indebted to Professor LIISI OTERMA and to Professor JUHANI KAKKURI for their help and interest in this work.

Table 1.

Errors when using the formula (2.7) for g_n.

z \ n	1	2	3	4	5	6	7	8	9
80°.00	3".30	-0".10	0".01	-0".00	0".00	0".00	0".00	0".00	0".00
81.00	4.34	-0.16	0.01	-0.00	0.00	0.00	0.00	0.00	0.00
82.00	5.85	-0.25	0.03	-0.00	0.00	0.00	0.00	0.00	0.00
83.00	8.10	-0.42	0.06	-0.01	0.00	-0.00	0.00	0.00	0.00
84.00	11.54	-0.73	0.12	-0.01	0.00	-0.00	0.00	0.00	0.00
85.00	16.95	-1.32	0.27	-0.04	0.01	-0.00	0.00	-0.00	0.00
86.00	25.64	-2.48	0.65	-0.12	0.02	-0.01	0.00	-0.00	0.00
87.00	39.39	-4.75	1.62	-0.37	0.08	-0.07	0.01	-0.01	0.01
87.50	48.51	-6.54	2.59	-0.69	0.16	-0.16	0.03	-0.02	0.02
88.00	58.56	-8.85	4.09	-1.26	0.30	-0.35	0.08	-0.06	0.05
88.50	67.56	-11.51	6.25	-2.28	0.55	-0.78	0.21	-0.16	0.17
89.00	70.37	-13.66	8.75	-3.84	0.93	-1.59	0.52	-0.36	0.49
89.25	66.25	-13.81	9.63	-4.65	1.13	-2.13	0.76	-0.50	0.79
89.50	55.53	-12.51	9.49	-5.07	1.25	-2.54	1.01	-0.63	1.13
89.75	34.95	-8.57	7.06	-4.19	1.05	-2.30	1.03	-0.59	1.23
90.00	0.00	0.00	0.00	0.00	0.00	0.00	0.00	0.00	0.00

Table 2.

Errors when using the formulae (3.6) for g_n.

z \ n	1	2	3	4	5	6	7	8	9
80°.00	-0".02	0".00	0".00	0".00	0".00	0".00	0".00	0".00	0".00
81.00	-0.03	0.00	0.00	0.00	0.00	0.00	0.00	0.00	0.00
82.00	-0.04	0.00	0.00	0.00	0.00	0.00	0.00	0.00	0.00
83.00	-0.07	0.00	0.00	0.00	0.00	0.00	0.00	0.00	0.00
84.00	-0.10	0.00	0.01	0.00	0.00	0.00	0.00	0.00	0.00
85.00	-0.15	0.00	0.01	0.00	0.00	0.00	-0.00	-0.00	0.00
86.00	-0.19	-0.00	0.04	0.01	0.00	-0.00	-0.00	-0.00	0.00
87.00	-0.19	-0.01	0.00	0.02	0.00	-0.01	-0.00	-0.00	-0.00
87.50	-0.13	-0.02	0.18	0.04	0.00	-0.01	-0.01	-0.01	-0.00
88.00	-0.01	-0.03	0.29	0.06	0.00	-0.03	-0.02	-0.01	-0.00
88.50	0.16	-0.04	0.41	0.08	-0.01	-0.07	-0.05	-0.03	-0.01
89.00	0.30	-0.03	0.47	0.08	-0.03	-0.12	-0.09	-0.06	-0.01
89.25	0.31	-0.03	0.42	0.06	-0.05	-0.13	-0.10	-0.07	-0.01
89.50	0.25	-0.02	0.31	0.04	-0.05	-0.12	-0.09	-0.06	-0.01
89.75	0.10	-0.01	0.12	0.01	-0.03	-0.06	-0.05	-0.03	-0.00
90.00	0.00	0.00	0.00	0.00	0.00	0.00	0.00	0.00	0.00

5. References

1. Henrici,P., Quotient-Difference Algorithms, p. 37 in 'Mathematical Methods for Digital Computers',Vol. II (Ed. Ralston and Wilf), Wiley & Sons, Inc., New York-London-Sydney, (1967).

2. Joshi,C.H., and Mueller,I.I., Review of Refraction Effects of the Atmosphere on Geodetic Measurements to Celestial Bodies, p. 83 in 'The Present State and Future of the Astronomical Refraction Investigations', Proc. of the Study Group on Astronomical Union Commission 8. (Ed. G.Teleki), Publ.Astr.Obs. of Belgrade No. 18, Belgrade (1974).

3. Mikkola,S., Computing Astronomical Refraction by Means of Continued Fractions. Rep. Finn. Geod. Inst. 78:6. Helsinki (1978).

4. Newcomb,S., A Compendium of Spherical Astronomy. The Macmillan Company, New York (1906). Reprinted by Dover Publications, Inc., New York (1960).

5. Oterma, L., Computing the Refraction for the Väisälä Astronomical Method of Triangulation, Astr.-Opt.Inst., Univ. of Turku, Informo 20, Turku (1960).

6. Wall,H.S., Analytic Theory of Continued Fractions. Van Nostrand, Princeton, New Jersey (1948).

DISCUSSION

B. Garfinkel: For what zenith distances is it possible to apply Mikkola's formulae?

J. Kakkuri: answered that he has no complete information on this subject.

B. Garfinkel, J.A. Hughes, K. Poder and J. Saastamoinen: discussed the possibilities of refraction calculation near to zenith distances of $90°$.

ON THE CHARACTERISTICS OF ASTRONOMICAL REFRACTION IN THE NORTHERN HEMISPHERE

C. Sugawa and N. Kikuchi
International Latitude Observatory of Mizusawa, 023 Japan

1. EXPONENTIAL REPRESENTATION OF ASTRONOMICAL REFRACTION

The astronomical refraction mainly depends on the vertical structure or the height profile of atmospheric density. Concerning the vertical structure of atmospheric density, various hypotheses were presented during the last century. Among them Newton's hypothesis of equal temperature and Ivory's one that temperature diminishes at a uniform rate with height appear to represent the height profile of temperature approximately in the lower part of the stratosphere and in the troposphere respectively.

Generalizing Newton's hypothesis, G. Teleki (1967) adopted the following exponential representation of refractive index (μ),

$$\mu - 1 = \varepsilon e^{-as}, \tag{1}$$

$$s = \frac{r}{r_0} - 1, \tag{2}$$

where r_0 is the geocentric radius of a station at the ground surface and r that at any height. The values of ε and a are to be determined from aerological data. Gladstone-Dale's law is adopted for the relation between refractive index (μ) and atmospheric density (ρ) as

$$\mu = 1 + c\rho, \tag{3}$$

where the value of c is 0.226. Combining (1) with (3), we can put

$$\rho = \varepsilon' e^{-z/H}, \tag{4}$$

where z is the height above the ground surface and H the height of homogeneous atmosphere or the scale height.

For the purpose of examining the fitness of the exponential representation of atmospheric density, we calculated the monthly mean values of atmospheric density for the year 1965 at 7 aerological stations in

Eastern Japan nearly along the meridian at Mizusawa (141°08'E), that is, Wakkanai (WA) (45°25'N, 141°41'E), Sapporo (SA) (43°03'N, 141°20'E), Akita (AK) (39°43'N, 140°06'E), Sendai (SE) (38°16'N, 140°50'E), Tateno (TA) (36°03'N, 140°08'E), Hachijojima (HA) (33°07'N,139°47'E), Torishima (TO) (30°29'N, 140°18'E), by using pressure and temperature data for 23 standard pressure levels (Sfs, 1000, 900, 850, 700, 600, 500, 400, 350, 300, 250, 175, 125, 100, 70, 50, 40, 30, 20, 15 mb).

The atmospheric density (ρ) is derived as

$$\rho = \frac{P}{RT}, \qquad (5)$$

where P denotes air pressure (mb), T absolute temperature ($273°.16 + t°C$) and R the gas constant for dry air ($2.8704 \cdot 10^6$). As the effect of water vapour is small for visible ray, dry air is treated thereafter.

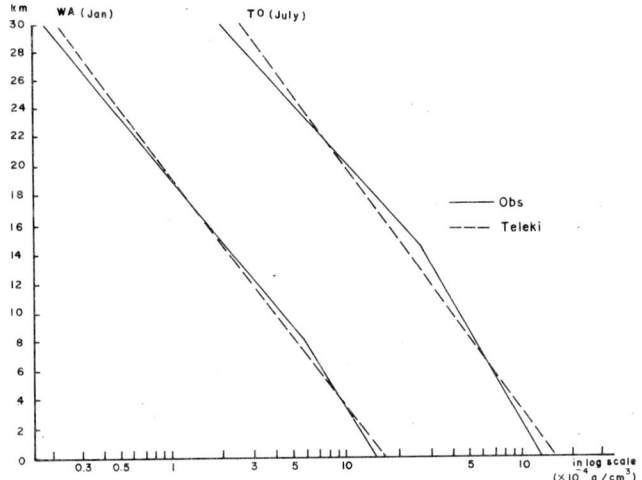

Figure 1. The vertical distribution of the atmospheric density at Wakkanai (Jan.) and Torishima (July) and the fitness with the exponential representations by Teleki.

In Figure 1. the vertical distribution of density at Wakkanai in January and at Torishima in July and the exponential representations by (4) are given in logarithmic scale. As can be seen in Figure 1, the vertical distribution of density in logarithmic scale clearly indicates a tendency of bending near the tropopause between the troposphere and the stratosphere. This tendency was already found by B.R. Bean and E.J. Dutton (1966) for radio wave refraction. In the troposphere the lapse rate of temperature is nearly constant (0°.6/100 m) and the lower part of the stratosphere consists of nearly isothermal layers. It seems, therefore, more reasonable to define the scale height separately for the troposphere and the stratosphere. We put the following representations for two spheres model.

$$\rho = \rho_0 e^{-z/H^*}, \qquad z^* \geq z \geq 0 \quad \text{(troposphere)} \quad (6)$$

$$\rho = \rho_m e^{-z/H^{**}}, \qquad z \geq z^* \quad \text{(stratosphere)} \quad (7)$$

where ρ_0, ρ_m, H^* and H^{**} are parameters to be determined from aerological observations. The height of a bending point, z^*, can be determined by the method of successive approximation so as to make coincidence the density representation from the surface in (6) with that in the stratosphere between 5 km (500 mb) and 16 km (100mb). At the same time H^*, H^{**} and ρ_m can be determined.

The height of the bending point (z^*) shows a distinct seasonal variation which is lower in winter and higher in summer. Its amplitude of the seasonal variation becomes larger with the northern latitudes. The height of the bending point is 6-7 km in winter and 11-12 km in summer at Wakkanai and Sapporo in Hokkaido. It is 7-8 km in winter and 12 km in summer at Akita and Sendai including Mizusawa. It is also about 11 km in winter and 13 km in summer at the southern stations, Hachijojima and Torishima. Generally, the height of the tropopause is highest, about 17 km, in the equatorial region, but it decreases till 9 km with higher latitudes. At the same station it is higher in summer and lower in winter. The height of z^* does not exactly coincide with that of the tropopause, but it appears to be very near the boundary between the troposphere and the stratosphere.

Then the scale height H^* shows a similar seasonal variation to that of z^*, lower in winter (9.2 km) and higher (9.6 km) in summer. The seasonal variation of the scale height appears to depend on a regional effect. On the other hand, the scale height H^{**} in the stratosphere gives an opposite seasonal variation, higher (6.5 km) in winter and lower (6.1 km) in summer. The seasonal variation of the scale height is quite opposite between the troposphere and the stratosphere.

The mean density in the stratosphere, ρ_m, shows a similar seasonal variation to the height of the tropopause, small in winter and large in summer.

2. NUMERICAL EXPERIMENTS FOR ASTRONOMICAL REFRACTION.

In Figure 2 the center of the Earth is denoted by C and the observation point at the ground surface O_1. The direction of the zenith is represented by Z. Assuming the Earth as a sphere, the concentric differential airstrata for every km from the surface to 30 km are taken as i-th layer (i = 1, 2, 3, - - -, 30). Putting observed zenith distance at the ground surface as z_1, and refractive index at the 1st layer as μ_1, we define horizontal and vertical directions, parallel to those at O_1, at O_2 where optical path passes the boundary between 1 and 2 layers, and the angle between the tangent to the concentric boundary and the horizontal direction. The angle θ_2 is taken positive clockwisely. The inclination of airstrata is taken as S_2, clockwisely positive. Let CO_2 be r_2 and CO_1 be r_1. Then we obtain for the triangle CO_1O_2 as

$$r_1 \sin(z_1 - \theta_1) = r_2 \sin(z_1 - \theta_2). \qquad (8)$$

From (8) we can derive as

$$\theta_2 = z_1 - \sin^{-1}\left\{\frac{r_1}{r_2} \sin(z_1 - \theta_1)\right\}. \qquad (9)$$

At the ground surface

$$S_1 = 0, \quad \theta_1 = 0.$$

Let the zenith distance at O_2 be z_2. From the fundamental relation for refraction we get

$$\mu_1 \sin(z_1 - \theta_2 - S_2) = \mu_2 \sin(z_2 - \theta_2 - S_2). \qquad (10)$$

From (10) we can derive z_2 as

$$z_2 = \theta_2 + S_2 + \sin^{-1}\left\{\frac{\mu_1}{\mu_2} \sin(z_1 - \theta_2 - S_2)\right\}. \qquad (11)$$

We proceed these procedures successively to the i-th layers (i=3, 4,- -, 30). Thus z_{30}, the last zenith distance, is corrected for refraction. We take the last zenith distance as the result of numerical experiment.

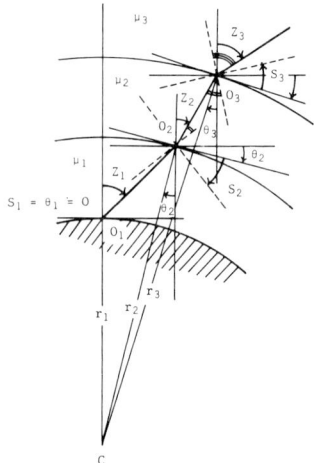

Figure 2

In Figure 3 the profile of the tilting of airstrata of equal density for one combination of two stations nearly along the meridian at Mizusawa, that is, Akita-Sendai (October). The profile derived from directly observed data show a typical character which is north up from the surface to about 10 km (the height of tropopause) and south up from 10 km to 25 km (lower part of stratosphere), as already found by Dines, Harzer, Sugawa and Teleki. It can be seen at once that two spheres model of the exponential representation appears to fit observed results approximately. It has been, therefore, proved that two spheres model is a more reasonable representation.

Moreover, two spheres model expresses the best approach to the actual seasonal variation of surface pressure.

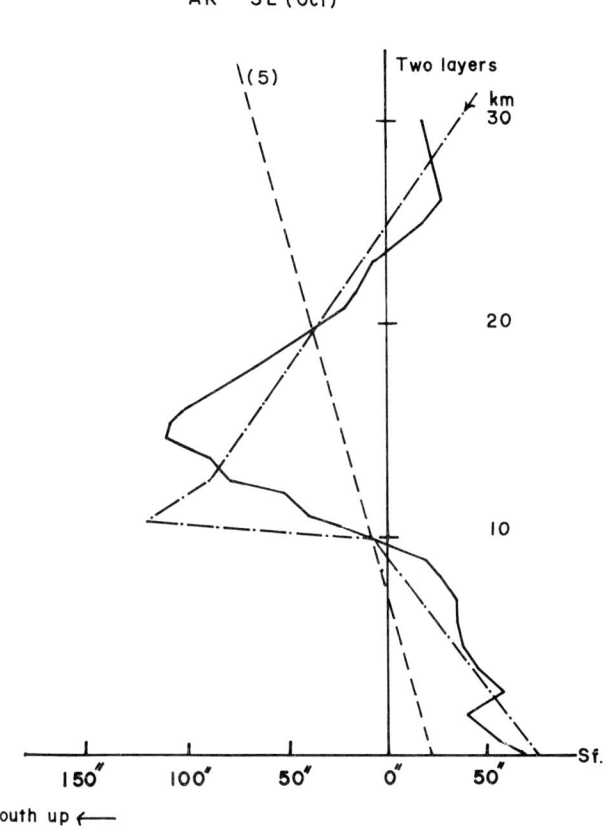

Figure 3 The vertical distribution of the inclination of airstrata
——— obs. —·—·— two spheres model — — — Teleki.

3. THE CHARACTERISTICS OF ASTRONOMICAL REFRACTION IN THE NORTHERN HEMISPHERE

Basic aerological data were obtained at the following standard pressure levels; the mean sea level, 1000, 850, 700, 500, 200, 100 and 30 mb respectively. The northern hemisphere is divided into a mesh with 10° in longitude and latitude ranging from 20° to 80°N.

The monthly mean values of aerological data were taken from the following aerological notes for every mesh.
Long Range Forecast, Technical Notes, No.6, Normal values at the heights of constant pressure level in the troposphere (100, 200, 300, 500, 700 and 850 mb) from 1951 to 1960, Japan Meteorological Agency, 1968.
Long Range Forecast, Technical Notes, No.10, Normal values at the heights of constant pressure level and anomalies at 30 mb from 1958 to 1966, Japan Meteorological Agency, 1970.
Normal of Atmospheric Pressure in the Northern Hemisphere (1909-1914 and 1924-1937) for Sfs and 1000 mb, Japan Meteorological Agency, 1966.

Five parameters of two spheres model in each mesh in the northern hemisphere have been calculated for each month. The global charts of the annual mean values on ρ_0, H^*, ρ_m, H^{**} and z^* are given in Figure 4(a), (b), (c), (d) and (e) respectively. For reference, the charts of the annual mean temperature and pressure at the mean sea level are shown in Figure 5(a), (b) respectively.

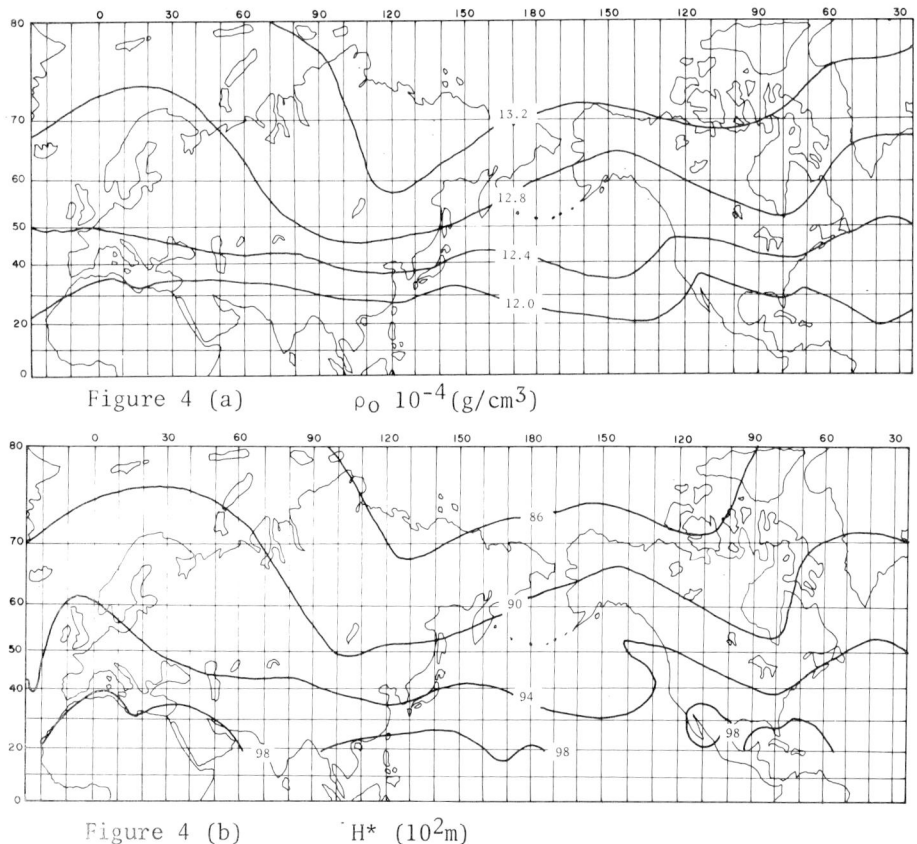

Figure 4 (a) ρ_0 10^{-4} (g/cm^3)

Figure 4 (b) H^* (10^2m)

ON THE CHARACTERISTICS OF ASTRONOMICAL REFRACTION

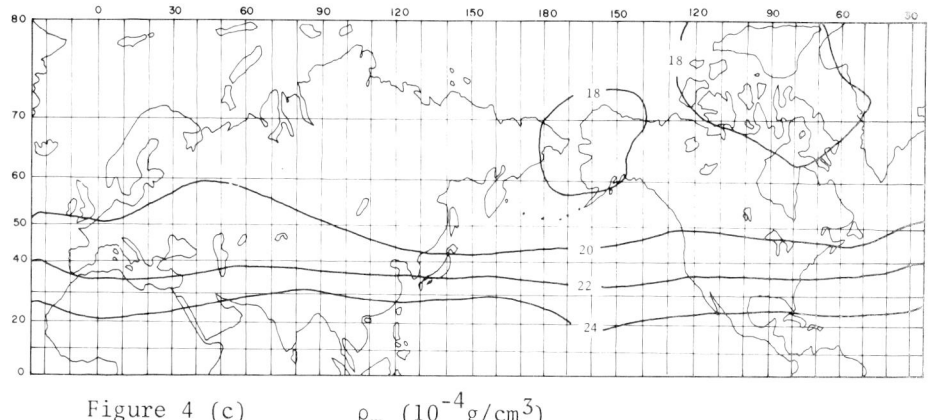

Figure 4 (c) ρ_m $(10^{-4} g/cm^3)$

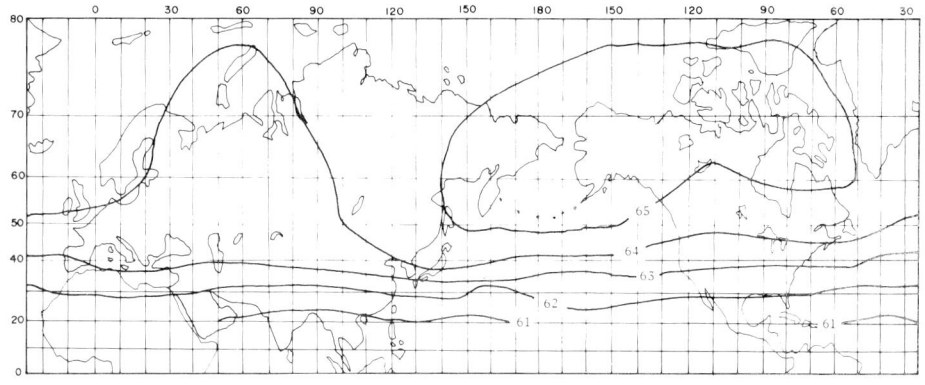

Figure 4 (d) H^{**} $(10^2 m)$

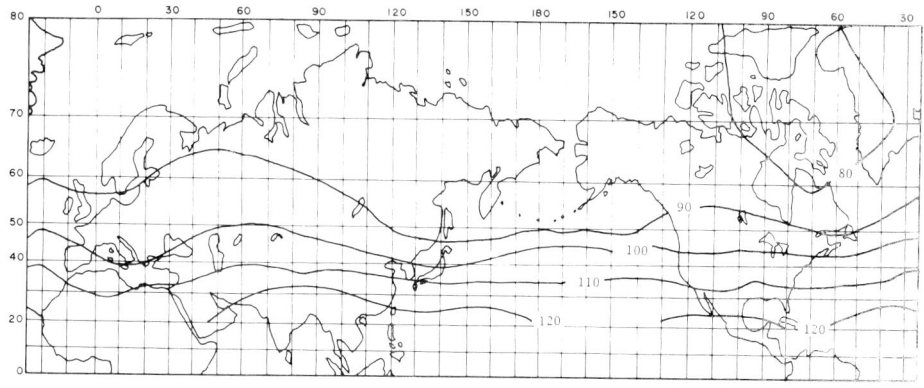

Figure 4 (e) Z^* $(10^2 m)$

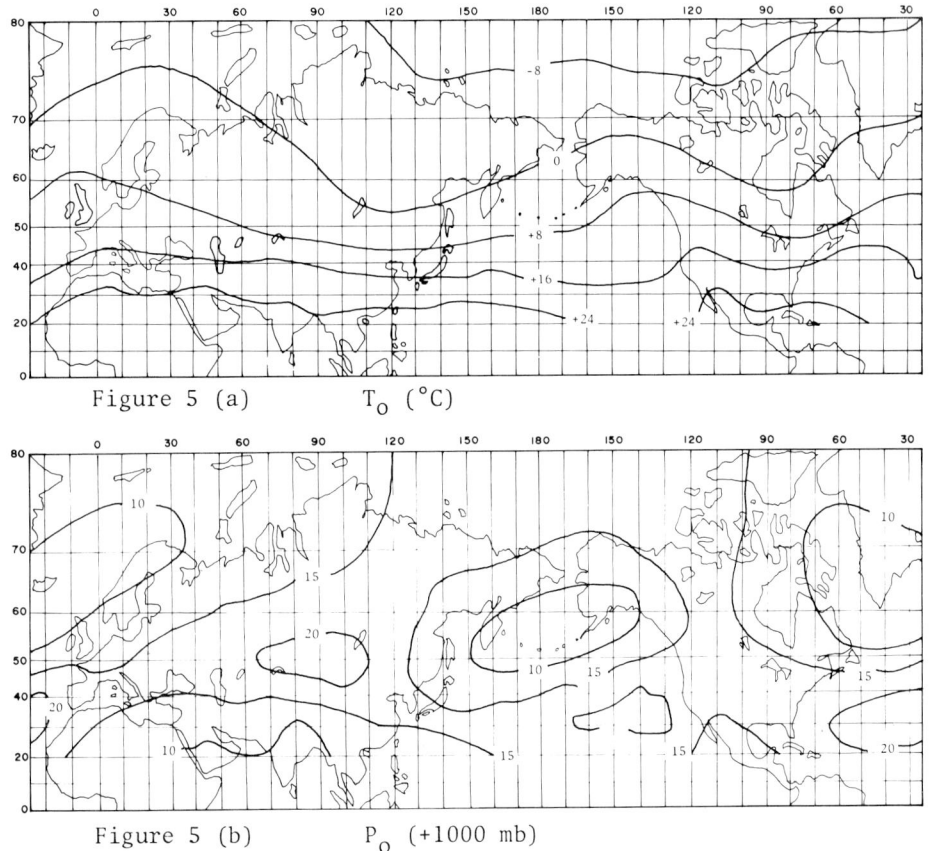

Figure 5 (a) T_o (°C)

Figure 5 (b) P_o (+1000 mb)

As can be readily seen in Figure 4(a), (b) and Figure 5(a), the world-wide distribution of surface density (ρ_o) and the scale height in the troposphere (H*) appear to be fairly similar to that of surface temperature.

The distribution of ρ_m corresponding to the equivalent density of the stratosphere shows two lower regions in Bering Sea and North West Territory in Canada in Figure 4(c). In the middle latitudes, the contour line of equal density appear to be nearly parallel to the equator. In Figure 4(d) the distribution of the scale height in the stratosphere (H**) appears to be also nearly parallel to the equator in the middle latitudes, but it shows some higher region in high latitudes. In figure 4(e) the distribution of the height of the bending point (z*) nearly corresponding to the height of the tropopause appears to give similar chart to that of ρ_m except Bering Sea region.

4. EXPERIMENTAL FORMULA FOR NORMAL OR PURE REFRACTION.

The fundamental integral of normal refraction is, as known well,

$$R = a\mu_o \sin z \int_1^{\mu_o} \frac{d\mu}{\mu(r^2 - a^2\sin^2 z)^{1/2}}. \qquad (12)$$

This integral can be developed approximately in the following principal terms.

$$R = (\mu_o - 1)\tan z + B \tan z(1 + \tan^2 z) = A \tan z + B \tan^3 z, \quad (13)$$

where

$$A = (\mu_o - 1) + B. \quad (14)$$

The value of B is nearly $-0\rlap{.}''0669$ under normal, 0°C and 760 mm Hg or 1013.25 mb. In the refraction table of "Connaissance des temp", A is put as

$$A = R_o \frac{\mu_o - 1}{\mu_s - 1} \quad (15)$$

where μ_s represents refractive index at the surface under normal condition. The value of R_o is adopted as $60\rlap{.}''154$.

Now the difference of the amount of refraction appears to depend on that of the scale height, as shown in Figure 6. Let ρ_o be constant but the scale height be different and the incident angle be equal in right and left sides. However, those at the uppermost layer are different due to the difference of the scale height in their way.

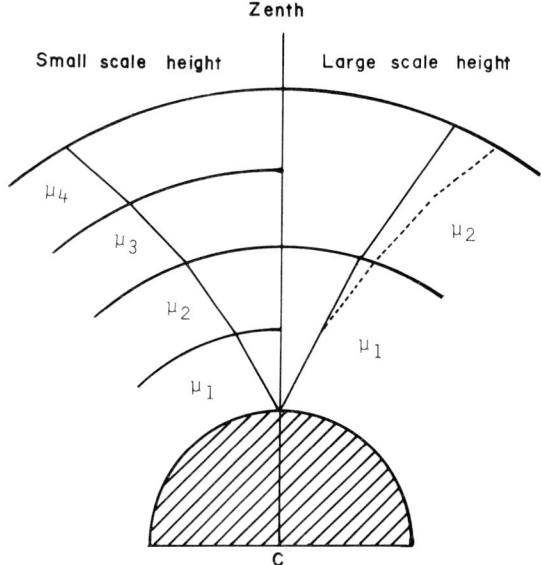

Figure 6

Putting R_o as the amount of refraction derived by numerical experiments for each mesh, we get

$$R_0 = R/(\frac{\rho_0}{\rho_S} \tan z), \qquad (16)$$

where R denotes equivalent refraction constant, ρ_S the standard value of density (0.00129 g/cm^3) and z a zenith distance. R can be approximated as

$$R_0 = \alpha + \beta \tan^2 z, \qquad (17)$$

where

$$\alpha = A + B \frac{H_0}{H_S},$$

$$\beta = C + D \frac{H_0}{H_S},$$

where H_0 is the scale height and H_S the standard scale height (7,999 m). From (17) we have

$$R = R_0 \frac{\rho_0}{\rho_S} \tan z$$

$$= (\alpha + \beta \tan^2 z) \frac{\rho_0}{\rho_S} \tan z$$

$$= (A + B \frac{H_0}{H_S}) \frac{\rho_0}{\rho_S} \tan z + (C + D \frac{H_0}{H_S}) \tan^3 z. \qquad (18)$$

As H_0 is proportional to T_0, we obtain

$$R = (A + B \frac{T_0}{T_S}) \frac{\rho_0}{\rho_S} \tan z + (C + D \frac{T_0}{T_S}) \frac{\rho_0}{\rho_S} \tan^3 z, \qquad (19)$$

where T_S is 237°16 k.
On the other hand, ρ_0, T_0 and P_0 are mutually related as

$$\frac{\rho_0}{\rho_S} = \frac{T_S}{P_S} \frac{P_0}{T_0} \qquad (20)$$

where P_S is 1013.25 mb.
Thus we can derive the following formula which contains P_0 as the whole atmospheric mass.

$$R = (A \frac{\rho_0}{\rho_S} + B \frac{\rho_0}{\rho_S}) \tan z + (C \frac{\rho_0}{\rho_S} + D \frac{P_0}{P_S}) \tan^3 z. \qquad (21)$$

From the values of astronomical refraction obtained by numerical experiments the following experimental formula was obtained by the method of least squares (number of equations of condition is 9,828).

$$R = 60\rlap{.}''23622 \frac{T_s}{P_s} \frac{P_o}{T_o} - 0\rlap{.}''06719 \frac{P_o}{P_s} \tan z$$
$$\pm 45 \qquad \qquad \pm 44$$

$$+ 0\rlap{.}''01249 \frac{T_s}{P_s} \frac{P_o}{T_o} - 0\rlap{.}''07502 \frac{P_o}{P_s} \tan^3 z. \qquad (22)$$
$$\pm 161 \qquad \qquad \pm 156$$

For reference the corresponding formula in "Connaissance des Temps" is shown as

$$R = 60\rlap{.}''154 \frac{T_s}{P_s} \frac{P_o}{T_o} \tan z - 0\rlap{.}''0669 \tan^3 z. \qquad (23)$$

Numerical experiments based on a single layer model of exponential representation were made and compared with (23). Using 600 samples in the range of ρ_o (0.0012 ~ 0.00137 g/cm^3), H_o (6 ~ 10 km), zenith distance (1° ~ 45°) and latitude (0° ~ 90°), numerical experiments were performed from surface to ∞ km.

Then the following formula was obtained.

$$R = (60\rlap{.}''24067 - 0\rlap{.}''07539 \frac{H_o}{H_s}) \frac{\rho_o}{\rho_s} \tan z$$
$$\pm 8 \qquad \pm 8$$

$$+ (0\rlap{.}''00804 - 0\rlap{.}''07415 \frac{H_o}{H_s}) \frac{\rho_o}{\rho_s} \tan^3 z. \qquad (24)$$
$$\pm 9 \qquad \pm 9$$

The latitude effect can be neglected according to (25). The value of $60\rlap{.}''24067$ exactly corresponds to $c\rho_s$ (c = 0.226). The differences between the values derived by (25) and those by numerical experiments remain within 10^{-5} of arc up to z = 45°.

5. ASTRONOMICAL REFRACTION IN A CASE OF THE TILTING OF AIRSTRATA OF EQUAL DENSITY.

The tilting of airstrata of equal density in a mesh with 10 in longitude and latitude was computed using five parameters in the vartical distribution from surface to 30 km. Tracing successively a refracted ray path of a fictitious star which is observed as zenith at the ground surface its zenith distance at 30 km is computed by numerical experiments. The results are given in Figure 7 (a), (b). For reference, the horizontal pressure gradients are shown in Figure 8 (a), (b). In these figures a positive sign indicates refracted directions towards south and east from the zenith. For the pressure gradient a positive sign indicates north and west up.

It has been accepted widely that the general tendency of the tilting of airstrata of equal density shows north up from the surface to about 9 km and south up about 10 km to about 30 km.

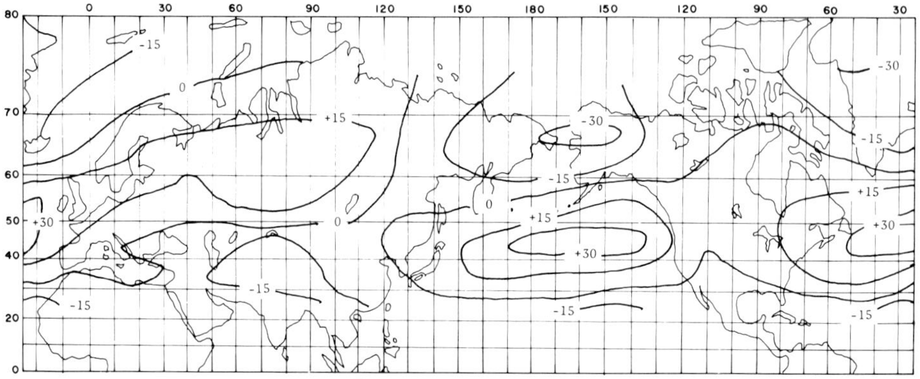

Figure 7 (a) Z_{NS} (unit: "10^{-4})

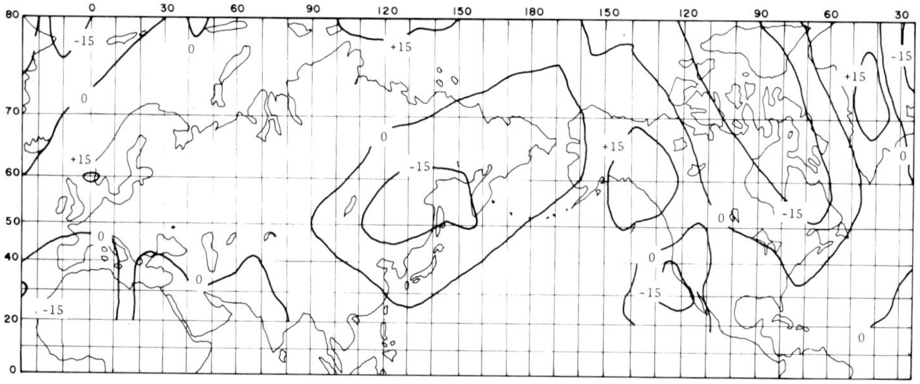

Figure 7 (b) Z_{EW} (unit: "10^{-4})

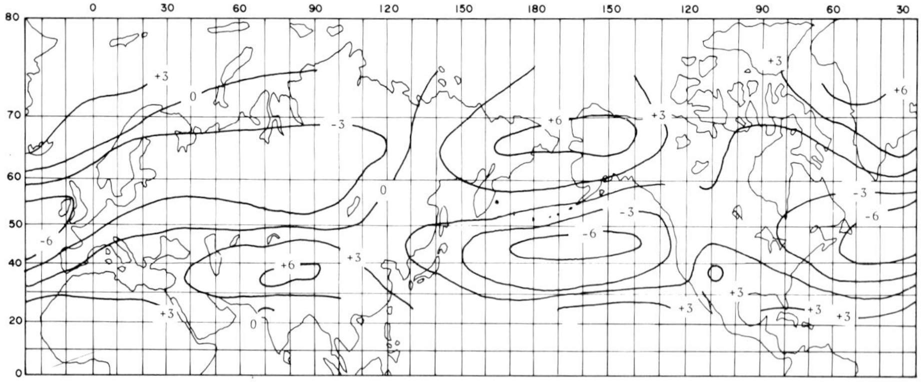

Figure 8 (a) P_{NS} ($P_N - P_S$) (unit: mb)

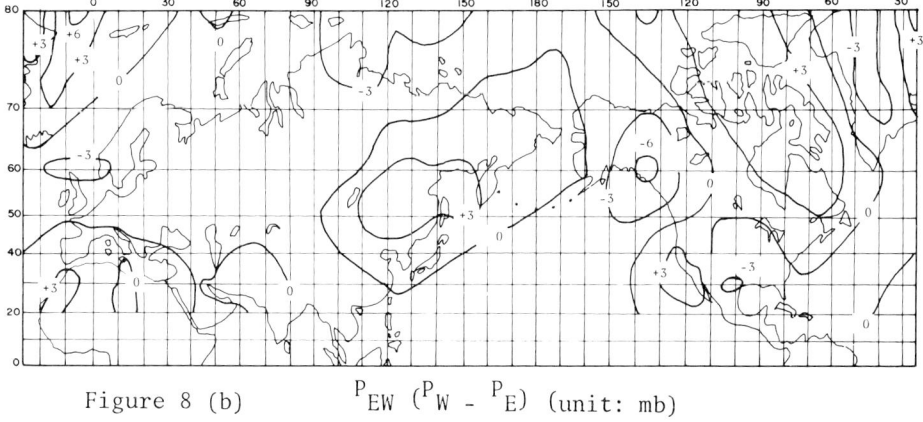

Figure 8 (b) P_{EW} ($P_W - P_E$) (unit: mb)

Then it has been discussed which is more important for the total effect, troposphere or stratosphere, but it has been confirmed from the chart of the annual mean zenith refraction that the total effect follows the horizontal gradient of P_0 as the total mass of the whole atmosphere. Namely, it is refracted towards higher pressure direction. The amount of refraction is about 0″.003 corresponding to the pressure difference of 7 mb in 10° of latitude (about 1,110 km). The correlation coefficients between Δz and ΔP are -0.971 for NS direction and -0.916 for EW direction when the value of P_0 is used for ΔP. They are also -0.994 for NS direction and -0.993 for EW direction when $P_0 = \int_0^\infty \rho(z) \, dz$ (g = 980.665 gal) is used for ΔP_0.

Assuming a linear relation, we put

$$z_0('') = u \times P_0 \text{ (mb/km)},$$

where $z_0('')$ is the amount of refraction of a zenith star in the second of arc and ΔP_0 the horizontal gradient of P_0 in the unit of mb/km.

$u = -0.583$ for NS direction,
$u = -0.477$ for EW direction.

The resultant relation is represented by the following formula

$$z_0('') = -0.48 \times P_0 \text{ (mb/km)}, \qquad (25)$$

where a negative sign is due to an opposite refracted direction with the tilt of airstrat in a polar coordinate system.

Now the following characteristics have been obtained for the case when the zenith distance is not zero. Taking a half sum of the values derived by numerical experiments for positive and negative zenith distances, 1°, 5°, 10°, 20°, 30° and 45° an experimental formula derived by using many samples can be approximated by

$$\Delta z = \Delta z_0 \sec^2 z, \qquad (26)$$

where Δz is an amount of refraction at a zenith distance of z. However, the expressions (25) and (26) can hold for z within 1″.

6. CONCLUDING NOTES.

It has been noticed that the exponential representation of refractive index gives some bending near the tropopause between the troposphere and the stratosphere. Therefore, we proposed the two spheres model as the improved exponential representation of refractive index.

The distribution of surface density (ρ_0) and scale height in the troposphere (H*) appears to be nearly similar to that of surface temperature. The distribution of ρ_m corresponding to the equivalent density in the stratosphere shows two lower regions, that is, in Bering Sea and North West Territory in Canada. In the middle latitudes the contour lines of equal density appear to be nearly parallel to the equator. The distribution of the scale height in the stratosphere appears to be also parallel to the equator in the middle latitudes, but it shows some higher region in high latitudes. The distribution of the height of bending point (z*) nearly corresponding to the height of the tropopause appears to give similar chart to that of ρ_m except Bering Sea region.

In this note two new methods of correction for astronomical refraction are proposed.
The effect of vertical gradient of atmospheric density has been so far neglected. However, it may be better to add a term for P_0 (total mass of the atmosphere), as given in (22). This effect is about 0″.001.
It may be possible to correct an anomalous refraction with an accuracy of 0″.1 using (25) and (26), where horizontal gradient of P_0 corresponds to a total integrated amount of the tilting of airstrata of equal density.

It is to be remarked that astronomical refraction is corrected by using P_0 common to two above-mentiond effects. The important element on astronomical refraction has been hitherto considered as air temperature at the surface (T_0), but the contribution of air pressure at the surface (P_0) should be anew recognized.

The above mentioned methods are based on the exponential distribution of the atmospheric density. If an exponential representation would be applicable for the general field of the atmosphere, the pressure difference 0.02 mb for a distance of 100 m in the scale height corresponding to that of about 3°.4 k in temperature may produce the effect of about 0″.08

in the amount of refraction from surface to 3 km for 20° of zenith distance even when there is no tilting of airstrata of equal density. However, the inversion of atmospheric temperature often occurs near the surface and several kilometers in height during clear nights. Moreover, considering the irregularity of the vertical distribution of atmospheric density due to the advection, it may be remained as a future problem to what extent our conception would enable us to correct anomalous refraction precisely.

We have usually called astronomical refraction with the tilting of airstrata of equal density as an anomalous refraction. However, it actually appears to occur every time and everywhere. Therefore, it would be more exact and convenient to call it as apparent refraction.

At the end of this note we express our hearty thanks to Prof. G. Teleki, a Chairman of Working Group on Astronomical Refraction (WGAR) of Commission 8 of the IAU for his kind advices and encouragements.

REFERENCES

Bean, B.R. and Dutton, E.J., 1966. "Radio Meteorology" pp,65-76.
Teleki, G., 1967. "Publ. Astron. Obs. Beograd," 13, pp, 1-147.

DISCUSSION

G. Teleki: stressed that Sugawa's and Kikuchi's investigation shows that for the calculation of new refraction tables we have to use a real atmosphere and not a mathematical spherical symmetric atmosphere.

B. Garfinkel: Is it possible to use also the polytropic atmosphere for this kind of investigation?

C. Sugawa: answered that it is possible. We have actually used the exponential representation for a two spheres model, which is a kind of the polytropic atmosphere.

ASTRONOMICAL REFRACTION AT THE MIZUSAWA LATITUDE OBSERVATORY

Shigetsugu TAKAGI and Yukio GOTO
International Latitude Observatory of Mizusawa, Japan

ABSTRACT

The formulation to calculate the precise refraction by means of the primitive equations by an electronic computer was made to be used for the calculation of the atmospheric refraction in the astrometry and the satellite geodesy with upper air data obtained with the atmospheric soundings.

This formulae were applied to calculate the astronomical refraction in the results of latitude observation at the Mizusawa Latitude Observatory. It was concluded that the results show lager refraction as compared with values due to Radau's and Pulkovo Refraction Tables.

The refraction calculated with the upper air data should be interpolated at each time of observation to be applied to the results of the astronomical observation.

To make an effective application of the formulae, the astronomical refractions were calculated on each clear night and the expressions to represent the actual astronomical refraction from the ground surface to the upper air were derived from these results.

It was found that these expressions can be given in a form

$$\log(n-1) = a + bh \quad \text{from ground to tropopause}$$

$$a' + b'(h-h_0) \quad \text{from tropopause to the upper air,}$$

where n is the refractivity.
The quantities a, b, a' and b' have almost same values during a night and these formulae can be used for the interpolation of the astronomical refraction at the time on the observation.

1. EXPERIMENTAL FORMULA

The astronomical refraction is one of the most important and difficult problem in the astrometry. We usually use the Refraction Table or experimental formulae to calculate the constitution of the atmosphere by means of the atmospheric factors measured at the observing site. Many attempts have been made to obtain the much more precise value of the astronomical refraction by means of the data of the actual atmospheric factors observed, for example, by the Radio Sonde. We attempted, in this report, to derive experimental formula to obtain the precise value of refraction for each astronomical observation.

According to the Fermat's principle, we derived the following primitive differential equations to compute the path of a ray (Takagi, 1974)

$$\frac{d\varphi}{dr} = \frac{N_\varphi \cos \varphi}{r \sqrt{n^2 r^2 \cos^2 \varphi - N_\theta^2 - N_\varphi^2 \cos^2 \varphi}},$$

$$\frac{d\theta}{dr} = \frac{N_\theta}{r \cos\varphi \sqrt{n^2 r^2 \cos^2 \varphi - N_\theta^2 - N_\varphi^2 \cos^2 \varphi}},$$

and

$$\frac{dN_\varphi}{dr} = \frac{\partial n}{\partial \varphi} \sqrt{1 + r^2 (\frac{d\varphi}{dr})^2 + r^2 \cos^2 \varphi (\frac{d\theta}{dr})^2} - \frac{nr^2 \cos\varphi \cdot \sin\varphi \cdot \theta_r^2}{\sqrt{1 + r^2 (\frac{d\varphi}{dr})^2 + r^2 \cos^2 \varphi (\frac{d\theta}{dr})^2}},$$

$$\frac{dN_\theta}{dr} = \frac{\partial n}{\partial \theta} \sqrt{1 + r^2 (\frac{d\varphi}{dr})^2 + r^2 \cos^2 \varphi (\frac{d\theta}{dr})^2}$$

where n is the refraction index and N_θ, N_φ and N_0 are defined by

$$N_\theta = nr^2 \cos^2\varphi \frac{d\theta}{dr} / N_0, \quad N_\varphi = nr^2 \frac{d\varphi}{dr} / N_0,$$

$$N_0 = \sqrt{1 + r^2 (\frac{d\varphi}{dr})^2 + r^2 \cos^2 \varphi (\frac{d\theta}{dr})^2}$$

having a polar coordinate system fixed to the Earth in which r is the radius of a point from the geocenter, θ the east longitude and φ the latitude. We applied these formulae to the actual data obtained in the vicinity of the Mizusawa Observatory and reduced them to the value at Mizusawa by the interpolation. We calculated the value of refraction for verious zenith distances in the previous paper (Takagi and Goto, 1975) and found the constant and the seasonal systematic differences between our value and these given by the Radau and the Pulkovo refraction tables. Moreover, our results shows that the refraction is not symmetric around the zenith. For instance, the difference between the refractions for north star and south star with zenith distance 20° amounts to 0".13 and shows slight seasonal variation with about 0".01 amplitude.

2. SOME ASPECTS OF REFRACTION INDEX

Difficulties in deriving the actual refraction are in the fact that we have hardly any data of the Radio Sonde at the time and at the site of the astronomical observation. There are many complex factors to determine the conditions of the atmosphere and these factors vary from time to time and from place to place. It is very difficult problem for the meteorologist to know detailed values of the meteorological factors from the data of Radio Sonde observation (Teleki 1974). We tried in this paper to devise a process to make interpolation of the atmospheric factors. We can see from our previous results (Takagi and Goto, 1975) that the refraction index can be expressed by a linear equations in three regions.

  a. from the surface to the top of the boundary layer
     (about 1.5 km height)   (Region I)

  b. from the top of the boundary layer to the tropopause,
     (Region II)

  c. from the tropopause to the top of the atmosphere. (Region III)

By assuming a formula

$$\log(n-1) = a + bh$$

where n ; refraction index
      h ; height (km)

we calculated the values a and b for the regions II and III for stations around the Mizusawa Observatory in a period from 1 January to 17 January, 1971. The values of a's will vary from time to time and from place to place, whereas the values of b's will not vary in a region with a small area. We will give the average values of a's and b's in this period for each station in Table 1 together with the standard deviation.

Table 1.

Station	Long. (E)	Lat. (N)	Region II a	b	S.D.	Region III a	b	S.D.
Sapporo	141.3	43.1	0.4814	-0.0503	0.0011	-0.0159	-0.0670	0.0008
Akita	140.1	39.7	0.4750	-0.0494	0.0015	-0.0688	-0.0672	0.0011
Sendai	140.9	38.3	0.4707	-0.0492	0.0015	-0.0029	-0.0668	0.0013
Wajima	136.9	37.4	0.4718	-0.0489	0.0012	-0.0024	-0.0673	0.0011

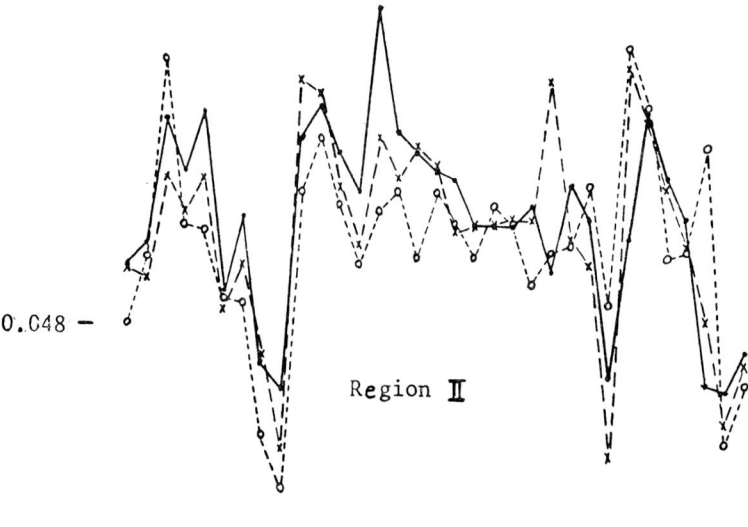

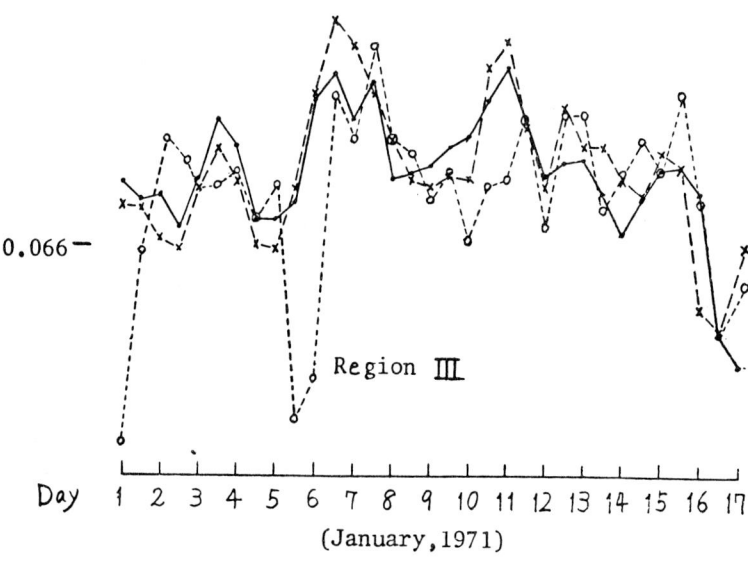

Fig. 1

The figure 1 shows the time variation of b's in the regions II and III at the same station. This figure shows us that the time variations in the three stations are very close and have almost the same tendency in the two regions, which reveals us the possibility to get the values of b's by the regional and time interpolation from the results obtained by the Radio Sonde observation. This problem should be solved with a numerical analysis which will need further investigation.

3. SUMMARY

In consideration of the results of our investigations we would like to suggest the following procedures to determine the precise refraction at any station.

  1) the atmospheric factors in the boundary layer should be observed at the observing site with a new device simultaneously with the astronomical or earth satellite observation.

  2) the atmospheric refraction above the top of the boundary layer can be interpolated from the data obtained at the stations around the observing site by assuming the simple formula of the linear variation of the refraction index.

Our investigations are not sufficient to detect the fact whether we can use the average value of b's for the both regions II and III obtained in many years or not. Our next investigations will be stressed to find a period when we can use the data to be applied for the interpolation at the time and at the site of the observation.

REFERENCES

Takagi, S.: 1974, Publ. Int. Latit. Obs. Mizusawa, 9, No. 2, 241.

Takagi, S. and Goto, Y.: 1975, Publ. Int. Latit. Obs. Mizusawa, 10, No. 1, pp. 41-51.

Teleki, G.: 1974, Publ. Obs. Astr. Beograd, No. 18, pp. 213-234.

DISCUSSION

J. Milewski: remarked that the variations of refraction table values mentioned in the presented paper show systematic changes, which can be connected with the change of relative composition of atmosphere layers, particulary with the permanent increasing of CO_2 component in the atmosphere. He proposed an investigation of the variations of atmosphere composition with time and its influence of different refraction table values elaborated at various period of time.

J. A. Hughes: The change of the index of refraction due to the presence

of CO_2 may be explicitly accounted for by using the expressions given
by Owens. This could be used for the proposed investigation.

G. Teleki: regret that the authors did not analyse the variations of
refractive indices at the boundary layer, which are dominant in the
calculation of refraction values, and that they did not compare their
results with similar investigations already published by Sugawa and
Kikuchi.

CHROMATIC REFRACTION IN THE VERTICAL CIRCLE OBSERVATION

B.K. Bagildinsky, S.P. Puliaev, E.G. Zhilinsky
Pulkovo Observatory of the USSR Academy of Sciences
Leningrad, USSR

In 1962 a photographic vertical circle (PVC) for determining absolute declinations was designed and constructed at the Pulkovo Observatory under the supervision of Prof. M.S. Zverev (Zverev, M.S., 1960). The Maksutov meniscus cassegrain system (D = 200 mm, f' = 2000 mm) was used for the first time in astrometry. Its construction is symmetric and it is automatically operated. In 1963-1966 the declination observation of bright and faint stars of the southern hemisphere was made with this instrument (Zverev, M.S. et al., 1966). At present after the reconstruction the PVC has been used again for determination of absolute declinations (Bagildinsky, B.K., Usanov, D.S., Smirnov, B.N., 1978).

Observations of the fundamental FK4 stars made in 1974-1976 show that the PVC instrumental system is sufficiently stable and ensures high precision as far as accidental errors are concerned (Bagildinsky, B.K., Zhilinsky, E.G., Shishkina, V.N., 1978). More than 200 stars at zenith distances $Z \leq 82°.5$ were observed. The mean square error of one observation is $\sigma_0 = \pm 0".18$ at the zenith and about $\pm 0".30$ at $Z = 50°$. Refraction was computed using the Pulkovo Refraction Tables for $\lambda_0 = 5753$ Å (Pulkovo Refraction Tables, 1956).

An analysis of observations showed the presence of chromatic refraction effects, particularly, at great zenith distances (Figure 1).

As seen from Figure 1 there is a dependence of the calculated deviations on Z, i.e. $\Delta\phi(Z)$; ϕ_i is the mean value of latitude from 5 - 8 observations of each star, ϕ_0 the accepted value of latitude. The results have been tentatively corrected for flexure (b = 0".95 sinZ). The division errors have not been taken into account. The conventional signs mark spectral types of the stars. It is also clear from Figure 1 that the early type stars (O,B,A) are most affected by chromatic refraction. The refraction effects are northwards and southwards asymmetric in relation to the zenith. Leningrad is situated to the north of the Pulkovo Observatory, so the northern part of the sky is more affected by the city lights and the atmosphere is less transparent. This circumstance seems to be the cause of the asymmetry of our results.

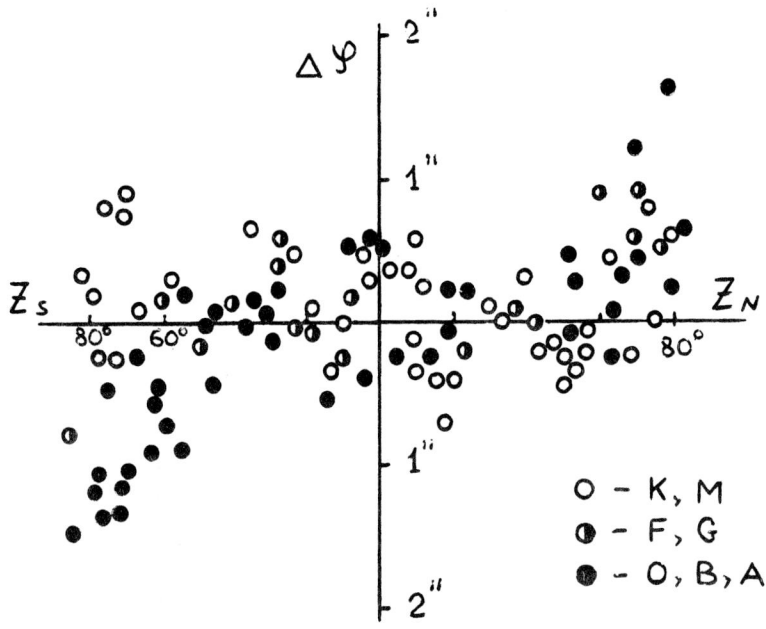

Figure 1

Observations were mostly made on panchromatic plates ORWO-NP27. Sometimes Ilford Ordinary No 30 were used.

For calculating chromatic refraction corrections spectral transparency of the atmosphere and optical system of the PVC and spectral sensibility of the ORWO-NP27 plates were taken into account. For calculating the effective wavelength λ_e the stars spectra beyond the atmosphere were also taken into account (Melnikov, O.A., 1957, Straisis, V., Svidezskene, Z., 1957). The λ_e values are presented in Figure 2. Corrections for chromatic refraction calculated as the differences between values of refraction for λ_e and $\lambda_0 = 5753$ Å at the same Z are given in the table.

For stars at $35° \leq Z \leq 60°$ we have a decrease in dispersion in latitudes ϕ_i and ϕ_0 a 1.3 fold gain in precision. For observations up to $Z \leq 82°.5$ a 1.7 fold increase was obtained. The results are preliminary and will be improved after the reduction of observations with the PVC.

To solve the chromatic refraction problem for meridian and vertical circles we intend to undertake a special investigation.

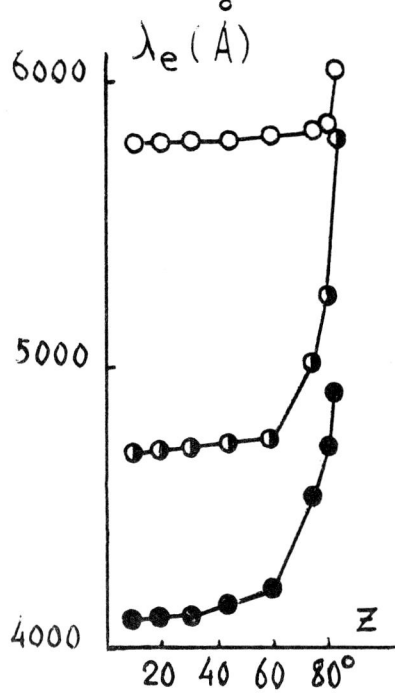

Figure 2

Z \ Sp	B5	G0	M0
10°	+ 0".17	+ 0".09	0".00
20°	0.36	0.19	− 0.01
30°	0.58	0.31	− 0.01
45°	0.95	0.50	− 0.02
60°	1.54	0.84	− 0.04
75°	2.34	1.16	− 0.11
80°	2.87	1.14	− 0.19
82°.5	2.83	− 0.09	− 0.64

Table of Chromatic Corrections

It is highly desirable that the programs of fundamental stars both faint and bright, consist of the K-M stars because the chromatic effects for these types of stars are minimum. The Soviet project of the catalogue of faint stars satisfies this requirement (Zverev, M.S., 1952). It seems advisable that the results of the observations of various spectral types stars be reduced to a unique system by applying corrections for chromatic refraction using argument $(\lambda_e - \lambda_0)$ where λ_0 is the wavelength for which refraction is calculated. For example, for λ_0 one can take $\lambda_v = 5430$ Å in accordance with the Vilnius photometric system (Straisis, V., Svidezskene, Z., 1957, Zdanavichus, K. et al., 1972) or $\lambda_v = 5500$ Å in the UBV system.

REFERENCES

Bagildinsky, B.K., Usanov, D.S., Smirnov, B.N.: 1978, Trans. of the 20th Astrometric Conference of the USSR, pp. 154-157.

Bagildinsky, B.K., Zhilinsky, E.G., Shishkina, V.N.: 1978, Trans. of the 20th Astrometric Conference of the USSR, pp. 55-58.

Melnikov, O.A.: 1957, Trans. of the Pulkovo Observ., No 157.

Pulkovo Refraction Tables, 1956, 4th ed., Moscow-Leningrad.

Straisis, V., Svidezskene, Z.: 1957, Bull. Vilnius Observ., No 35.

Zdanavichus, K. et al.: 1972, Bull. Vilnius Observ., No 34.

Zverev, M.S.: 1952, Catalogue of Faint Stars as an Astrometric Problem, Moscow.

Zverev, M.S.: 1960, Trans. of the Pulkovo Observ., No 166.

Zverev, M.S. et al.: 1966, Trans. of the Pulkovo Observ., No 181.

DISCUSSION ABOUT REFRACTION CORRECTIONS IN STAR CATALOGUE WORK AND GEODETIC ASTRONOMY

Chairman: J.A. Hughes

J.A. Hughes; gave, as an introduction, a summary of the problems involving refraction which are encountered in fundamental astrometry (star catalog programs):
As is well known, the values calculated on the basis of commonly used refraction tables (or theories) are not sufficiently accurate, and hence it is necessary to analyze circumpolar observations in order to deduce corrections. This procedure is only marginally acceptable since a single number (derived from many different nights and from only one part of the sky) is propagated, on a more or less ad hoc basis, over the entire sky. Thus the main problem is how to get information regarding the systematic local or regional effects, and regarding their variations as well.

B. Garfinkel: asked for information on the limitation of the Space Telescope, which will be outside of the earth's atmosphere.

J.A. Hughes: The astrometric limitations of the Space Telescope derive from the fact that it can measure very small differential angles only. The prime task of fundamental astrometry involves measuring large arcs in an absolute instrumental system.

G. Teleki; discussed several questions:
1) He underlined the importance of the connection and comparison of the old and new astrometrical observational series and for this task the refractional influences must be known with higher accuracy than nowadays.
2) Related to the refractional influences there are several philosophical and theoretical considerations, but the astrometry needs the practical instructions for the calculation of these values. Therefore it is necessary to translate our words into deeds, and to introduce nothing but new methods for the refraction calculations which are closer to the reality.
3) It is much to be wished to devise such observing methods of determining latitudes, declinations etc., which would enable us to reduce or even eliminate the refraction influences.

K. Ramsayer: For the calculation of refractional influences we must use the true atmosphere instead of a model atmosphere. On the basis of measurements of pressure and temperature with balloons we can calculate the refractive index variations and can construct the topography of the layers of equal refractive index. From these layers we could calculate by numerical integration a better value of the refraction. However, the difficulty is that the distances between the meteorological stations are too large and that the stations are in most cases too far away from the observation place.

J.A. Hughes: Agreed completely with Ramsayer's remarks, and mentioned the possible use of LIDAR to measure the isopycnic tilts. Whether or not this method could be used to absolutely measure the density of the atmosphere remains to be seen.

K. Poder: added his opinion to Ramsayer's idea, and mentioned that H. Moritz has shown that the refraction correction is a kind of conformal mapping. In this case we need several mathematical relations between purely refractional measurements and also astrometrical or geodetical observations.

B. Garfinkel: stressed the importance of refraction values calculation on the basis of modern theories, based on polytropic atmospheres, and not on the last century theories, which used the exponential distribution of density (temperature). In his opinion it would be better, instead of the application of actual atmosphere, to determine the adequate polytropic atmosphere using the balloon observations and to check this calculation against what we know.

C. Sugawa: informed about his intention to compare the formulae for the calculation of refraction for radio and optical waves in the troposphere.

TWO COLOR REFRACTOMETRY, PRECISION STELLAR CATALOGS, AND THE ROLE OF ANOMALOUS REFRACTION

Douglas G. Currie
University of Maryland

I. INTRODUCTION

Atmospheric refraction, with its non-predictable variations, is one of the dominant sources of error in the precise determination of the apparent position of a star. It appears that this will be the dominant error source for the new generation of instrumentation now being installed for the determination of stellar catalog and Universal Time (i.e., the USNO 65-cm PZT). The effects of anomalous refraction measurements of stellar position propagate to the various quantities derived from this data. This includes the compilation of stellar catalogs, the determination of Universal Time, and variations in the latitude.

A unique instrument, the Two-Color Refractometer, can for the first time, directly measure the total deviation which the light of a star suffers due to refraction and thus can be used in a measurement program to evaluate the various aspects of anomalous refraction. In addition, this system may be used to make fundamental astrometric observations which are free from the deleterious effects of refraction.

A. Atmospheric Refraction

There are two aspects of refraction errors which are useful to discuss separately. The first of these effects are the short-term errors or the "random error" in an individual measurement. These errors affect an individual measurement so that it may be significantly different from measures made before and after it on the same night. This type of error may be of the order of 0.15 arc-seconds or larger. When we later discuss all of the errors more generally in terms of the power spectra of the anomalous zenith refraction, the random error is characterized by the magnitude of the power spectra in the domain having periods of a few minutes.

The many individual measurements may be analyzed and the results averaged together for an entire night. This procedure should result in an improved determination of the quantities derived from the stellar position measurements. The improvement should be parameterized by a factor proportional to $1/\sqrt{n}$. This will occur if the errors of each separate measurement are independent. However, if there are phenomena which affect the anomalous refraction which have long periods, then the determination of the averaged quantities may now show this improvement. Again, discussing the question in terms of the power spectra, one will obtain the improvement in performance only if the power spectra vanishes for periods longer than a few minutes. We now address the effects of the power spectra for periods greater than the time which is required for a single observation. The systematic error to be expected may be derived from the power spectra with the period in question.

B. Derivations of Star Catalogs

We wish not to consider the question of the improvement of the star catalogs. This will require one to both improve the measurement of the single observation and reduce the systematic errors obtained when averaging data over long periods. The former will improve the basic random error and the latter will insure that the result of many measurements will reduce this diminished random error in proportion to $1/\sqrt{n}$. In general, the stellar catalog measurement using the Transit Circle is not affected by three types of error. These are: the anomalous refraction, the encoder errors, and the lack of stability of the personal equations. It is difficult to separate the effects of these errors prior to the measurements of anomalous refraction. Therefore, we shall address the remaining remarks to measurements made with the Photographic Zenith Tube (PZT). For the PZT, the errors of encoders and lack of reproducibility of personal equation are far less significant than in the Transit Circle. In addition, to first order one is independent of normal refraction and only sensitive to the anomalous refraction. In this discussion, we refer to the "normal refraction" as that theoretical quantity derived from the zenith distance of the star, and the local pressure, temperature, and humidity. The anomalous refraction is the difference between this normal or calculated refraction and the actual measured refraction.

In the case of the PZT, the normal random error is of the order of 0.18 arc-seconds. This is the internal precision for one night and does not include the effects of longer period phenomena which is probably significantly smaller. It is more difficult to estimate the errors with a longer period. These effects, due to phenomena like the heat island effect of Washington, D. C., the seasonal variation of the heat island effect, and long term weather effects may lead to violation of the horizontal uniformity which is assumed in the normal data reduction procedure. The derivation of the averaged measures for one night at the USNO may be of the order of 0.06 arc-seconds for periods between one and one hundred days.

C. Two-Color Refractometer

In order to address the role of anomalous refraction in these determinations of star position, we suggest the development and use of the Two-Color Refractometer (TCR). The TCR is a system which measures the magnitude of the refraction due to the atmosphere when observing a star. This measurement is not affected by problems of an encoder, a star catalog, or variations in personal equation. It is a measurement which isolates the refraction alone.

The primary subsystem of the Two Color Refractometer is the Quadrant Sensor System (QSS). The QSS is an operating system which has been developed at the University of Maryland. The QSS is an "eyepiece" which permits one to determine the apparent star position with great accuracy. This determination of the apparent direction to the star is performed by observing in two different colors. Using a knowledge of the index of the refraction and the dispersion of the air, one may compute from this data the instantaneous magnitude of the angle of refraction. Using the local temperature, pressure, and humidity, one may then compute the normal refraction. The difference between the normal angle of refraction and the measured angle of refraction is the anomalous refraction. The latter will be done by exactly the same calculation which is normally used by USNO for the reduction of transit circle observations.

In general, the anomalous refraction will depend upon both zenith distance and time. Let us for the moment neglect the time variation. Then a number of stars may be observed "at one instant" (actually a period of about thirty minutes). The measured value of the anomalous refraction may then be expressed in a power series function about the zenith in terms of "east" and "north" direction. The first term, which will be a constant term, which might be due to the use of erroneous values of local pressure, temperature and humidity, for the computation of the normal refractions. Any linear terms which appear would be due to gradients in the atmosphere. Quadratic terms should be small if the theory is proper and the constant term is not large.

II. OPERATION AND PERFORMANCE OF QUADRANT SENSOR SYSTEM

In this section we consider an ideal Quadrant Sensor System (QSS) and discuss the expected performance of such a system. The objective of the first part of this section is to define the type of sensor system which would be most desirable and to provide a model for the performance analysis which will appear later in this section. The realization of the QSS system will be discussed in Section IV.

The QSS system is based upon a special photosensor, the "Quadrant Photo Sensor" produce by the Electronic Vision Company, a division of

Science Applications, Inc. The Quadrant Photo Sensor (QPS) consists of a quadrant silicon diode which is mounted in an evacuated envelope. Photons are converted to photoelectrons at the photocathode and accelerated electrostatically to an energy of 15 KeV. They are then electrostatically imaged onto the quadrant diode. The overall mechanical configuration of the Quadrant Photo Sensor (QPS) is illustrated in Figure 1.

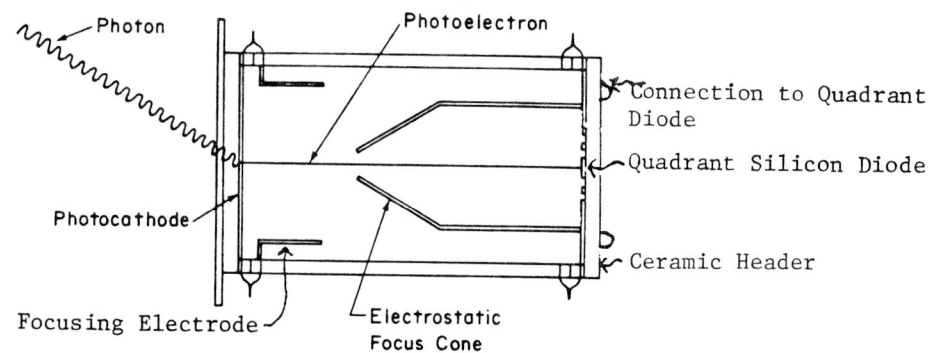

Schematic Diagram of Quadrant Photo Sensor
Figure 1

The photoelectron, which is accelerated to 15 KV, produces entensive ionization within the silicon in the photodiode. The output of a photodiode which is bombarded by a single photoelectron is a charge packet which consists, at the normal operating voltage, of about 2,000 electrons. This pulse signal generated by this charge packet is amplified, shaped, and finally detected with a level discriminator. In order to operate such a system, we wish to provide functions indicated in Figure 2.

Thus Figure 2 illustrates the photon detection system, which provides a high-level pulse for the arrive of each photoelectron. By accumulating these counts, one would have available the intergal number of counts for any interval. However, we wish to use such a system in several applications where a greater variety of outputs are required. The above system will be used as the basis for the calculations presented in Section II. At present, however, we shall now continue this discussion to describe the full Quadrant Sensor System requirements. The subsystem described in Figure 2 is connected to a data processing subsystem.

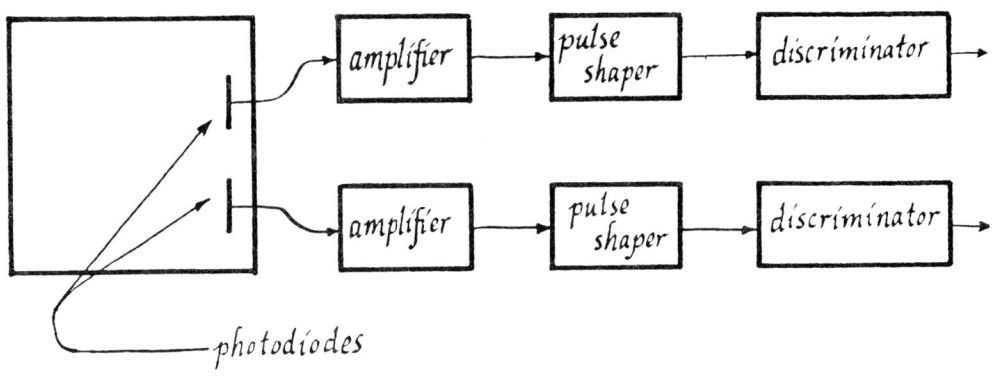

QSS Detection Subsystem for Single Photoelectrons
Two of the Four Channels are Shown
Figure 2

Thus Figure 2 illustrates the photon detection system which provides a high-level pulse for the arrival of each photoelectron. By accumulating these counts, one would have available the intergal number of counts for any interval. However, we wish to use such a system in several applications where a greater variety of outputs are required. The above system will be used as the basis for the calculations presented in Section II. At present, however, we shall now continue this discussion to describe in Figure 2 is connected to data processing subsystem.

The QSS has been designed both to provide the input for a telescope used for tracking of stars, but also to detect laser returns from satellites and track on these returns. However, the latter function is not used in this application and will not further be considered.

When operating in the photon sensing mode, the system is expected to produce an output signal which is proportional either to the intergal number of counts or the count rate smoothed in some fashion. Two outputs of this type, which are the difference between the counts from each axis, are available (X high pass, Y high pass). In addition, the output is then filtered with a low pass filter having a time constant of about 1 second. This is available as $X_{low\ pass}$ and $Y_{low\ pass}$. The high speed outputs will be used as the input or an external subsystem to stabilize the image with a high frequency response. The low frequency output will provide the signals required to move the telescope. The signals described above are available in analog form in order to operate the analog servo-loops. They are also multiplexed onto an A-to-D Converter to provide outputs which may be used for computer control and recording recording. Thus this unit is capable of both stabilizing the image, and through the photon counts, provide interval number of counts in the various channels.

In this discussion, we shall evaluate some of the numerical performance parameters for the quadrant sensor. In this section we shall presume the use of a quadrant photon counting sensor. We will assume that the photocathode has a uniform quantum efficiency and that the electronics system imposes no rate limits. The description of the actual sensor and its limitations appears in a later section.

A. Photon Noise in Direct Mode

In this section we shall compute the photon noise which affects the determination of the centroid of an image by a Quadrant Sensing System. For pedigogical ease, we hall first consider a stellar image which is square, so the brightness has the form indicated in Figure 3.

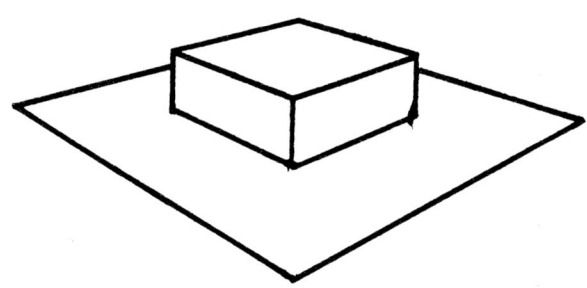

Brightness of Stellar Image
Figure 3

The difference between the performance criteria for this approximation to the stellar image (by a square pillbox) and a more realistic approximation, i.e., a Gaussian or a Lorentzian form, will be discussed later. The more realistic forms will result in about a 2% modification of the numerical results obtained for the squre pillbox image.

The results shall be expressed in terms of the total number of photoelectrons which arrive on all four quadrants (denoted by the symbol N) during a time interval denoted by the symbol T. The full width of the image described in Figure 3 in denoted d, which might typically have a value of two arc-seconds for a reasonable telescope used at a good astronomical site. The one dimensional projection of such an image has the form indicated in Figure 4.

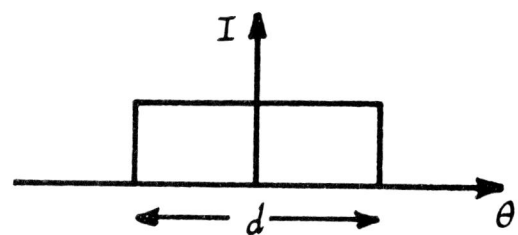

One Dimensional Projection of Square Pillbox Image
Figure 4

The number of photoelectrons per linear arc-second at the center of the square image, denoted by $\tilde{N}$, is given by

$$\tilde{N} = N/d = N/(d_o\sqrt{3}) \tag{II.A.1}$$

where d_o is the "rms diameter". This quantity is obtained by doubling the rms radius, which is the deviation of the light intensity distrition over the image plane. We may reexpress $\tilde{N}$ in the form

$$\tilde{N} = (\eta N)/(2d_o) \tag{II.A.2}$$

where we have defined the symbol η to represent a numerical coefficient which is related to the shape of the image. Thus from the above expressions, we see that for a square image

$$\eta_s = 2/\sqrt{3} = 1.547 \tag{II.A.3}$$

Now we consider an image with a more realistic profile, i.e., a Guassian. The one dimension projection of such an image is indicated in Figure 5.

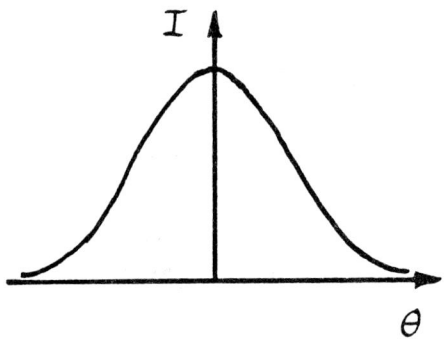

One Dimensional Projection of Gaussian Image
Figure 5

The intensity $\tilde{N}$, expressed as the number of photoelectrons/arc-seconds, has the form

$$\tilde{N}(x) = \frac{\sqrt{2}\, N}{\sqrt{\pi}\, d_o} e^{-2x^2/d_o^2} \qquad\qquad (II.A.4)$$

Thus at origin

$$\tilde{N} = \tilde{N}(o) = \frac{\sqrt{2}\, N}{\sqrt{\pi}\, d_o} = \eta_g\, N/2d_o \qquad\qquad (II.A.5)$$

Where we have again used the coefficient as η. Thus for the Gaussian we have

$$\eta_g = (2\sqrt{2})/\sqrt{\pi} = 1.5958 \qquad\qquad (II.A.6)$$

Thus in general, the value of η will be dictated by the model for the seeing disk, or from a more pragmatic viewpoint, from an actual measurement of the size and profile of seeing disk.

We may later take into account a large deflection or error in image position by expressing η as a function of the offset of the image. Thus if there is an average offset of Δ, then $\eta(\Delta)$ will have a smaller value. In this way we will be able to address some problems which will arise in the two color application. Thus for a Guassian image

$$\eta(\Delta) = \frac{2\sqrt{2}}{\sqrt{\pi}} e^{-2\Delta^2/d_o^2} \qquad (II.A.7)$$

The quantity $\tilde{N}$ is physically related to the change in signal or differential count rate motion of the image. Thus if we apply a sinusoidal offset or a "dither" to the position by an optical device, we will see a similar variation in $\tilde{N}$. This will obviously be related to the profile, as expressed by η. On the other hand, if we assume a profile, we may invert this relation and use the measured value of $\tilde{N}$ which is obtained by applying a dither to the servo-system to evaluate the seeing disk diameter, using the expression

$$d_o = \eta N / 2\tilde{N} \qquad (II.A.8)$$

If the image is nominally centered, then a small image offset which is parameterized by the value of will result in a difference in count rates which is given by:

$$\Delta N = N_L - N_R = (N_{Lo} + \tilde{N} \cdot \Theta) - (N_{Ro} - \tilde{N} \cdot \Theta) \qquad (II.A.9)$$

We will describe the one-dimensional case. If the image was initially centered and the quantum efficiencies of the effective areas of the photocathode are equal then $N_{Lo} = N_{Ro}$ and the above equation becomes

$$\Delta N = 2\tilde{N}\Theta = \eta N\Theta/d_o \qquad (II.A.10)$$

which may be reexpressed as

$$\Delta N/N = \eta \Theta/d_o \qquad (II.A.11)$$

or using ΔN to compute the offset angle

$$\Theta = (\Delta N/N) d_o/\eta \qquad (II.A.12)$$

We may now define a relative angle which is expressed in terms of the image diameter, so we have

$$\alpha/d_o = \Delta N/N\eta \tag{II.A.13}$$

This expresses the angular offset due to an inequality in counts as a dimensionless angle.

Now let us determine the standard deviation of the angular position due to shot noise, i.e., the statistical error in the determination of the centroid due to the photon statistics. In each pair of quadrants (presuming we sum the photocounts from both quadrants of each side to determine the error) the mean value is

$$\overline{N}_L = \overline{N}_R = N/2 \tag{II.A.14}$$

Thus for large value of N, the statistical noise in the value of each of these is

$$\delta N_L = \delta N_R = \sqrt{N/2} \tag{II.A.15}$$

Since the statistical noise in each of these quantities are independent, we have

$$\delta N = \sqrt{(\delta N_L)^2 + (\delta N_R)^2} = \sqrt{N} \tag{II.A.16}$$

inserting this value of δN into the above expression for yields an equivalent angular offset due to the noise. Thus we have

$$\alpha = 1/\eta \sqrt{N} \tag{II.A.17}$$

At this point, these considerations will be extended to include the effect of uniform sky background on the performance. Thus the presumed brightness now takes the form

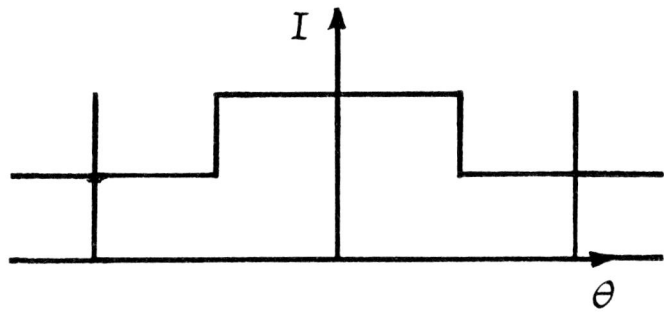

Brightness Distribution for Sky Background
With Boundard of Photosensitive Region Marked
Figure 6

where N_b is the total number of counts which appear in the four quadrants due to the brightness of the backgound sky. This is equal to the product of the field of view (in arc-seconds), the background illumination (in photoelectrons/arc-second/sec), and the integration time (in seconds). The product is expressed in photoelectrons. The image of the star is characterized as in the previous discussion. In this treatment, one simply adds the background to the image, so the mean number of counts on each side is given by

$$N_L = N_R = (N + N_b)/2 \qquad (II.A.18)$$

Thus the photon shot noise is one of the pairs of quadrants (again for relatively large signals) is given by:

$$\delta N_L = \sqrt{N_L} = \sqrt{(N + N_b)/2} \qquad (II.A.19)$$

and the values for N_R and δN_R have similar expressions. Now if we take the difference of these two independent qualities and use this quantity to determine the apparent error in pointing angles, with the presumption that the photon statistics on the two sides are independent and the sky illumination is spacially uniform, we have

$$\Delta N = \sqrt{N + N_b} \qquad (II.A.20)$$

Thus for α we have

$$\alpha = \sqrt{N + N_b}/\eta N = \sqrt{1 + (N_b/N)}/ \eta \sqrt{N} \qquad (II.A.21)$$

where we have used the total counts for the noise in the numeration, and the counts in the image for the signal which appears in the denominator. We have two types of behaviour of interest, the first of which consists of the case when the background is negligible. For this case the error becomes

$$\alpha = 1/\eta \sqrt{N} \qquad (II.A.22)$$

as discussed earlier. This is illustrated for the case of a Gaussian image in Figure 5 by the solid line.

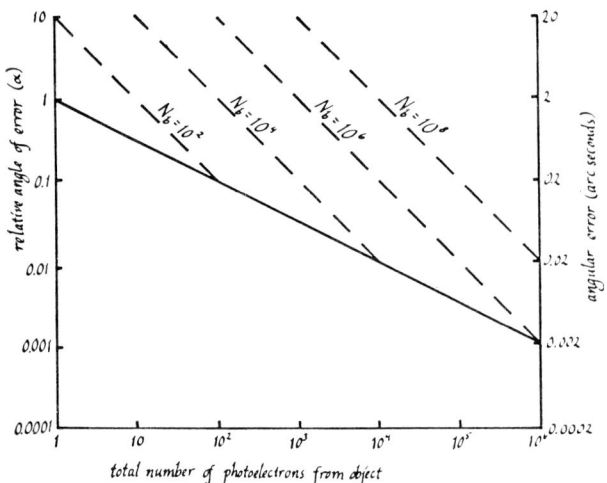

Figure 7

The other extreme case is when there is a high level of background illumination. This might represent observations conducted in the daytime with a blue sky. In this case we have:

$$\alpha = \sqrt{N_b/N} / \eta N = \sqrt{N_b} / \eta N \qquad (II.A.23)$$

This may be summarized in graphical form by the dashed line in Figure 7.

Thus Figure 7 indicated the performance that one may expect to obtain with the quadrant sensor. The expected error in the relative angle and the expected error in arc-seconds, presuming a seeing disc diameter of two arc-seconds are related to the total number of photoelectrons received during the observation period.

We may now tranform this relationship to describe the behavior with respect to stellar brightness and telescope aperture. Let us, for example, consider a 48-inch telescope and photoelectron count rates which are caused by a particular stellar brightness for a G-2 star (or solar reflection from a satellite). We assume a quadi- S-20 photo-cathode as measured for a tube of this type and six mirrors with reflectivities of 0.7, 0.7, 0.86, 0.86, 0.86, and 0.86. Thus for the case of a Guassian image with negligible sky background we have the relation indicated in Figure 8.

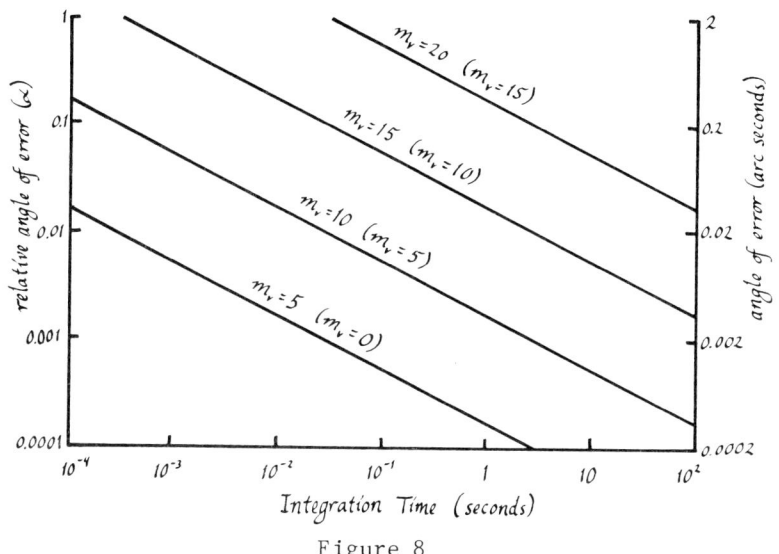

Figure 8

This indicated the ability to go to rather faint objects and still have very small angular errors due to the photon counting. The numbers in parentheses indicate the results for a 4.8-inch aperture.

III. TWO COLOR APPLICATIONS

The Quadrant Sensor System, as a component of the Two-Color Refractometer system may be used to obtain high accuracy measurements in absolute astrometry. In particular, it will determine the deviation in the apparent star position caused by the earth's atmosphere. As star light enters the earth's atmosphere, the deviation of index of refraction of the atmosphere from that of the vacuum results in a change in the direction of propagation. For a flat, horizontally-stratified atmosphere, this change in direction is a known function of the index of refraction of the air in the immediate vicinity of the telescope (the local index of refraction). This may be determined from the local temperature, pressure and humidity. However, if there are horizontal inhomogeneities, this change of propogation direction can not be predicted from a measurement of local surface conditions. In this section, we address a procedure to determine the magnitude of this change in direction (the "angle of refraction") from measurements made on the light of the star using the Quadrant Sensor System.

We take advantage of the fact that the magnitude of the angle of refraction depends upon wavelength, due to the dispersion of the index of refraction of air. Thus by measuring the apparent position of the star as viewed in red and in blue light, the difference in apparent position may be evaluated. This dispersion angle may be used with a knowledge of the dispersion in the index of refraction to determine the angle of refraction. Thus using the dispersion angle between the red and blue images, and the known dispersion of the air, we may determine the deviation of the direction of propogation of the starlight due to the atmosphere.

In order to illustrate the operation of the Two-Color Refractometer (TCR) let us consider the basic relations governing the refraction. The symbol Θ_λ describes the deviation of a ray of wavelength λ from its initial direction and ϕ is the true zenith distance of the star. The index of refraction of the air at the wavelength λ is denoted by η_λ. Thus from Snell's law, we have

$$\Theta_\lambda = (\eta_\lambda - 1)\phi/\eta_\lambda \quad . \tag{III.1}$$

Since η_λ is very close to unity, we shall approximate this expression by:

$$\Theta_\lambda = (\eta_\lambda - 1)\phi \tag{III.2}$$

where the definition of these quantities may be seen in the following figure

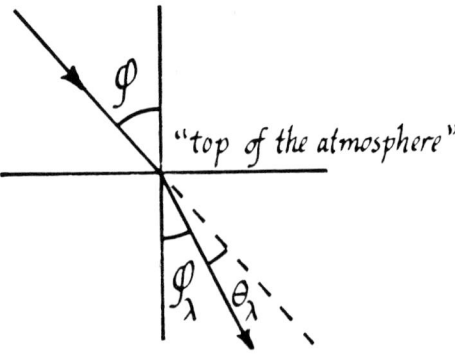

Effect of Refraction on Starlight
Figure 9

While the figure illustrates a simple idealized case, a little reflection indicates that equations apply to the more general case of a continuously variable, non stratified, spherical atmosphere. For the calculation made in the later part of this section, we use the following expression for the index of refraction for dry air

$$(n_\lambda - 1) \times 10^{-6} = 64.328 + 29498.1 \, (146 - \lambda_0^{-2})^{-1} + 255.4 \, (41 - \lambda_0^{-2})^{-1}$$

(III.3)

This applies for a temperature of 15°C, a pressure of 760 mm and no water vapor. This expression is not the optimal form for expressing the index of refraction when reducing the data for the operational system, but it is entirely satisfactory for the system analysis and error analysis presented in this section.

In order to permit the independent variation of the wavelength selected for obervation within a reasonable error analysis, we shall use the blue and red measurements and a parameter which depends on the wavelengths used to determine the magnitude for the angle of refraction at a reference wavelength chosen to be the center of the visual band (5600Å). Thus for the wavelengths at which we will measure the relative angular position, we use:

$$\Theta_R = (n_R - 1) \phi \qquad (III.4)$$

$$\Theta_B = (n_B - 1) \phi \qquad (III.5)$$

We will then use these expressions to determine the zenith distance ϕ, which we will relate to the angle of refraction in the green by the relation

$$\Theta_V = (n_V - 1) \phi \qquad (III.6)$$

Thus we will determine the angle of refraction in the visual band. This particular equation does not appear to represent the optimal procedure for minimizing the propogation of experimental errors but we shall use it as a useful tool for system analysis. Thus we obtain an expression of the form:

$$\Theta_V = \Omega \, (\Theta_B - \Theta_R) \qquad (III.7)$$

where

$$\Omega = (n_V - 1)/[(n_B - 1) - (n_R - 1)] \qquad (III.8)$$

If we measure the difference in apparent angular position of the blue and the red image, this angular difference, multiplied by the quantity Ω, will yield the angle of refraction suffered by the visible light. This is the value for local conditions which are 15°C and a pressure of 760 mm Hg.

In order to provide a general frame of reference, let us consider the numerical values for the refraction and the dispersion angles. Let us consider the value of for various choices of the limiting wavelengths. We have

Effective Blue Wavelength	Effective Red Wavelength	Ω
3500Å	6000Å	30.36
3300Å	6000Å	25.27
3500Å	7000Å	26.89

Relation Between Selected Wavelengths and Ω Multiplier
Table 1

Since the eventual error is proportional to the value of Ω, we would prefer the smallest possible value. From Table 1 we see that the value of Ω decreases and thus the error multiplier decreases as the spread between the wavelengths increase. This dependence is particularly sensitive to the wavelength of the blue observation where a change of 200Å provides as much improvement in Ω as a change of 1000Å in the wavelength of the red observation. However, for various technical reasons we will accept the first case as a reasonable, practical limit.

Let us now evaluate the sources of error in this determination. Differentiating the expression for Θ_V, we have:

$$\delta\Theta_V = \frac{\delta\Omega}{\Omega} \Theta_V + \Omega (\delta\Theta_B - \delta\Theta_R) \qquad (III.9)$$

We now presume the uncertainty in the measurement of the angular position of the image in the red and in the blue have the same magnitude, and are independent. Concerning the former, it assumes similar filter widths which will probably not be the case, but it is a useful point for discussion. Thus for the part of the variation due to uncertainty in angular measurement we have

$$\delta\Theta_V \quad \sqrt{2} \; \Omega \; \delta\Theta \qquad (III.10)$$

where

$$\delta\Theta = \delta\Theta_B = \delta\Theta_R \qquad (III.11)$$

We may now use the relations between the total number of counts and angular errors to express the accuracy of the determination of $\delta\Theta_V$ as a function of the total number of photoelectrons. In Figure 10 we have also indicated the various system operating points obtained when the QSS is running at the maximum count rate (the normal procedure) for various intervals to time. These results include the effect of the UV filter and the red filter.

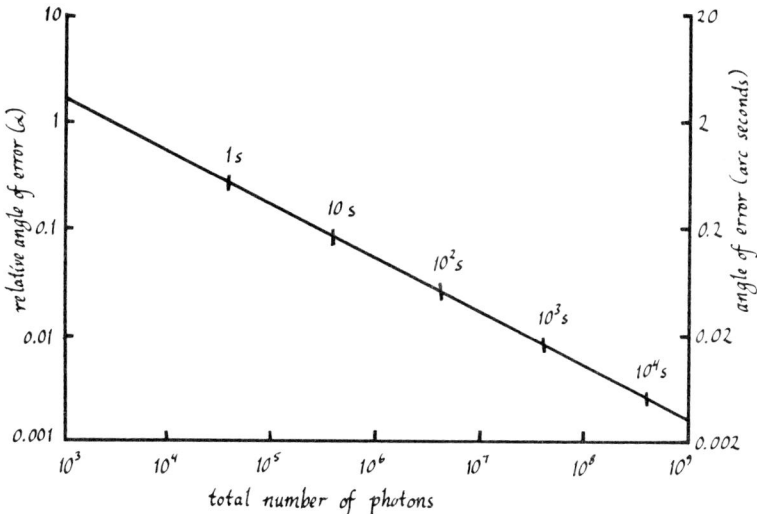

Relation Between Total Number of Counts and Error in
Anomalous Refraction for TCR
Figure 10

We now consider the other source of error which appeared in the Equation for $\delta\Theta_V$. This is the possible variation in the quantity Ω. It is presumed that Ω is calculated for the local value of refraction (i.e. using pressure and temperature). The predominant remaining reason for an controlled uncertainty in $\delta\Theta_V$ is due to the variable water vapor content in the atmosphere. In order to evaluate this phenomenon, the following table is presented. This presents the value of Ω and the change produced in Ω due to precipitable water for various pairs of effective wavelengths. We use the mean water vapor which is expressed in the terms of precipitable water vapor (for a flat atmosphere).

Effective Wavelength for Blue Filter	Effective Wavelength for Red Filter	Ω for no Water	Precipitable H_2O	Ω with Water Vapor	$\delta\Theta/\Omega$
3500	6000	30.3657	10 mm	30.1885	0.0058
3300	6000	25.2696	10 mm	25.1232	0.0058
3500	7000	26.8927	10 mm	26.7351	0.0059
3500	6000	30.3479	1 mm	--	0.00065

The Effect of Water Vapor on Systematic Errors
Table 2

Thus we see that the water vapor correction is approximately proportional to the amount of water and generally independent of the selection of the effective wavelength of the filters.

Let us now consider a numerical example. If we observe a star at zenith distance of 30° as one might with an astrolabe, then using the expression for the refraction under standard temperature and pressure

$$\Theta_V = 58\overset{"}{.}2 \tan Z \qquad (III.12)$$

which applies to a flat atmosphere we have, at 30°, a deflection of about 34". If we wish to be able to determine this correction with an accuracy of $0\overset{"}{.}05$, then we need

$$\frac{\delta\Theta_V}{\Theta_V} = \frac{\delta\Omega}{\Omega} = \frac{0\overset{"}{.}05}{33\overset{"}{.}6} = 0.0015 \qquad (III.13)$$

This means one to know the water content to 2.3 mm of precipitable water.

However, this requirement is not particularly difficult since commercially available microwave ratiometer are accurate to about 1 or 2 mm of precipitable water. In addition, a system attached to the TCR using near IR measurements in and out of a water absorption band should be significantly more accurate and less complicated.

A. Some Practical Considerations

We now consider a few practicle aspects of the design of the TCR. The full discussion of the system will be considered at a later time.

We first consider the procedure for sampling the light at two different colors. One might attempt to measure the position of the red and blue images at the same time with two different quadrant sensor. The parameters which enter Figure 6 are based upon this assumption. However, a more practical approach to this problem is to use a single quadrant sensor. The detailed reasons will be considered in a separate discussion. We shall, briefly, discuss this approach and how we should use it to maximize the accuracy of the evaluation. Thus we note that due to the fact that the refraction in the atmosphere has significant power at various frequencies, one might wish to sample the two different colors continuously and at the same time. If this is done, then both the power at high frequency and the power at low frequency (the image motion) are properly tracked and do not affect the accuracy of the measurement.

However, the mechanical stability requirements strongly suggest the use of a single detector. If one switched the two colors onto the one sensor at a rate of several hundred times a second, one may show that the residual error is expected to be random and has a value of less than 0.01 arc-seconds for the measurement lasting several minutes.

The remaining uncertainty is in the determination of the water vapor. From the previous discussion, and typical values of water vapor to be expected, this effect may be totally ignored for the initial measurements out to a distance of 15° from the zenith.

B. Expected Accuracy for Measurements of Zenith Refraction

We now discuss the accuracy which we would expect to obtain with measurements using this instrument. These will be expressed in terms of the measurement errors to be expected on the power spectral density, and will be based upon use of the 48-inch telescope at the Goddard Optical Research Facility (GORF). However, approximately the same numbers apply to the set of measurements which would be performed on the 36-inch telescope. For these evaluations, we will presume the use of the present circuit for the quadrant sensor which has a maximum counting rate of 10 kilohertz/quadrant. This will permit the observation of 8th magnitude stars with the proper count rate. Since we will limit our measurements to a region that is between 5 and 10 degrees, it is worth noting that we would statistically expect to have about 80 stars within a five degree circle of the zenith.

In general, the relationship between the stellar magnitude, the telescope aperture and the expected photon statistical error for a ten minute observation is expressed in Figure 11.

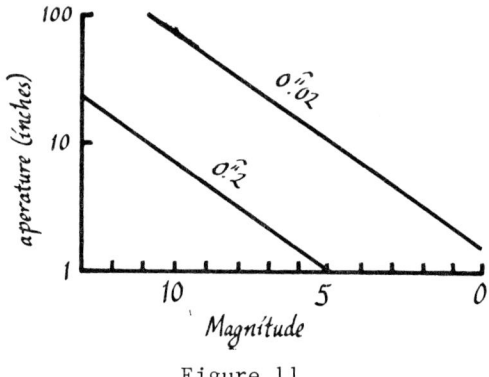

Figure 11

Combining the results of Figure 10 and Figure 11, we see that we may trade off between the aperture size of the telescope with the observing time to use a small telescope and still have a reasonable number of stars to observe.

Now presuming this count rate, we can evaluate the power spectrum of the system errors which we expect. To facilitate this present discussion, we will not consider the true power spectrum but indicate the standard deviation of the system error for a measurement of a given length. The overall results are thus summarized in Figure 12. The system error caused by the photon statistical noise is illustrated by the broad solid line. For periods shorter than one second, one may increase the centroid of the white light image and rely on the smoothness of the telescope tracking to determine the refraction (called image in this domain) rather than the two-color extraction. This is expected, with some study of the records and knowledge of the telescope, to give about $0\rlap{.}''2$. If these measurements were conducted on the 100-inch telescope at Mt. Wilson, which has a very smooth and regular drive, and were conducted on a night of low wind, this system error could probably be reduced to 0.1 arc-seconds. For periods of more than 10 seconds, however, one would expect that errors in the telescope drive will become too large. In order to study the behavior for longer characteristic times, we consider a limit for a single observation of twenty minutes which results in an error of 0.01 arc-seconds. Thus one might expect to obtain an increase in accuracy by about a factor of three, but this will mean the observation of new stars. The feasible accuracy in this region must be more carefully explored. For the longer period term, one averages the data obtained on successive nights to obtain these results. However, since we do not expect to observe for 24 hours a day, there is a break in this curve as one goes to the dashed portion. Presuming an observation period of two hours for a night, we obtain the dashed curve. At about 0.03 arc-seconds one expects to run into systematic errors. It seems possible that these may be calibrated to the 0.001 or 0.002 arc-second level. Accuracies beyond this cannot currently be projected for the first generation instrument.

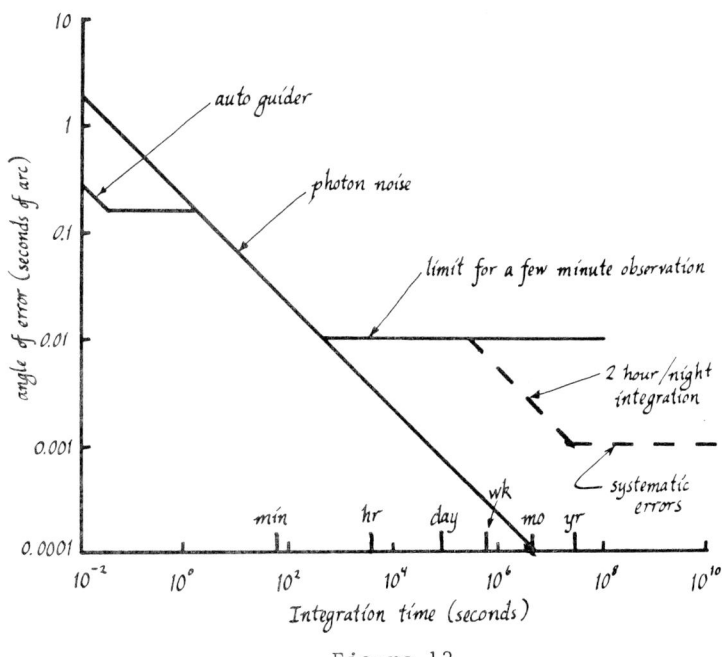

Figure 12

IV. IMPLEMENTATION OF THE QUADRANT SENSOR SYSTEM

In this section we describe the Quadrant Sensor System which has been developed within the Amplitude Interferometer Program at the University of Maryland. This system is the practical realization of the hypothetical QSS which was described in Section II. The basic element of the system consists of two standardized units, the photosensor head and the control unit which have been used in three different astronomical system. The quadrant sensor system is basically a special system which detects light in four channels. These are closely placed and are mechanically and geometrically rigid. In order to satisfy the system requirements, a special device has been obtained from Electronic Vision Company and an electronic system has been developed at the University of Maryland to make these observations. The requirements on this optical/electronic system demand a very high precision, electronically, optically, and mechanically, as well as a considerable amount of generality. The electronic system will permit the quadrant sensor to interface to a variety of data processing system, i.e., computers or hardwired processors.

The Photosensor head contains the Quadrant Photosil. The structure has already been indicated in Figure 4 and is shown here

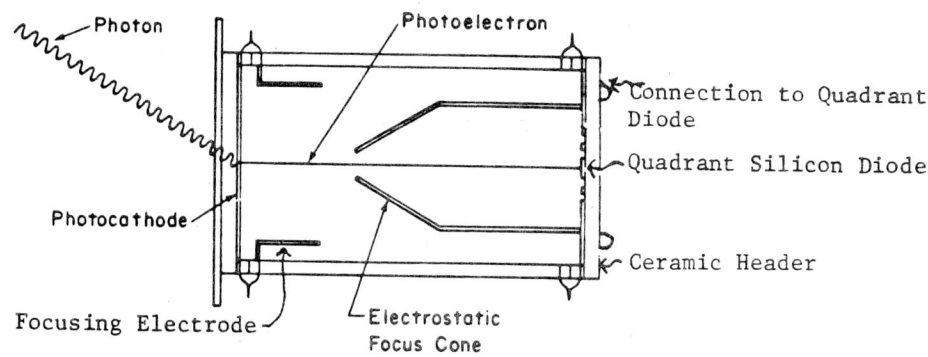

Schematic Diagram of Quadrant Photo Sensor
Figure 13

as discussed in the first section, the circuitry following the Quadrant Photosil is indicated in Figure 14

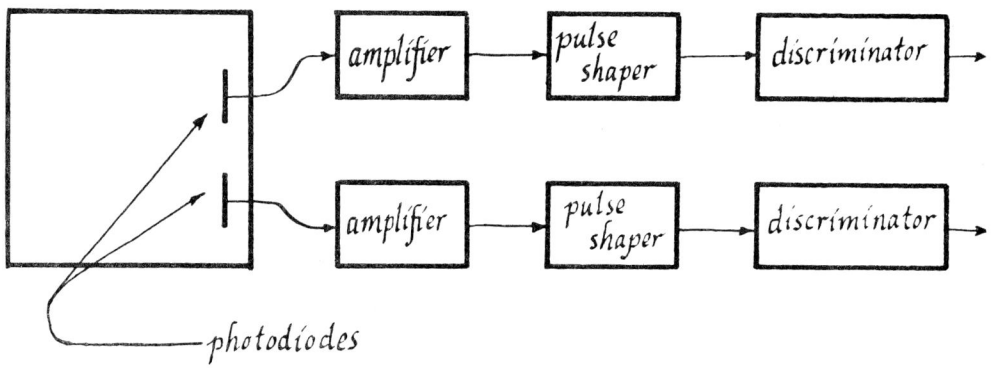

QSS Detection System for Single Photoelectrons
Two of the Four Channels are Shown
Figure 14

This results in a pulse height distribution which permits a straight-forward discriminator. The preamplifier is similar to that photon counting Digicon system developed by E. Beaver at UCSD. The pulse height distribution at 16 KV is shown in the following figure.

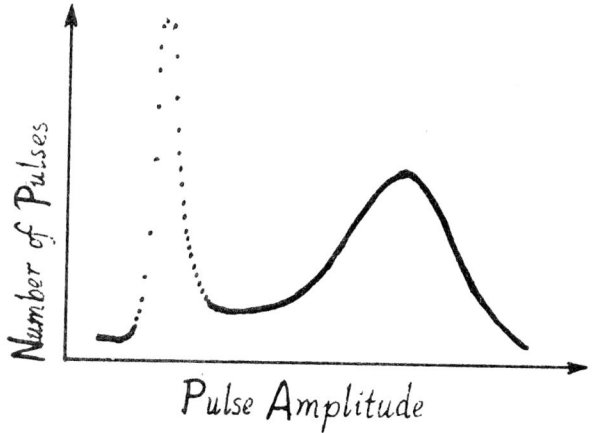

Figure 15

This system is expected to operate in environments with high levels of external EMI. This is a serious problem since the signals are smaller than conventional PMT's by a factor of about 1000 to 10,000.

Special shielding and grounding procedures are requried in order to prevent external TVI from interfering with the discrimination process. This has been provided for the input high voltage leads by the use of solid aluminum shields. The amplifiers are specially isolated and grounded. Even so, the pickup is noticable under certain circumstances.

The preamplifier in the QSS head provides a signal which is relatively immune to external problems. This is then sent by four short cables to the QSS Control Unit. Here the signal is reamplified and sent to discriminators. These discriminators may be adjusted independently for each of the four channels. These are normally set at about the point indicated by the arrow in Figure 15. The shape of this pulse height distribution leads to a relative immunity from effects due to the variation one expects in operating voltages and circuits in the field.

In order to interface to a computer and provide data for the closure of a servo-loop, the four signals are multiplexed and then applied to an A-to-D converter. The output is an eight-bit word. The operation acts as an asynchronous peripheral which supplied data when asked. Thus generalized system has been used in several applications. We shall now describe how it has already been used.

A. Application in Single Aperture Amplitude Interferometer

The first application of the quadrant sensor system was to provide image stabilization in Single Aperture Amplitude Interferometry (SAAI). In this system it operates in only one axis. It incorporates a device which provides high frequency stabilization with a frequency response up to about 60 hertz. The gain is set to permit the utilization of the stabilization to this level. The expected noise performance was discussed in the previous section.

As an indication of the proper performance of this system in Amplitude Interferometry, the measurements of the fringe visibility which are performed when this system is operating are significantly improved. Special procedures have been developed, both for the proper alignment of this system and to permit the simultaneous use of the manual guiding of the telescope. This is required at present since the QSS has not yet been interfaced to the large telescopes due to the necessity of performing the coordinate transformation without a computer.

B. 48-inch Telescope at the Goddard Space Flight Center

The Quadrant Sensor System (QSS) has been used to make an Automatic Guider System (AGS) to control the 48-inch telescope at the GORF. In this case, the output from the high frequency error signal is multiplexed and digitized and fed directly to the computer. There this data on the error in pointing is processed and filtered in a digital form. This system has succeeded in providing precise pointing information and also has been useful in detecting difficulties in the encoder system. Instabilities in the computer processing of the data and the telescope setup drive can cause oscillation under some circumstances. However, the computer procedure for handling the data is now being reprogrammed to handle this properly.

In addition, the GSFC system also provides for the capability to sense the offset of lasers from a satellite reflection. To this end, the capability is provided to gate the photocathode off while the outgoing laser pulse is near the telescope with a high voltage pulse applied to a grid of a modified Quadrant Photosensor. The sensor system can detect the arrival of the laser return. The signal on each diode is amplified, and then compressed for greater dynamic range, and stores before it is presented to the multiplexer. This sequence is activated by the receipt of a photo pulse which has a value which is greater than three photoelectrons. Thus this gives the error in pointing with respect to the laser return. This system has been implemented and tested in a laboratory but the required computer programs for the 48-inch telescope and the required telescope time are not yet available in order to test the laser tracking system on the telescope.

DISCUSSION

G. Teleki: How will the observed stars, with different spectral type, influence your measurements?

D.G. Currie: answered, that because of the different spectral type of observed stars he expects some systematical effects (it is to be hoped that these effects can be determined), but the main problem is connected with the colouring of the atmosphere at different zenith distances.

J. Milewski: What a magnitude of variations (in time) of the dispersion do you expect?

D.G. Currie: answered, that he expects the variations of hundredths of arc second.

B. Garfinkel: Are you able to separate the vertical and azimuthal refractions? Do you intend to determine also the time dependence effecting only the azimuthal component of refraction or both of them?

D.G. Currie: answered the first question in the affirmative. Connected with the second question, he said that the total dependence will be examined as a function of different parameters.

PATH LENGTH VARIATIONS DUE TO CHANGES IN TROPOSPHERIC
REFRACTION

Allen Joel Anderson
Department of Solid Earth Physics
Box 556
S-751 22 UPPSALA Sweden

Abstract

An experiment is described in which microwave Doppler is used to determine very small changes in path length to spacecraft tracked by the Deep Space Tracking Network (DSN). The experiment was carried out to test the detection capabilities of the DSN system to gravitational radiation of very low frequency (10^{-2}-10^{-4} Hz). In this work spectral analysis of Doppler variations were performed for periods over 4 hours and more.

These results indicated that one of major sources of noise was due to rapid variations in tropospheric refraction. The results obtained a differential path length variation, $\Delta L/L$, of 1 part in 10^{14} for periods between 100 and 1 000 seconds.

Doppler spectra are shown and a general discussion of the experiment is given.

1. INTRODUCTION

In experiments first begun in 1969 attempts have been made to measure very low frequency gravitational radiation wave pulses and isotropic cosmic background using Doppler data from the Nasa Deep Space Tracking Network (Anderson, 1971, 1974, 1977). Because the system has certain unique response functions to gravitational radiation, it is possible to spectrally filter the data to improve the sensitivity of the measurement and uniquely determine the result. The two frequency X/S band systems presently envisioned allow for an elimination of space plasma effects, but do not allow for corrections of variations in tropospheric refraction. It therefore became essential to understand these tropospheric effects and to provide means of measuring and modelling them correctly.

Work has begun towards these goals. In a paper by Moran and Penfield, 1976, an evaluation of the use of early water vapor radiometers to determine their capability to measure tropospheric path length propogation was made. Comparison with radiosonde profile measurements gave a rms of 2 - 5 cm absolute errors while those of the radiometers gave a rms of 1.5 cm in the correlation between measured brightness temperature and the calculated wet path length at zenith.

2. TROPOSPHERIC PATH LENGTH VARIATIONS

Variations of tropospheric path length propagation were studied by Hargrave and Shaw, 1978, using the 5 km baseline of the Cambridge Radio Telescope. Interpreting their data in terms of time variations indicated that path length differences amounting to 0.8 cm over periods of 5 000 seconds were typical (Callahan, 1978). It was also pointed out, however, that under favorable conditions, such as dense fog, and under other conditions where atmospheric convection is limited, measured path length variations were reduced by as much as 20 times this value. In other words, there is large variability in the tropospheric path length variation at different times under different conditions.

Figure 1 shows the type of data analysis used for the present study. Here is plotted the spectra of Doppler variations as received from Pioneer 9 under good conditions over an interval of 4.5 hours. Using the relationship for path length variation:

$$\Delta L/L = \frac{\lambda_{sb} \cdot A_s \cdot T_s}{4 \pi c R} \tag{1}$$

where:

λ_{sb} = microwave Doppler wave length = 13 cm.

A_s = amplitude of spectral value (Hz).

T_s = period of spectral value (sec).

c = velocity of light (cm/sec).

R = light time to space craft (sec).

We obtain an estimate of total path length variation over any spectral period available from the analysis.

A flat spectrum is generally observed in the interval between 100-1 000 seconds, indicating that the path length error is proportional to T in this range. Beyond 1 000 seconds the spectrum rises, indicating

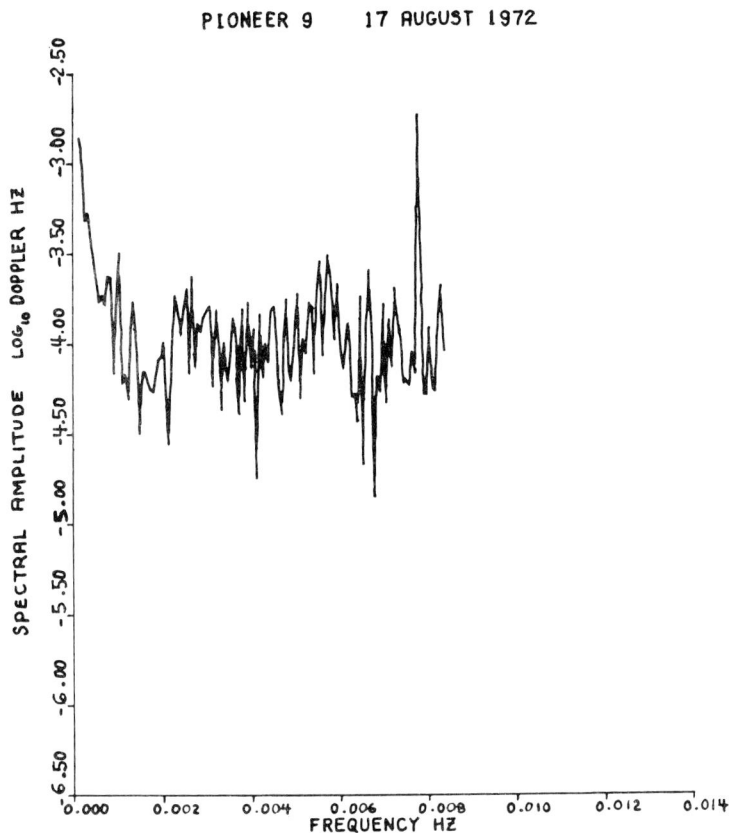

Figure 1. Amplitude Spectrum of Doppler variations from the Pioneer 9 spacecraft as tracked by DSN (Anderson, 1976).

that path error becomes proportional to T^2. The sharp peak in the Pioneer spectra at 129 seconds is a periodic offset of the spacecraft's antenna of 0.24 cm due to spacecraft rotation.

A comparison of results on measured tropospheric path length variation is given in Figure 2. The absolute variation ΔL (cm) is given for typical measured conditions and most favorable measured conditions. The measured results of Doppler variations from Pioneer 9 are included. The total effect on differential path length variation $\Delta L/L$ is given for a spacecraft at 10 AU. distance.

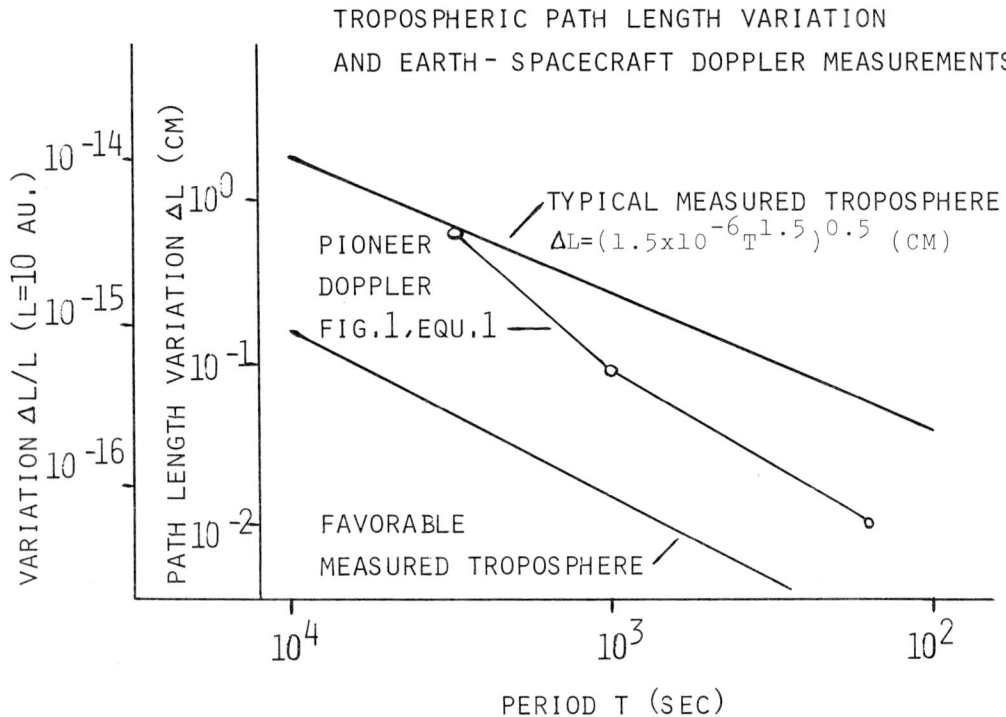

Figure 2. Tropospheric path length variability as measured by Thompson, et al, 1975, and Hargrave and Shaw, 1978 compared with Doppler path length variations.

3. CONCLUSIONS

The method of Doppler tracking of spacecraft provides a means of monitoring tropospheric path length variations. Under favorable tropospheric conditions and/or by monitoring these variations, measurements of the total path length variation, $\Delta L/L$, to less than 1 part in 10^{15} can be obtained when tracking spacecraft at a distance of 5 AU and beyond.

REFERENCES

1. Anderson, A. J. (1971) Probability of long period gravitational radiation, Nature, 229, 547-548.

2. Anderson, A. J. (1974) A study of the anomalous effects in the spectra of Doppler from the DSN, JPL MEM, 391-560.

3. Anderson, A. J. (1977) Detection of gravitational waves by spacecraft Doppler data, in Experimental Gravitation, 235-246.

4. Callahan. P. S. (1978) A first look at errors from small scale tropospheric fluctuations. JPL IOM 315, 1-272.

5. Hargrave, P. L. and Shaw, L. J. (1978) Large-scale tropospheric irregularities and their effect on Radio Astronomy seeing. Mon. Not. Roy. Astr. Soc., 182, 233.

6. Moran, J. M. and H. Penfield (1976) Test and evaluation of water vaper radiometers and determination of their capability to measure tropospheric propagation path length. Final Report, Smithsonian Astrophysical Obs., Cambridge, Mass.

7. Thompson, M. C., Wood, L. E., Janes, H. B., and Smith, D. (1975) Phase and Amplitude Scintillations in the 10 to 40 GHz Band. IEEE Trans. on Ant. and Prop. 23, 190.

DISCUSSION

K. Poder: You are transmitting over a 500 second transmission line path, which must be a long one, and even a very small gradient will give a correction. Of course, if you are lucky that this gradient does not change within your measuring time, you are safe. Obviously, the direct refraction can more or less be connected by using the dispersion, but the other one is beyond correction by dispersion. The only thing you have to know is to get an idea of how the gradients of the refractive index is perpendicular to the propagation path. It is a second order effect, but it goes up with the third power of distance. For all geodetic measurements, it is really one of the most limiting factors.

A.J. Anderson: The estimates on the induced pathlength correction, that is, the total amplitude of the tropospheric corrections, the dry and the wet components, give upwards of 2.6 meters on the total pathlength, at an azimuth angle of, say, $60°$. Then the ionospheric corrections give you another 2 to 3 meters. Finally you have to consider the total ion density value of the solar system path, which changes very much with the sunspot cycle, but totally you can get upwards of 10 meters of pathlength correction, due to the bending of the ray from the spacecraft to the point on the earth, so we are dealing with a total of 10 meters refractive value due to three different sources.

D.G. Currie: If the spacecraft are close enough in the sky, the refraction would be similar, going to the earth's atmosphere. If they would be ideally in the same direction, you would then be measuring the distance between them at high accuracy with no tropospheric (wet or dry component).

A.J. Anderson: Yes, that is possible. However, the constraints on the DSN-network have been, so far, that one has not been able to track simultaneously two spacecraft of this type. One has the hope to do that, however. We presently can track two spacecraft, being distant in the sky, which are tracked at separate sites. Let me give you an example: One can track two spacecraft simultaneously at two sites at Goldstone. Generally the angles have been very large between spacecraft and they have not been relevant for anything in which we would be interested in, however with some additional equipment there will be simultaneous tracking at Goldstone in the early 1980's of the two Voyager spacecraft and they will be very close to each other in the sky, and it is hoped to carry out that experiment you have mentioned.

DIURNAL CHANGES OF RADIO REFRACTION

Wilhelm J. Altenhoff
Max-Planck-Institut für Radioastronomie
Auf dem Hügel 69
5300 Bonn, West Germany

ABSTRACT

Pathlength variations due to refraction changes in the troposphere may impose a severe limitation to VLB experiments (like random clock drifts). In connection with a VLB experiment Moran and Penfield (1976) analysed surface values of water vapor density, data of several hundred radiosonde launches and of measurements of sky brightness near the water vapor line at $\lambda = 1.3$ cm. They found that the surface values allow to estimate the pathlength to an accuracy of 5 cm in summer and 2 cm in winter. Sky brightness data give a prediction accuracy of 1.5 cm for all weather conditions, but for cloudfree conditions the accuracy was 0.3 cm.

Three different methods were used in radio astronomy to determine the water vapor content:

1) IR absorption measurements in the direction of the sun (Wesseling et al., 1974),

2) relative emission measurements near the water vapor line at $\lambda = 1.3$ cm (Moran et al., 1976),

3) absolute measurements of emission in the range of 100 to 1000 GHz (Hills et al., 1978).

The question, which line and which method is best, is still unanswered. In principle the absorption measurements may be preferred, since they give exactly the amount of water vapor in the line of sight, but this method depends on the sun. Method 2 may turn out to be most useful, since it allows to map the water vapor emission in any direction at any time; the exact calibration may give some problem. Method 3 seems to be too involved to be practical for routine pathlengths corrections.

REFERENCES

Hills, R.E. et al.: 1978, "Absolute Measurements of Atmospheric Emission and Absorption in the Range 100 to 1000 GHz", paper presented at the Third International Conference on Submillimeter Waves, Guildford, U.K.

Moran, J.M., Penfield, H.: 1976, "Test and Evaluation of Water Vapor Radiometers". Contract NAS5-20975, final report.

Wesseling, K.H. et al.: 1974, Radioscience, 9, 349.

DISCUSSION

P.V. Angus-Leppan: asked for a clarification of the use of the 1.3 cm wavelength measurements for water vapor content.

W.J. Altenhoff: repeated, that Moran and Penfield studied the radiation from the water vapor line of $\lambda = 1.3$ cm. Using aerosonde measurements of water vapor content, correlated to the radiation results, they conclude, that it is possible to predict wet pathlengths to within 0.3 cm. The speaker, however, regarded this accuracy as overestimated.

J.C. de Munck: asked if any attempts had been made to determine the humidity of the air by acoustical means. By combination of measurements of acoustical waves and radar waves, it should be possible to map the humidity distribution in the atmosphere.

W.J. Altenhoff: answered that it would be possible, but that he was not aware of any such attempt. In his opinion, the absorption measurements were the most reliable ones for VLBI.

J.A. Hughes: declared as his opinion, that acoustical means can map the water vapor distribution, but since in the case of radio interferometry one is interested in the instantaneous line of "sight" integrated effect, he doubted that acoustical methods could improve upon the direct radiation measurements. The acoustical methods do work however, so he agreed with de Munck, that they could be used if sufficient time and space resolution were available.

USE OF METEOROLOGICAL MEASUREMENTS FOR COMPUTING REFRACTIONAL EFFECTS - A REVIEW

P.V. Angus-Leppan
University of New South Wales, Sydney, Australia

ABSTRACT

Corrections which are dependent on the atmosphere include the first and second velocity corrections and the curvature correction in EDM, and the refraction correction for trigonometric heighting. Generalised correction formulae have been developed to make use of variable values of the refractivity N and the coefficient of refraction k. There are no universally applicable values of these parameters, so atmospheric models of the temperature gradient and, for microwaves, the humidity gradient, are needed to represent the very significant variations. The models should take into account variations due to meteorological factors, surface conditions and the height above the surface. Many models for the vertical gradients have been produced. That of Brocks (1948) is very important and has been widely used and developed by researchers. More recently Monin and Obukhov, basing their work on the physics of the lower atmosphere and dimensional analysis, have included equations for the vertical gradients in their turbulence theory. The Turbulent Transfer Model, which embodies later results in this theory, is currently being refined and developed for geodetic applications.

1. INTRODUCTION

The atmospheric effects on geodetic observations place a limit on their accuracy. The simple approaches - using endpoint meteorological measurements for EDM reductions and the coefficient of refraction in trigonometric heighting - are effective for surveys of low or intermediate accuracy. When higher accuracies are needed, improvements can be found by adopting an atmospheric model which corresponds more closely to the real atmosphere, or using atmospheric dispersion, by measuring with two or more colours. The dispersion method will not be discussed in this paper, beyond noting that there are practical difficulties involved, and that even in this method some atmospheric modelling is essential. In a modified form, the so-called parallel measurement, the practical difficulties are avoided by using existing lightwave and microwave instruments (KUNTZ and MÖLLER, 1971) but the model adopted for the relationship between temperature and humidity is critical in the reduction.

The atmospheric model is required to give the temperature gradient and, if microwaves are involved, also the humidity gradient. All the required corrections can be derived, using these gradients and meteorological measurements at the geodetic stations. The model must take into account the large number of factors which affect the gradients. No progress will come from trying to find the static, universal model, for example the best average value of the temperature gradient or the best value of the coefficient of refraction. The atmospheric model must be multi-factored to take into account the variations in the atmosphere, which are very large and real. For example, long term temperature measurements (BEST et al., 1952; FLOWER, 1937) show the average error due to observing temperature at 1.5 m instead at the height of the line, say 50 m above the surface. In the month of April (spring) the error, in parts per million (ppm) varies from -3 in the day to +5 at night, at Rye, in Southern England. At Ismailia, Egypt, the average daily variation is -6 to +6 ppm. In mid-summer the errors are similar, but in winter they are quite different, with lower averages, but a far larger variation from day to day (ANGUS-LEPPAN, 1967).

The variations in k, the coefficient of refraction, are even more remarkable. Table I shows values at heights of 1.5, 7.5 and 75 m for the same month, April.

Table I   COEFFICIENT OF REFRACTION k

Minimum (day) and maximum (night) values for Spring (April) at Rye, England, and Ismailia, Egypt.

Height above surface	Rye		Ismailia	
	Day	Night	Day	Night
1.5 m	-1.6	2.5	-3.7	3.1
7.5	-0.2	0.8	-0.7	0.9
75	0.14	0.25	0.18	0.35

Ref. ANGUS-LEPPAN, 1967

Even at 7.5 m the values bear little resemblance to the standard value of k = 0.13, being -0.2 at Rye and -0.7 at Ismailia. Only at 75 m do the daytime values approach 0.13.

2. EDM REFRACTION CORRECTIONS - CONVENTIONAL FORM

The three corrections to EDM which are related to atmospheric refractive index are:
- the first velocity correction $\quad C_1 = d(n_{ref}-n)$ (1)
- the second velocity correction $\quad C_2 = (-k+k^2)d^3/12R^2$ (2)
- the correction for curvature of the $\quad C_3 = -k^2d^3/24R^2$ (3)
  wave path

(RUEGER, 1978)

Here, d is the measured distance,
n_{ref} is the reference refractive index adopted in the particular instrument type,
n is the refractive index along the wave path,
k is the coefficient of refraction, defined as the ratio of the radius of curvature of the ellipsoid to the radius of curvature of the wave path,
R is the radius of curvature of the ellipsoid.

For n and k, values appropriate to the EDM carrier wave are needed. For light waves the group refractive index and the refractive index under atmospheric conditions are calculated according to the well-known formulae of BARREL and SEARS (1939), while the equally familiar formula of ESSEN and FROOME (1951) is used for the atmospheric refractive index of microwaves. The coefficients of refraction normally adopted are 0.13 for lightwaves and 0.25 for microwaves (but see the discussion in BRUNNER, 1977 A).

3. EDM CORRECTIONS BASED ON EIKONAL EQUATION

Correction formulae (1) - (3) are suitable when n and k are assumed to be constant. When variations in n and k are taken into account, more flexible formulae are needed. A suitable general formula has been derived by MORITZ (1967) and developed by BRUNNER and ANGUS-LEPPAN (1976). It has the advantages that it combines all three corrections and that its integration is along the chord between the end points and not along the unknown wave path.

The basis of the derivation is the <u>eikonal</u> equation:

$$(\text{grad } d)^2 = n^2 , \qquad (4)$$

a differential equation which corresponds to Fermat's principle, namely that the path followed by electromagnetic waves between two points is that which takes the minimum time. The correction formula is:

$$\Delta s = d-s = 10^{-6} \int_0^s N \, dX - \frac{10^{-12}}{2} \cos^2\beta \int_0^s \frac{[\int_0^X \frac{dN}{dz} \xi \, d\xi]^2}{X^2} \, dX \qquad (5)$$

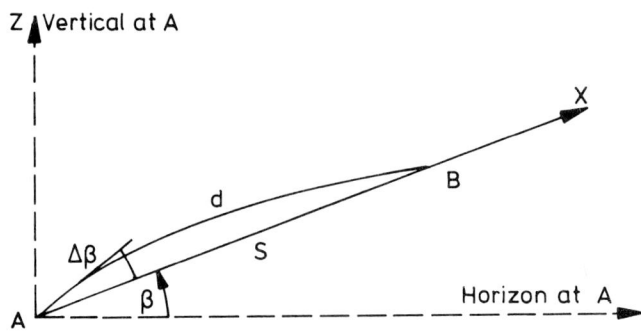

Figure 1   Wave path d and chord s

where   s   is the chord length AB,
        N   is the refractivity = $10^6(n-1)$,
        X   is distance along the chord,
        $\frac{dN}{dz}$ is the vertical gradient of refractivity,
        β   is the vertical angle,
        ξ   is an integration variable along the chord.

See Figure 1. The integrations are along the chord AB.

The determination of N and the refractivity gradients dN/dz will be discussed later. These will be values at discrete points. Rather than expressing them as functions of X for (5), numerical integration is used. Simpson's Rule is appropriate for the main integrals. However, for the minor integration with respect to ξ the Trapezoidal Rule is more suitable, as the number of ordinates is sometimes odd and sometimes even, and Simpson's Rule requires an odd number of ordinates; also, the correction is small.

If for example it is decided to divide the line into 6 sections, there will be 7 values of N: N_1, N_2 ... N_7, where N_1 is the value at A and N_7 the value at B. The first integral is a straightforward application of Simpson's Rule. Before applying Simpson's Rule in the second term, each ordinate must be formed by integration and squaring. For example at the second ordinate X = s/6 and {(dN/dz)ξ} must be integrated from 0 to s/6, applying the Trapezoidal Rule.

4. TRIGONOMETRICAL HEIGHTS WITH VARYING k

It has been shown that if a line, length S, is divided into sections s_1, s_2, s_3 ... s_n with refraction coefficients k_1, k_2, k_3 ... k_n, then with some minor restrictions, the trigonometrical refraction correction is given, to sufficient accuracy, by:

$$\Delta z = \frac{1}{2R} \{k_1 s_1 (2S-s_1) + k_2 s_2 (2S-2s_1-s_2) + k_3 s_3 (2S-2s_1-2s_2-s_3) \ldots$$
$$\ldots + k_n s_n^2\} \qquad (6)$$

The sections $s_1, s_2 \ldots s_n$ need not necessarily be equal.

More rigorously, the vertical angle of refraction $\Delta\beta$, the angle between the tangent to the light ray and the chord AB (Figure 1), can also be derived from the eikonal equation:

$$\Delta\beta = \frac{10^{-6}}{S} \cos\beta \int_0^S \frac{dN}{dz} (S-X) dX \qquad (7)$$

5. ATMOSPHERIC MODELS-GENERAL

Meteorological measurements at the end points of an EDM line can be expected to give representative values if observations are from towers and the line traverses over flat terrain with even surface conditions. However tower observations do not always give a higher accuracy, as noted by JONES (1971) from observations by the Geodetic Survey of Canada. The need for an atmospheric model can also be avoided by taking direct measurements along the line. Meteorological observations from planes set to fly along the EDM wave path are reported by MEADE (1969) and PRESCOTT and SAVAGE (1974). Such measurements have not been adopted more generally, probably because of the expense and the difficulties of organisation. Other possible problems are the low flight altitude required and uncertainties in the friction heating effect and time lag of the temperature sensor. The use of tethered balloons or kites for intermediate measurements has also been advocated, for example by MITTER (1962). In practice, the balloons and kites are troublesome and not compatible with sensitive electrical apparatus.

Many different atmospheric models have been suggested for refraction corrections. In order to provide a good representation, the model needs to take into account, directly or indirectly, variations arising from:
- daily and annual cycles of sun and temperature,
- amount of cloud cover,
- height above the surface,
- wind velocity,
- roughness and heat properties of the surface.

The first two factors might be covered by measurements of radiation.

Generally the models are based on measurements at special meteorological stations. The stations are carefully chosen in the midst of large areas of flat terrain with even surface conditions. In contrast, geodetic stations cannot be chosen for their meteorological suitability. The models could be systematically different from conditions above the typical geodetic station, which is on a hilltop. The models leave out the important factors of topography, uneven surface conditions and the effects of movements of masses of air (advection). While it is true that schematic representations of temperature structure under typical geodetic lines are implied or shown by many writers, for example SCHÄDLICH (1975), ANGUS-LEPPAN (1967) and MAIER (1977), they are somewhat conjectural and may not be generally applicable.

It can be concluded that, at present, the horizontal variations of meteorological conditions are inadequately represented. It will be worthwhile in the future to undertake research towards the development of suitable models.

6. REFRACTION IN TRIGONOMETRIC HEIGHTS

The coefficient of refraction k has a long history. In the celebrated geodetic survey in Lapland, 1736-7, Maupertius made calculations which implied the use of the coefficient. In the 19th century Jordan compiled values from the surveys of various countries and derived a mean value which is still valid today. There is an extensive literature on investigations to find the value of k. In using it, we are assuming that the atmospheric gradient of refractivity is constant, and, in practice, that the temperature gradient is also constant.

Refraction posed an early challenge in determining the heights of the high Himalayan peaks. From the foothills they appeared to rise and fall 150 m per day. J. de Graaff-Hunter, at the time Mathematical adviser to the Survey of India, made investigations which enabled the refraction correction to be calculated with some precision (de GRAAFF-HUNTER, 1913).

In another field, precise levelling, the investigations of KUKKAMÄKI (1938) were important. He adopted a function for the temperature gradient,

$$dT/dz = a\, z^b \tag{9}$$

in which a is the temperature gradient at height 1 m and the exponent b was determined from a series of temperature observations by BEST (1935).

A fundamental contribution was that of BROCKS (1948). He adopted the same function as Kukkamäki, equation (9), for the gradient in a layer up to about 30 m, above which the gradient is adiabatic, $-1^\circ C/100$ m, for a few hundred meters. For daytime conditions, the index b in (9) is generally equal to -1. Based on long series of temperature measurements in France, Germany, England and Egypt, Brocks deduced values of the parameters a and b for various heights, times of day and seasons of the year. These parameters also define the values of k for corresponding conditions.

The work of Brocks has been very influential in refraction research. Numerous investigators have adopted his functions as the basis of their atmospheric models. However, this influence did not extend into general practice. Brocks' work should have put an end to the concept of a single fixed value for k. For each observation it was possible to look up an appropriate value of k. Since 1948 many more sets of temperature observations have become available, enabling modified values to be determined for different localities and conditions. Unfortunately this did not happen, at least not in the English-speaking parts of the world, and the search for the elusive k continued.

At about the same time a group from the Institut Géographique National was investigating refraction in the French Alps. Their report, in 1953, introduced a model for atmospheric temperature as a function of height and time of day (LEVALLOIS and d'AUTUME, 1953). Fixed parameters in the function were determined from earlier radiosonde observations at a nearby weather station, and variable parameters from the temperatures at the geodetic station throughout the day of observation.

ANGUS-LEPPAN (1967) tried using the same model in flat terrain but found it needed modifications. His investigations resulted in a comprehensive function to model temperature gradients, but it required the determination of 10 constants.

Recent developments in technology have made it easier to measure temperature gradients, but the use of direct temperature gradient measurements in calculating refraction has not been successful. This is because of the large fluctuations in the temperature ($1 - 2^0$), and because local measurements are not representative of the line. Reciprocal vertical angles have been used for a long time, but it has only recently been realised that simultaneous reciprocal observations over short lines (1 - 2 km) have an accuracy almost up to that of third order levelling (BRUNNER, 1974).

Another model proposed is the Heat Balance-Turbulent Transfer Model (TTM). Unlike the earlier, basically empirical, models, the TTM has a basis in atmospheric physics and turbulence theory. It will be described in the following section on EDM. It is also applicable in vertical refraction. BRUNNER (1977) reports results in which the coefficient of refraction, which varied from -1.0 to +0.6 during a day, is calculated with a standard devation of $\pm$ 0.13 by the TTM method. The method also represents rapid changes in k which occur when shadow moves over or away from the line.

7. ATMOSPHERIC MODELS IN EDM

In the 1960's the emphasis in refraction research shifted to EDM, where the basic problem is how to determine the representative temperature and humidity for a line, given the values at the end points. As early as 1961 it was realized that better meteorological data was the key to improved accuracy in EDM (RINNER, 1961). An effective model for temperature and humidity in the lowest atmospheric layers should provide the solution, but it has not been easy to find.

Many functions have been tried. Brock's function, equation (9) has been applied, for example, by ADLUNG (1963), FELLETSCHIN (1978), LANG (1969), MAIER (1977), SCHÄDLICH (1975) and others. BRETTERBAUER (1966) used a non-linear representation of the refractivity-height relationsship to deduce the mean refractivity. KUNTZ (1970) developed a parabolic profile, termed the "spherical-parabolic-model". Generally the results have been inconclusive. SCHÄDLICH (1975), for example, claims that his "topographic-atmospheric reduction" gives good agreement with test data, but provides no evidence, while PARM (1967), representing his meteorological parameters as logarithmic functions of height, finds after comparison with normal reductions, that the technique does not give the expected improvement.

A useful review of research on microwave EDM is given by LANG (1969). He notes particularly that there is disagreement on three matters:
- the most favourable time of day for EDM observations,
- the determination of the representative refractivity,
- the minimum height of the line above the surface.

A number of writers have noted systematic differences between geodimeter (lightwave) and tellurometer (microwave) measurements, and between day and night measurements. Responsibility for these has been traced to systematic errors in the representative refractivity, by Bretterbauer and others. Extensive test measurements in Canada are described by JONES (1971). These were compared, and the systematic differences deduced. Applying the temperature and humidity observations of BEST et al. (1952), the expected systematic errors were calculated, and were found to be in good agreement with those from measured distances. Perhaps surprisingly, there has been no general agreement to apply corrections for such systematic errors, nor to adopt a model based on observations such as those of Best.

In 1962 Saastamoinen proposed the use of zenith distances, measured with the EDM, to improve the refraction correction. The concept of using the values of k to provide integrated information on the line, is appealing. The theory was fully developed by KUNTZ (1970) in conjunction with his spherical-parabolic model. Practical tests were conducted by DIECHL and REINHART (1969), who concluded that the use of zenith distances at one end only was not favourable, whereas reciprocal zenith distances gave the same result as the normal meteorological reduction, that is, no improvement.

The Heat Balance-Turbulent Transfer Model, put forward by Angus-Leppan and Webb (ANGUS-LEPPAN, 1971; ANGUS-LEPPAN and WEBB, 1971) differs from most other models in being based on atmospheric physics and also in being highly sensitive to a wide range of meteorological conditions at the time of observations. It is based on theories of turbulent transfer in the boundary layer to which contributions were made by Monin, Obukhov, Priestley, Panovsky, Webb, Deacon and many others.

Although the TTM has a different origin, it has some parallels with Brocks' model. A distinction has to be drawn between stable and unstable thermal stratification in the atmosphere. For unstable (daytime) conditions there are three strata, defined in terms of a stability parameter L, the Obukhov length. L fluctuates rapidly with time, varying as the cube of the wind speed and inversely as the upward heat flow due to turbulence. On the typical breezy summer day the lowest stratum might average up to 1 m, and the middle stratum, 1 to 30 m. In the upper stratum the temperature gradient is $-1^o/100$ m as in Brocks' model. The parameter b is -1 in all cases except the middle stratum where it takes the value -1.33. The parameter a is different in each stratum, being a function of meteorological parameters, including those which determine turbulent heat flow (sun and sky radiation, cloudiness, heat properties of the surface, surface moisture etc.), wind velocity, surface roughness and air temperature. BRUNNER and FRASER (1977 A) have compared values from the model against test data, with very favourable results.

8. CONCLUSIONS

Effective atmospheric models for temperature and humidity are badly needed. Perhaps geodesists have not been collaborating sufficiently closely with meteorologists, and have not made sufficient use of existing meteorological data. Even if changes in measurement techniques do away with the need to determine refractivity along an EDM line, studies of the atmosphere will remain important in geodesy, since the atmosphere is the medium in which the geodesists make their observations. There is another aspect of geodesy-meteorology collaboration which should be borne in mind. Geodetic observations can be valuable in meteorology. They are very efficient in determining spatially integrated values of meteorological parameters.

9. ACKNOWLEDGEMENTS

The author is currently holder of a Fellowship under the Alexander von Humboldt Foundation, held at the Geodetic Institute of the University of Karlsruhe. Grateful acknowledgement is made to the Foundation for the opportunity offered by the Fellowship, and to Professor Kuntz and colleagues in the Geodetic Institute.

10. BIBLIOGRAPHY

ADLUNG, A. (1963): Über die Ausbreitung von Zentimeter- und Dezimeter-
wellen in der bodennahen Luftschicht. Abh. d. Meteorol. u. Hydrol.
Dienstes der DDR, Nr. 79 Bd. IX, Akademie-Verlag, Berlin.

ANGUS-LEPPAN, P.V. (1961): A Study of Refraction in the Lower Atmosphere.
Empire Survey Review. No. 120, pp 62-69; No. 121, pp 107-119;
No 122, pp 166-177.

ANGUS-LEPPAN, P.V. (1967): A Mathematical Model for Temperatures in the Lower Atmosphere, and its Application in Refraction Calculations. Österreichische Zeitschrift für Vermessungswesen, Sonderheft 25, pp 219-227 + suppl.

ANGUS-LEPPAN, P.V. (1971): Meteorological Physics applied to the Calculation of Refraction Corrections. Proceeding of conference of Commonwealth Survey Officiers, Cambridge, Paper No. B5, 9 pp.

ANGUS-LEPPAN, P.V. and WEBB, E.K. (1971): Turbulent Heat Transfer and Atmospheric Refraction. General Assembly IUGG Travaux. IAG Sec. I, 15 pp.

BARRELL, H. and SEARS, T.E. (1939): The Refraction Dispersion of Air for the Visible Spectrum. Phil. Trans. Roy.Society, Series A, London, p. 238.

BEST, A.C. (1935): Transfer of Heat and Momentum in the lowest Layers of the Atmosphere. Met. Office Geophys. Mem. No. 65, HMSO London.

BEST, A.C., KNIGHTING, E., PEDLOW, R.H. and STORMONTH, K. (1952): Temperature and Humidity Gradient in the first 100 m over South-East England. Met. Off. Geophys. Mem. No. 89, HMSO London. (Reprinted 1962).

BRETTERBAUER, K. (1966): Die Ausbreitung von Mikrowellen in einem Atmosphärischen Modell. Allgemeine Vermessungs-Nachrichten, 8, pp 313-318.

BROCKS, K. (1948): Über den täglichen und jährlichen Gang der Höhenabhängigkeit der Temperatur in den unteren 300 Metern der Atmosphäre und ihrem Zusammenhang mit der Konvection. Berichte d. Deutschen Wetterdienstes in der U.S.-Zone, Nr. 5, Bad Kissingen.

BRUNNER, F.K. (1974): Trigonometrisches Nivellement - Geometrisches Nivellement. Österreichische Zeitschrift für Vermessungswesen, 62, pp. 49-60.

BRUNNER, F.K. (1977): Experimental Determination of the Coefficients of Refraction from Heat Flux Measurements. Proc. Int. Symp. on EDM and Atm. Refr., Wageningen, pp 245-255.

BRUNNER, F.K. (1977 A): On the Refraction Coefficient of Microwaves. Bulletin Geodesique, 51, pp 257-264.

BRUNNER, F.K. and ANGUS-LEPPAN, P.V. (1976): On the Significance of Meteorological Parameters for Terrestrial Refraction. Unisurv G, 25, pp 95-108, University of New South Wales, Sydney.

BRUNNER, F.K. and FRASER, C.S. (1977): An Atmospheric Turbulent Turbulent Transfer Model for EDM Reduction. Proc. Int. Symp. on EDM and Atm. Refrn., Wageningen, pp 304-315.

BRUNNER, F.K. and FRASER, C.S. (1977 A): Application of the Atmospheric Turbulent Transfer Model for the Reduction of EDM. Unisurv G, 27, pp 1-21. University of New South Wales, Sydney.

DE GRAAFF-HUNTER, J. (1913): Survey India Professional Paper. Dehra Dun.

DIECHL, K. and REINHART, E. (1969): Zur Bestimmung des Brechungsindexes bei der elektrooptischen Entfernungsmessung. Allgemeine Vermessungsnachrichten, 7, pp 269-281.

ESSEN, L. and FROOME, K.D. (1951): The Refractive Indices and Dielectric Constants of Air and its Principal Constituents at 24000 mc/s. Proc. Phys. Soc., (London), B-64, pp 862-875.

FELLETSCHIN, V. (1978): Analyse und Steigerung der Genauigkeit bei elektronischen Entfernungsmessungen mit Licht- und Mikrowellen im Testnetz Karlsruhe. Deutsche Geodätische Kommission, C-246, München.

FLOWER, W.D. (1937): An Investigation into the variation of the Lapse Rate of Temperature in the Atmosphere near the ground at Ismailia, Egypt. Met. Office Geophys. Mem., No. 71, HMSO London.

JONES, H.E. (1971): Systematic Errors in Tellurometer and Geodimeter measurements. Canadian Surveyer, 25, pp 406-423.

KUKKAMÄKI, T.J. (1938): Über die nivellitische Refraktion. Publication of the Finnish Geodetic Institute, 25, Helsinki.

KUNTZ, E. (1970): Messung von Zenitdistanzen in der elektrooptischen Entfernungsmessung. Allgemeine Vermessungs-Nachrichten, 77, pp 41-56.

KUNTZ, E. and MÖLLER, D. (1971): Gleichzeitige elektronische Entfernungsmessung mit Licht und Mikrowellen. Allgemeine Vermessungs-Nachrichten, 78, pp 254-266.

LANG, H. (1969): Über den Einfluß der bodennahen Luftschicht auf die Mikrowellen-Entfernungsmessung. Arbeiten aus dem Verm. u. Kartenwesen der DDR, 19, Leipzig.

LEVALLOIS, J.J. and MASSON-D'AUTUME, G. (1953): "Etude sur le Refraction Géodésique et le Nivellement Barométrique." Institut Géographique National, Paris.

MEADE, B.K. (1969): Corrections for Refractive Index as applied to Electro-Optical Distance Measurements. IAG Symposium on Electromagnetic Distance Measurement and Atmospheric Refraction, Boulder, Colorado.

MAIER, U. (1977): Genauigkeitsuntersuchungen zur elektrooptischen Messung langer Strecken. Thesis, Geodetic Institute, University of Karlsruhe.

MITTER, J. (1962): Über die Bestimmbarkeit der Ausbreitungsgeschwindigkeit der Trägerwellen bei elektrischen Entfernungsmessungen. Allgemeine Vermessungs-Nachrichten, 5, pp 139-159.

MORITZ, H. (1967): Application of the Conformal Theory of Refraction. Österreichische Zeitschrift für Vermessungswesen, Sonderband 25, pp 323-334. (IAG Symposium on 'Recent research on atmospherical refraction for geodetical purposes').

PARM, T. (1967): Investigations on Refractional Corrections in Tellurometer measurements. Österreichische Zeitschrift für Vermessungswesen, Sonderband 25.

PRESCOTT, W.H. and SAVAGE, T.C. (1974): Precision of Geodolite Distance Measurements. IAG Symposium on Terrestrial EDM and Atmospheric Effects on Angular Measurements, Stockholm.

RINNER, K. (1961): Über Schranken für die geodätische Anwendung der elektronischen Entfernungsmessung. Deutsche Geodätische Kommission, B-95, München.

RUEGER, T.M. (1978): Introduction to Electronic Distance Measurements. Monograph No. 7, School of Surveying, University of New South Wales, Sydney.

SAASTAMOINEN, T. (1962): The effect of Path Curvature of light waves on the Refractive Index: application to Electronic Distance Measurement. Canadian Surveyor XVI, 2.

SCHÄDLICH, M. (1975): Die topographisch-atmosphärische Reduktion des mittleren Brechungsindex. Vermessungs-Technik, 2/75, pp 68 -71.

WEBB, E.K. (1969): The Temperature Structure of the Lower Atmosphere. Proc. of REF-EDM Conference, University of New South Wales, Sydney, pp 1-9.

DISCUSSION

L. Hradilek: As far as the trigonometric leveling is concerned, is it satisfactory to measure the temperature gradient at the observation station only, or is it necessary to have some information about the gradients over the whole line of sight?

P.V. Angus-Leppan: Meteorologists can tell us quite a lot about the vertical gradients but not so much about variations in the horizontal direction. The typical site for meteorological research is specially selected with very flat terrain and very even surface conditions. In special circumstances, such as a mountain top, there may be information on the variations in the horizontal. In general not enough is known about the changes which will occur in vertical gradients as one moves from point to point in the horizontal.

L. Hradilek: When starting our refraction research, we measured at the station all basic meteorological data, especially the temperature grad-

ient by two independent methods. When introducing these values into
the formulas of Jordan, we obtained a result which was 30 times larger
than the actual variation in refraction determined by precise theodolite measurements.

P.V. Angus-Leppan: In general the direct measurements of temperature
gradients have been very disappointing, because they are representative only of a single point. It is very difficult to get representative values over the length of a line whose height above the surface
varies.

T.J. Kukkamäki: In Finland we observed daily periodic variation in
the lateral direction of a 5 km long sight running rather low along
a steep sideward slope. The difference between day and night reached
6 seconds of arc. We made efforts to derive this lateral refraction
from vertical temperature gradient observations carried out simultaneously on the spot. Assuming that the isothermic layers ran parallel
to the sloping ground surface, we calculated the horizontal component
of the temperature gradient and on that basis its effect on the lateral
direction of the sight. The correlation between the observed and the
calculated refractions was good. (T.J. Kukkamäki: On lateral refraction
in triangulation, Bull. Géod. No 11, 1949).

P.V. Angus-Leppan: There is a line in the Australian Geodetic Survey,
where the angle varied by 13 seconds, in a daily cycle. It lies along
a low bluff some distance from the sea. It is easy to measure the variation in lateral refraction, but very difficult to measure the absolute
value.

T.J. Kukkamäki: Also in Finland we have carried out observations on
lateral refraction along the seashore. The variation between day and
night directions of a 30 km long sight was 2 seconds of arc. The sight
ran parallel to the shoreline. The horizontal temperature gradient was
determined from the temperature recordings at two stations, one on a
small island and the other on land, situated symmetrically on either
side of the sight. Correlation between the observed and the calculated
refraction was as good here as on the slope.

B. Garfinkel: In making the calculation of terrestrial refraction, it
is crucial that the choose of a good physical model. You suggested
Brock's model with the parameters a and b. Now these parameters are
presumably subject to fluctuations with time, depending on the time
of the day and perhaps the season of the year. Instead of using formulas with constant parameters for such a dependence it would be practical to measure at the sight three values of temperatures along some
vertical direction, and from these values calculate those parameters
a and b for the particular time of the observation.

P.V. Angus-Leppan: I do not recommend Brock's model any more. It was
good at the time, but the turbulent transfer model is much better. It
is based upon theoretical and empirical research in the boundary layer

by meteorologists, and it comes close to a true representation of the physical processes that take place: the heat-balance at the surface and the transfer of heat upwards, taking into account the various factors such as the radiation intensity, wind and surface conditions.

B. Garfinkel: I have a question regarding that other model. Is that the one that involves the potential temperature, the temperature which you denote with letter θ ? You suggest there are some difficulties in using such an equation, because the potential temperature is not directly measurable in the field. Is that correct?

P.V. Angus-Leppan: No, it is just more convenient to use θ rather than T in the theory.

B. Garfinkel: What about the practise?

P.V. Angus-Leppan: One can convert from one temperature parameter to the other without difficulty.

B. Garfinkel: Do you have any reference to this other model, which involves the potential temperature?

P.V. Angus-Leppan: There are papers by Webb listed in the references to my paper.

STATISTICS APPLIED TO THE ESTIMATION OF THE INFLUENCE OF THE ENVIRONMENT ON RESULTS OF OBSERVATIONS

M.K. Szacherska and Z. Wiśniewski
Institute for Geodesy and Photogrammetry,
Agricultural and Technical Academy at Olsztyn
Poland

ABSTRACT

The assumption of a generalized model of the composition of random and systematic errors of measurements can constitute a basis for the estimation of the influence of the environment on the results of geodetic observations in the course of their numerical elaboration. The method of the estimation of systematic errors on the basis of the analysis of an empiric distribution of the results of observations has been established on the assumption of such a model, which permits also to estimate systematic errors in the course of the adjustment of observations, these errors being treated as additional unknowns in the system.

1. INTRODUCTION

Statistical methods are nowadays frequently used for the analysis of the precision of geodetic measurements and the estimation of random and systematic observation disturbances of different kind. The possibility of applying mathematical statistics to the estimation of the influence of the environment and atmospheric conditions on the results of observations is widely discussed (Tengström, E., 1975), (Proceedings, 1977). Investigations deal with the application of statistics to the elaboration of mathematical models describing the character and the development of the physical phenomena as well as their influence on the results of geodetic observations. In other researches statistical methods are utilized for the numerical elaboration of the results of observations and simultaneous estimation of systematic disturbances caused by the influence of the environment.

As basis for the numerical elaboration of observation results we can adopt a generalized model of the composition of random and systematic errors of measurements (Szacherska, M.K., 1974), (Szacherska, M.K., 1977). Let us assume, that a set of errors $\{Z\}_c$ has been obtained as the result of a number of measurements N. We can use then as theoretical model of the composition of errors (Szacherska, M.K., 1977) the general formula

$$(z_c)_i = \sum_{l=1}^{n_1} (z_r)_{l,k} + \sum_{j=1}^{n_2} (z_s)_{j,i} \quad , \qquad (1.1)$$

for $\quad i = 1, 2 \ldots N$,

where
- k — is the random variable with uniform distribution (the interval being $0 < k \leq N$),
- n_1 — denotes the number of random disturbances,
- n_2 — the number of systematic disturbances,
- n — is the number of all the disturbances of the analysed measurements $(n = n_1 + n_2)$.

In formula (1.1) the symbol

$$(z_r)_l \in R_l(a_l, s_l) \quad , \qquad l = 1, 2 \ldots n_1$$

denotes random errors characterized by the probability distribution $R_l(a_l, s_l)$ with the expected values a_l and the standard deviations s_l. If we assume, that the random components of the mixture are of elementary character and their number $n_1 \to \infty$, the first sum in formula (1.1) can be replaced by $(z_G)_k$, $0 < k \leq N$, that is a random error with the Gaussian distribution.

The systematic errors are expressed by the formula

$$(z_s)_{j,i} = \phi_j(t_i) \quad , \qquad \begin{matrix} j = 1, 2 \ldots n_2 \\ i = 1, 2 \ldots N \end{matrix} \quad , \qquad (1.2)$$

where
- t — denotes any measurement parameter,
- $\phi(t)$ — the relation between alterations of this parameter and the value of the occuring systematic errors.

The generalized model of the composition of errors — briefly recapitulated here — has been justified and discussed in previous publications (Szacherska, M.K., 1973), (Szacherska, M.K., 1974), particularly in a guest lecture (1977) at the Geodetic Institute of Uppsala University. These publications deal also with the method of estimating the precision on the basis of an analysis of the empiric distribution of observations, founded on the assumption of a generalized model of the composition of errors. This assumption is also the basis for the method of the estimation of systematic errors in the course of the adjustment of observations (Szacherska, M.K., Wiśniewski, Z., 1977).

2. ADJUSTMENT OF OBSERVATIONS WITH ESTIMATION OF SYSTEMATIC ERRORS

The traditional methods of adjustment are based on the assumption that systematic errors have been previously eliminated. Practice shows however, that this assumption is not always realised, particularly when disturbances caused by the influence of the environment and atmospheric conditions have occured. The method of the adjustment of observations with simultaneous estimation of the influence of systematic disturbances is founded on the assumption of a generalized model of the composition of errors described by formula (1.1).

If we assume, that observations u_i of a certain parameter of the geodetic network have been performed, we can express the corresponding true value by the equation

$$U = u_i + (z_G)_i + \sum_{j=1}^{n_2} \phi_j(t_i), \quad i = 1, 2 \ldots N. \tag{2.1}$$

Equation (2.1) shows, that the results of observations u_i and their environment described by the functions $\phi_j(t_i)$ are taken into consideration when determining the most probable value of a parameter. We can adopt the principle described by equation (2.1) for the adjustment of a geodetic network, thus making use of the geometric conditions of the network and the physical conditions of the observations.

With regard to the chosen method of adjustment, formula (2.1) can be utilized in order to establish a system of observation equations or conditional equations. The principle of establishing observation equations can be expressed by the formula

$$U_h = u_{h,i_h} + (z_G)_{h,i_h} + \sum_{j=1}^{n_{2_h}} \phi_{h_j}(t_{i_h}) = f_h(x_1, x_2 \ldots x_p),$$

$$i_h = 1, 2 \ldots N_h \tag{2.2}$$

$$h = 1, 2 \ldots M$$

where M - denotes the number of observed parameters,

U_h, $u_{h,i_h} \in \{u\}_h$, N_h - the true value, an observation result and the number of measurements of one of the observed parameters,

$\{x\}$, p - the set of determined parameters and their number.

The number of observation equations, which can be established in conformity with the proposed principle is

$$r_o = \sum_{h=1}^{M} N_h$$

Systematic components of the mixture of errors are additional unknowns in the system of equations; the total number of necessary observations will be consequently determined by the sum $(p + q)$. The symbol q denotes the number of additional unknowns, which depends from the type of the observations and the kind of disturbances.

Applying formula (2.1) we can establish also a system of conditional equations, the construction of which is described by the relation

$$F_d[u_{1,i_1} + (z_G)_{1,i_1} + \sum_{j=1}^{n_{2_1}} \phi_{1_j}(t_{i_1}), \ldots, u_{M,i_M} + (z_G)_{M,i_M} +$$

$$+ \sum_{j=1}^{n_{2_M}} \phi_{M_j}(t_{i_M})] = c_d \qquad (2.3)$$

$$d = 1, 2 \ldots r_c .$$

The true value of a parameter U_h can be defined on the basis of a chosen observation of this parameter $u_{h,i_h} \in \{u\}_h$. Taking into account the geometric scheme of the network and different combinations of observations $u_{h,i_h} \in \{u\}_h$ in an equation of the type (2.3), we can form a system of conditional equations, the number of which

$$r_c = \sum_{h=1}^{M} N_h - (p + q)$$

will depend from the number of all the observations and the number of fundamental and additional unknowns of the system.

The proposed scheme of the adjustment of observations allows the joined determination of the parameters of the network and of the systematic errors treated as additional unknowns in the system. A fundamental problem is the correct determination of the character of disturbances of the measurement process and the choice of the function $\phi_j(t_i)$.

The choice of the function $\phi_j(t_i)$ ought to be based on a thorough knowledge of the measurement process and of the disturbances of this process. Informations can be obtained also by means of an analysis of the empiric distribution of observation errors (Szacherska, M.K., 1973), (Szacherska, M.K., 1977). The computations are generally carried out in several variants with different assumptions of the function $\phi_j(t_i)$, that is for different models of the disturbances of the measurement process. The conformity of the distribution of corrections after the adjustment with the theoretical distribution is decisive for the choice of the optimal variant of the solution.

3. APPLICATION OF THE METHOD TO THE ESTIMATION OF THE INFLUENCE OF THE ENVIRONMENT ON THE RESULTS OF OBSERVATIONS

The method of estimating systematic errors in the course of the adjustment of the network is completed by the algorithm of the computations and an Algol-programme. With their help control tests have been made at the Institute for Geodesy and Photogrammetry of the Agricultural and Technical Academy at Olsztyn. The analysis of the precision levelling network of the Silesian Coal-Basin as well as of the trigonometric levelling network in the region of Cracow shows the considerable influence exerted by the environment and the atmospheric conditions on the results of the observations.

At the elaboration of the precision levelling network of the Silesian Coal-Basin (1519 sections forming 79 lines of I and II order levelling) a model of the composition of errors has been applied, in which the characteristic errors of observers (6 observers), errors connected with the type of the instruments (Wild N 3, Opton Ni 1) and the influence of the refraction were treated as systematic components. The separation of the systematic errors contributed to increase the real precision of the determinations, proved by a reduction of the mean square error amounting to 30% of its value (Szacherska, M.K., Wiśniewski, Z., 1977).

Another interesting example is furnished by the analysis of the triangulation network situated in the submontane region east of Cracow. We have chosen for the analysis a section of the network consisting of 110 triangles, for all sides of which reciprocal observations of the vertical angles have been performed. The measurements have been carried out in 1973. The vertical angles have been measured in three series by seven observers using a Wild T 3 theodolite, the length of the sides - with a tellurometre CA 100.

Since in the chosen section of the network we had for all the sides of the triangles the results of reciprocal measurements of the vertical angles, we could in the analysis make use of conditions furnished by the differences of the results of reciprocal measurements of the altitude difference and the misclosure of altitude in triangles of the network. In the analysis we utilized all the informations concerning the conditions of the measurements noted in the reports of the observers, that is date and time, weather and temperature.

According to theory (Levallois, J.J., 1969), refraction and deflection of the vertical exert an influence on the results of measurements of the vertical angles. Examples of the determination of adequate values in the course of the adjustment of observations have been discussed in literature (Hradilek, L., 1972), (Makowska, A., Zorski, Z., 1976). The precision of any measurements depends of course also from the characteristic features of the observer and the qualities of the instrument.

Considering the principles of the determination of altitude differences by the method of trigonometric levelling, we ought to include in the composition of errors, apart from the random component, systematic errors connected with the influence of changeable atmospheric conditions, the position of the stations, as well as errors of the observer and the instrument. In conformity with formula (1.1) we can express

$$z_{c_i} = z_{G_i} + \phi_1(\{A\}_i) + \phi_2(\{P\}_i) + z_r, \qquad (3.1)$$

where $\{A\}_i$ and $\{P\}_i$ - denote sets of informations concerning the measurement conditions and the position of the stations for the i-observation, z_r is the error of the r-observer.

We have assumed such a general model of the composition of errors for the analysis of the discussed triangulation network. According to the adopted scheme of elaboration (Szacherska, M.K., Wiśniewski, Z., 1977) the fundamental task consisted in the choice of the functions $\phi_1(\{A\}_i)$ and $\phi_2(\{P\}_i)$.

When establishing the function $\phi_2(\{P\}_i)$ we have used well known relations (Levallois, J.J., 1969). The analysis of the influence of the environment has been made in several variants based on different assumptions of the function $\phi_1(\{A\}_i)$. The choice of the variant was limited by the informations contained in the set $\{A\}_i$, for we have not utilized specially planned experimental observations, only the results of typical field works.

We have carried out the computations for each of the variants in two stages. At the first stage unknown systematic disturbances have been computed on the basis of a system of conditional equations (type I) established for each side, the differences between the results of reciprocal observations being the free terms. At the second stage conditional equations (type II) of geometric figures and altitude differences between constant points have been introduced into the system. The differences between the results of these two stages were insignificant when compared with the differences between the results obtained in the particular variants.

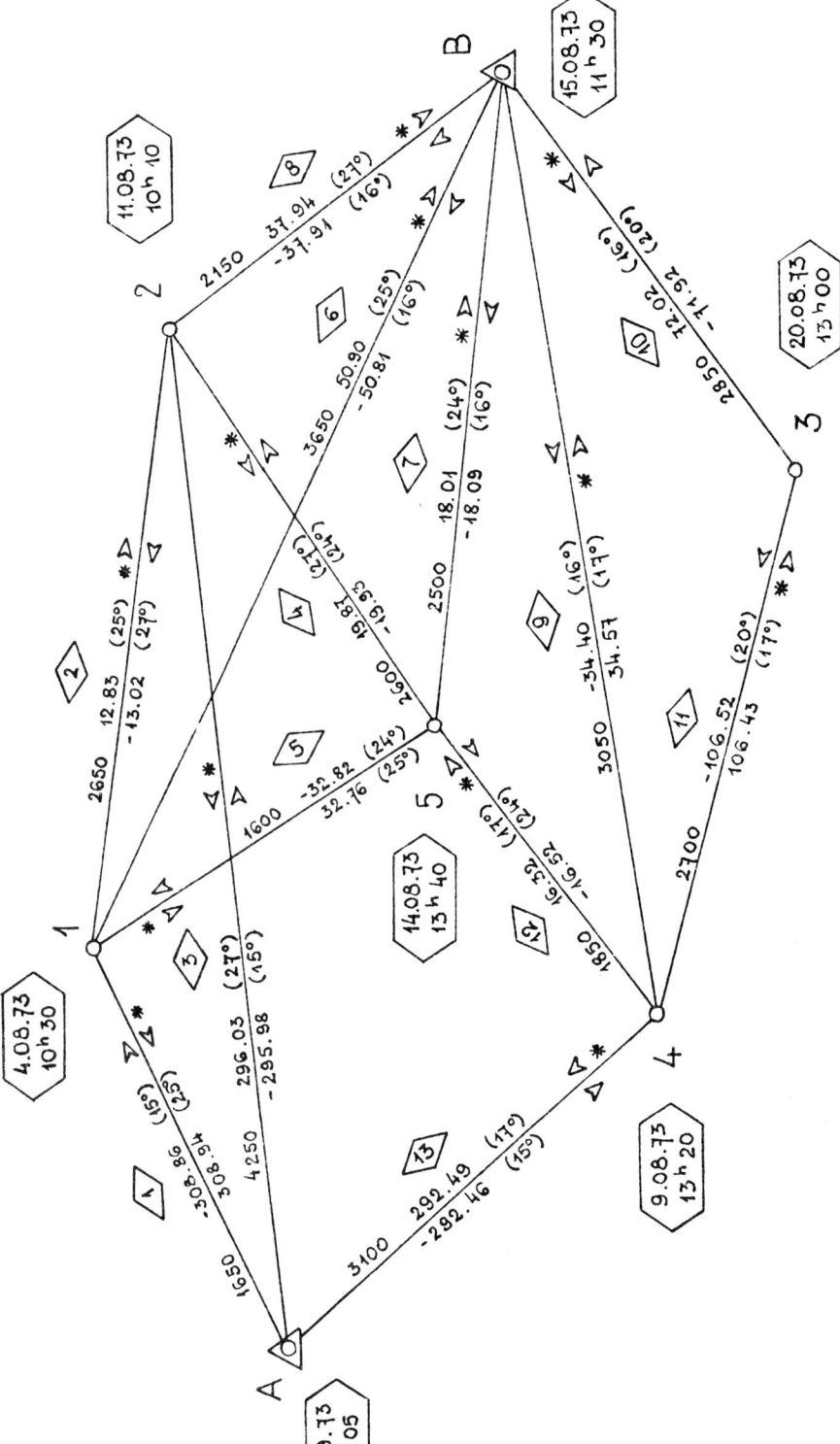

Fig. 1. Scheme of a section of the network with results of reciprocal observations.

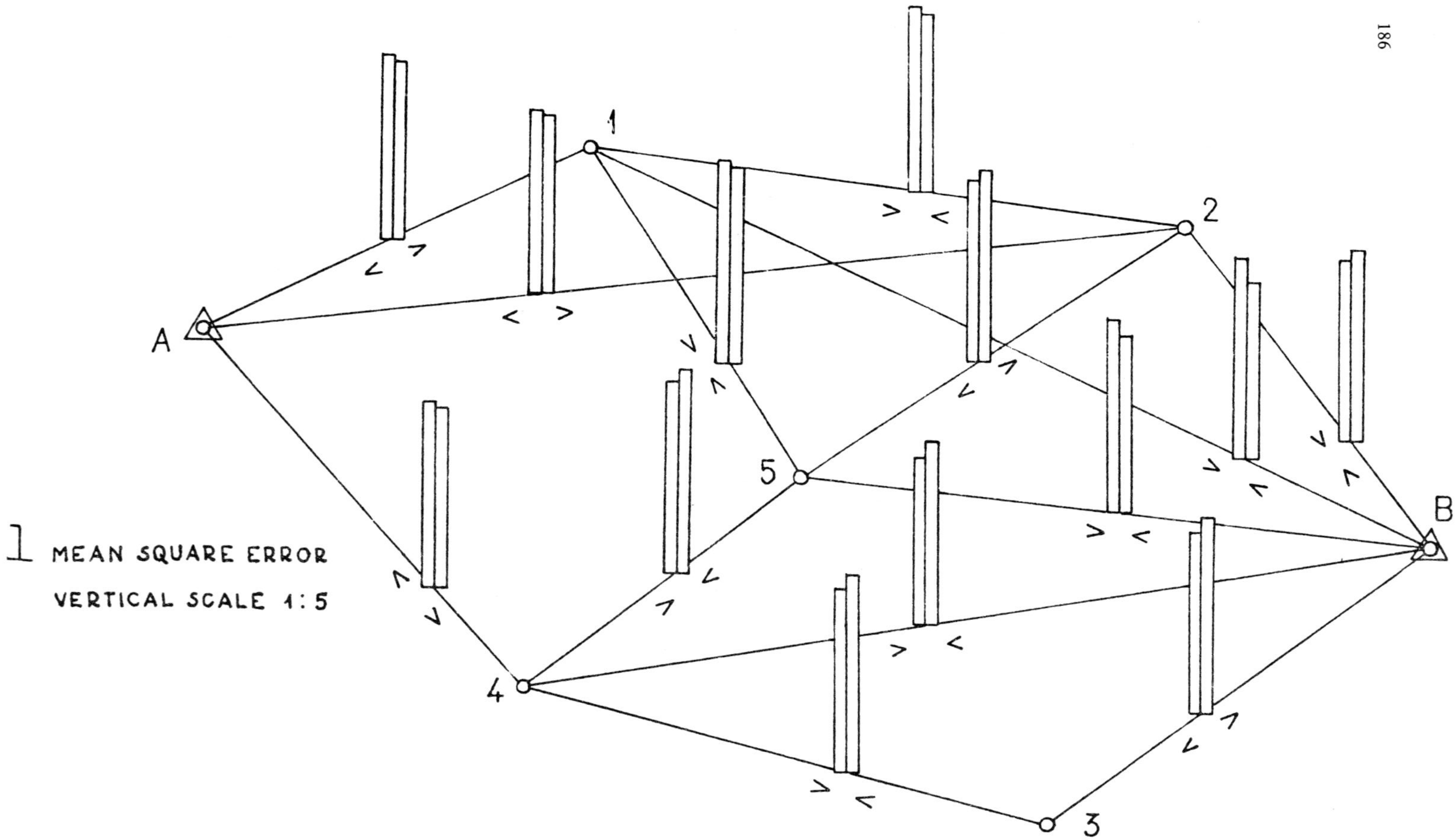
A
B
1
2
3
4
5
1 MEAN SQUARE ERROR
VERTICAL SCALE 1:5

Fig. 2. Systematic components of mixture of errors connected with the influence of observation conditions.

The conformity of the distribution of the random components of the mixture of errors with the model of Gauss and the minimum standard deviation was the criterion of the choice of the variant. We have chosen as optimal the variant of the solution, in which systematic errors resulting from the influence of the environment are described by the linear function of the temperature and the circular function of the time of the observations.

The presented scheme of the analysis is illustrated by an example of a section of a network (fig. 1). In this system 13 conditional equations of type I and 8 independent conditional equations of type II can be formed. The directions of the observations, the results of which have been used for the computation of the free terms of the equations of type II, are indicated on fig. 1.

Systematic errors caused by the influence of the observation conditions, determined for the optimal variant by the adjustment of the system, are shown by fig. 2. The above mentioned components of the mixture of errors of altitude differences, determined on the basis of single measurements, are indicated on this figure. The differences of these values illustrate the corresponding components of the mixture of errors in reciprocal measurements. The reliability of the obtained results is confirmed by the value of the mean square error given at the bottom of the figure.

4. CONCLUSIONS

The method of adjustment presented here is a proposition to apply statistics to the estimation of the influence of the environment, of the atmospheric conditions and refraction on the results of observations. The method has been conceived in the first place as a means of elaborating typical field observations and has been tested in practice on such examples.

It is not necessary, that the observations destined for the analysis should follow a special programme and the observations of each of the parameters need not be more numerous than recommended for typical field works. It is however very important, to register as detailed informations as possible about the conditions in which each measurement was carried out. It results from the assumptions of the method, that each of the parameters ought to be measured at least twice in different atmospheric conditions; this principle is however always observed in practice. It is advisable, that each of the observations should be carried out in atmospheric conditions differing from each other as much as possible. A considerable density of sights in the network is advantageous.

Since the systematic errors are treated as additional unknowns, the total number of unknowns in the system increases, thus weakening the system. This can be prevented by diminishing the number of unknowns by means of a joined determination of the influence of different disturbances. A strengthening of the system can be achieved also by carrying out additional observations, which should be double.

A fundamental question is the correct choice of the functions describing the influence of the systematic disturbances of observations on their results. It seems advisable to elaborate some scientifically justified variants together with recommendations for the practical realization, concerning the indispensable additional informations about the conditions of the measurements. Utilizing the results of typical field works, it will thus be possible to complete the effects of scientific experiments.

REFERENCES

Hradilek, L.: 1972, "Refraction in trigonometric and three-dimensional terrestrial networks", Canadian Surveyor, 26, pp. 59-70.

Levallois, J.J.: 1969. Géodésie Générale, I, Paris.

Makowska, A., Zorski, Z.: 1976, "Analiza wyrównania trygonometrycznej sieci wysokościowej z uwzględnieniem odchyleń pionów i refrakcji" (Analyse de la compensation d'un réseau trigonométrique des altitudes en tenant compte des déviations de la verticale et de la réfraction), Geodezja i Kartografia, 25, pp. 3-16.

Proceedings of the International Symposium on Electromagnetic Distance Measurement and the Influence of Refraction: 1977, Wageningen.

Szacherska, M.K.: 1973, "Metoda analizy rozkładu błędów obserwacji geodezyjnych" (Méthode de l'analyse de la répartition des erreurs d'observations géodésiques), Geodezja i Kartografia, 22, pp. 255-275.

Szacherska, M.K.: 1974, "Model kompozycji błędów pomiarów geodezyjnych" (Modèle de la composition des erreurs de mesures géodésiques), Geodezja i Kartografia, 23, pp. 21-51.

Szacherska, M.K.: 1977, "Method of analysis of the composition of random and systematic observation errors", Guest-lecture at the Geodetic Institute of Uppsala University, Uppsala, Sweden.

Szacherska, M.K., Wiśniewski, Z.: 1977, "Evaluation des erreurs systématiques au cours de la compensation des observations", Invited paper for XV International Congress of Surveyors, Proceedings of FIG Congress, Stockholm, vol. 5, pp. 229-238.

Tengström, E.: 1975, Report of the work of IAG Special Study Group No. I.23, Atmospheric effects on angular measurements, 1971-1975, 16th Assembly of IUGG, Grenoble.

DISCUSSION

K. Poder: expressed his high appreciation of the research done by Szacherska et al, which has started a new theoretical cooperation of great interest to all scientists in Europe, not least in Scandinavia, where also Uppsala University plays a role of unifying all efforts of statistical approaches in geodesy.

EVALUATION OF REFRACTION BY GEODETIC MEASUREMENTS

Ludvík Hradilek
Faculty of Science, Charles University, Prague

Abstract: The merits of two refraction methods are discussed with the result that the line refraction method is appropriate for elaboration of trigonometric leveling traverses whereas the station refraction method is comparatively more convenient for trigonometric and three-dimensional networks of a larger extent especially those designed in high mountain regions.

Problem of the superiority of the methods

The following two approaches can be made for the evaluation of refraction by geodetic measurements
   i) The line refraction method which determines a special coefficient of refraction for each line by reciprocal zenith distances.
   ii) The station refraction method estimating one coefficient of refraction for each observation station by the adjustment of vertical angles, inclined distances and other observables.

Advantages of one method over the other were discussed by professor Ramsayer at the IAG Symposium held at Stockholm in 1974. He found the line refraction method as superior to the station refraction method because the distances between the stations of the traverse were nearly equal and therefore the separation of station refraction from the elevations of the stations was very poor. The statement was indeed correct. However, the examples as given were unfavorable for the station refraction method because the trigonometric leveling traverses which are commonly observed for vertical angles between the neighboring points or a network consisting of equilateral triangles are not relevant for testing of the station refraction method. Also a three-dimensional net without a certain number of slope distances which are inclined over $15°$ is not relevant for the complete checking of three-dimensional procedures. Infact the station refraction method requires both shorter and longer lines of sight radiating from each station and the inclined distances over 15° are necessary in three-dimensional nets for the determination of vertical scale and also for the elimination of systematic errors due to the refraction.When designed with respect to the evaluation of station refraction, the three-dimensional and trigonometric nets can separate the station refraction from the elevation of the station and yield nearly the same accuracy in vertical coordinates as in horizontal ones (Hradilek 1977). When additional observations of vertical angles for longer lines are performed, the station refraction may be estimated even in trigonometric leveling traverse. However, such an observation procedure may have a practical significance for traverses with longer lines of sight (over 3 km) which are designed in high mountain areas without an adjustment to the spirit leveling (Blažek and Hradilek 1978).

Superiority of the line refraction method

Results obtained by Ramsayer (1978), Brunner (1974) and others have indicated that the line refraction method corresponds better to the nature of trigonometric leveling traverses and it yields more realistic mean refraction value over the distance than the station refraction method.Application of the line refraction method by trigonometric leveling traverses in Czechoslovakia resulted in an accuracy which is comparable to that of a lower order spirit leveling.Special sighting targets fitted to the telescopes of the theodolites were used for simultaneous reciprocal pointings.Such an observation procedure eliminates the eccentricity errors and provides an immediate information on the mean value of refraction over the distance.

Superiority of the station refraction method

The three-dimensional and trigonometric leveling nets of a larger extent observed for simultaneous reciprocal zenith distances may prove to be uneconomical especially in high mountain regions.The station refraction methods corresponds better to the nature of such nets and it has additional power for separating refraction from the deflections of the vertical and determining the latter at most of the stations (about 70%) as proved by the elaboration of eight three-dimensional and trigonometric leveling networks (Hradilek 1972,1973).In this regard,the following conditions for the design,observation and elaboration of such nets should be fulfilled

i) All the lines of sight radiating from the same station should differ significantly both in their azimuths and their lengths and they sould be observed within one hour.The observations are repeated at least twice after intervals of two hours or more.The observation stations should be situated at the mountain peaks or at the observation towers at least 15 m above the ground.

ii) The changes in refraction by the repeated angular measurements are used for testing of the station refraction model and refining the latter for each individual line of sight as and when necessary (Hradilek 1973).

Conclusions

Above studies indicate that the line refraction method is superior to the station refraction approach when elaborating trigonometric leveling traverses whereas the station refraction method is more convenient for trigonometric and three-dimensional networks of a larger extent, especially those designed in high mountain regions.

References

Blažek,R.and Hradilek,L.:1978,Proceedings IAU Symp.89,Uppsala.
Brunner,K.:1974,Oest.Zeitschr.f.Verm.und Photogr.62,pp.49-60.
Hradilek,L.:1972,Canadian Surveyor 26 ,pp.59-70.
Hradilek,L.:1973,Zeitschr.f.Verm.98,pp.243-252.
Hradilek,L.:1977,Zeitschr.f.Verm.102,pp.57-63.
Ramsayer,K.:1974,Proceedings IAG Symp.Stockholm,Vol.5.
Ramsayer,K.:1978,Proceedings IAU Symp.No 89, Uppsala.

INVESTIGATION ON REFRACTION IN TRIGONOMETRICAL LEVELING TRAVERSES

Radim Blažek
Technical University, Prague
Ludvík Hradilek
Charles University, Prague

Abstract: When investigating refraction in trigonometrical traverses, additional observations for vertical angles have made it possible to evaluate the coefficient of refraction for each station. The improvement in elevation accuracy was substantial in traverses with longer distances (about 4 km) between stations.

1. DOUBLE TRAVERSE FOR EVALUATION OF REFRACTION

At the IAG Symposia in Stockholm 1974 and in Wageningen 1977 one of the authors presented a method for evaluation of refraction by the adjustment of three-dimensional nets and illustrated an application of such nets for the determination of crustal movements in the western Carpathians. He outlined further investigation on refraction in trigonometric leveling traverses (Hradilek, 1974, 1977).

To obtain more information about refraction, trigonometric leveling traverses were designed as double traverses, i.e. two vertical angles were observed both foresight and backsight at most of the stations (Fig.1). Additional observations for vertical angles made it possible to calculate one coefficient of refraction for each station.

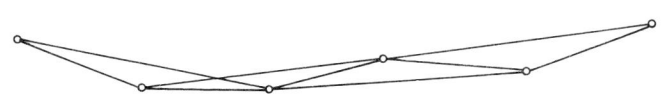

Figure 1. A double traverse.

2. FIELD EXPERIMENTS

2.1. A double traverse with distances of about 200 – 300 m between the neighboring stations was observed by forced centering equipment along the highway Praha-Brno (Fig.2).

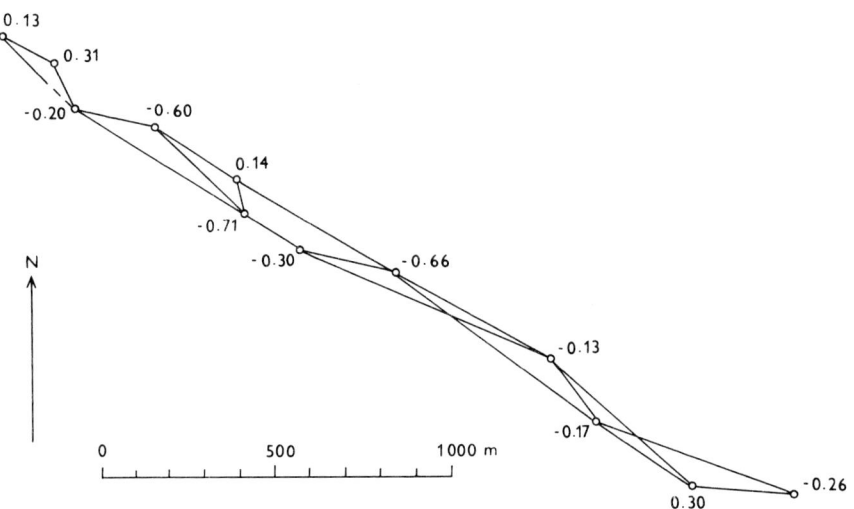

Figure 2. The double traverse situated along the highway Praha-Brno. The numerical quantities indicate the values of coefficients of refraction calculated at each station.

Following refraction models were used for evaluation of the traverse:
 a) a constant value $k = 0.13$ for all stations,
 b) one unknown value k for all stations,
 c) one unknown value k_i for each station P_i.

In Table 1 are given the results A, B of the adjustments; A being an average value of the standard deviation of adjusted trigonometric elevations, B denotes the mean square difference between the trigonometric and spirit leveling elevations.

Table 1. The mean square difference between the elevations determined by double traverse and by spirit leveling along the highway Praha-Brno.

Refraction model	a	b	c
Results A	6.5 mm	4.2	4.3
Results B	2.5	3.0	3.8

The coefficients of refraction calculated under the procedure c) are given in Fig.2.

The results of the adjustment indicate:
  i) the choices of the refraction model a), b), c), respectively, are immaterial in trigonometric leveling traverses with distances of about 300 m between the neighboring stations,
  ii) the precision of trigonometric leveling is comparable to that of a lower order spirit leveling.

2.2. A double traverse with distances of about 500 - 1000 m between the neighboring stations was observed by forced centering equipment in the Žiar Valley (western Carpathians,Fig.3). The values of refraction coefficients calculated under the assumption c) are given in Fig.3.The largest elevation difference in the traverse reached 900 m and was checked by the results of a three-dimensional triangulation with a discrepancy of 30 mm.

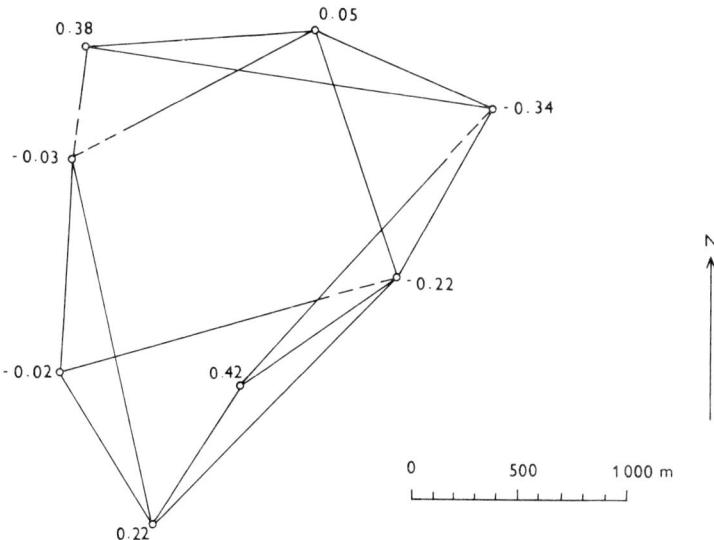

Figure 3. The double traverse observed in the Žiar Valley.The numerical quantities indicate the values of the coefficients of refraction calculated at each station.

2.3. A double traverse with distances of about 4 km between the neighboring stations was designed in the western Carpathians (Fig.4).The values observed were taken from the observation material of a three-dimensional network.Eight computing procedures,denoted as d),...k) were used for the elaboration of the traverse:
  d) The adjustment of a single (i.e.normal) traverse with vertical angles observed (not simultaneously) between the neighboring stations only.The refraction model was chosen

according to b), the deflections of the vertical were neglected. The elevation of the point No 16 was taken as a fixed value into the adjustment.

   e) The same procedure as in d) except two fixed points No 16 and No 4.

   f) and g) The same procedure as under d) and e), respectively, except the deflections of the vertical. The latter had been known before the adjustment and reduced the vertical angles to the spheroid.

   h) The adjustment of the double traverse (Fig. 4) with all the lines of sight observed for vertical angles (not simultaneous observations). All the vertical angles were reduced to the spheroid except that at the station 13. The components of deflections of the vertical at the point No 13 were introduced as unknown parameters and calculated by the adjustment of the traverse. The refraction model was chosen according to the assumption c). The elevation of point No 16 was considered a fixed value.

   i) The same procedure as under h) except two fixed elevations of points No 16 and No 4.

   j) and k) The same procedure as under h) and i), respectively, except the deflection of the vertical at the point No 13. The deflection had been known before the adjustment, and the vertical angles at all stations were reduced to the spheroid.

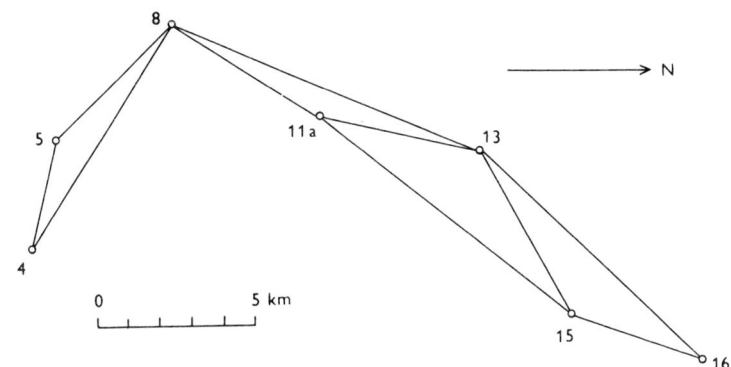

Figure 4. The double traverse designed in the western Carpathians.

The results of the adjustment procedures d),...k) are given in Tables 2 and 3. They were checked by a three-dimensional network surrounding the traverse. The accuracy of the network was estimated to be about 20 mm in elevations and 1.5 seconds of arc in deflections of the vertical. The results given in Table 3 indicate the well known distinction between the traverses with one and two fixed end points, respectively.

Table 2. The coefficients of refraction calculated by the adjustment of the double traverse designed in the western Carpathians.

Point No	Adjustment procedures d,...k. Refraction models (b),(c)							
	d (b)	e (b)	f (b)	g (b)	h (c)	i (c)	j (c)	k (c)
4	0.114	0.113	0.131	0.130	0.126	0.124	0.126	0.123
5	0.114	0.113	0.131	0.130	0.131	0.131	0.132	0.131
8	0.114	0.113	0.131	0.130	0.139	0.139	0.139	0.140
11a	0.114	0.113	0.131	0.130	0.127	0.122	0.125	0.123
13	0.114	0.113	0.131	0.130	0.131	0.130	0.130	0.128
15	0.114	0.113	0.131	0.130	0.117	0.123	0.119	0.121
16	0.114	0.113	0.131	0.130	0.175	0.179	0.176	0.179

Table 3. The differences between the elevations determined by the double traverse and by the three-dimensional triangulation in the western Carpathians (MSE denotes the mean square difference).

Point No	Adjustment procedure d),...k)							
	d	e	f	g	h	i	j	k
4	145 mm	0	93	0	61	0	48	0
5	117	20	51	36	26	33	12	31
8	97	17	36	37	32	21	18	16
11a	49	49	75	13	57	3	41	10
13	80	10	34	11	15	20	8	16
15	34	18	15	5	26	27	25	31
16	0	0	0	0	0	0	0	0
MSE	88	22	53	21	37	20	27	19

The stronger design of the double traverse diminish this distinction significantly. The deflections of the vertical calculated by the procedures h) and i) at the station No 13 differed by 2 seconds of arc from its check value.

3. CONCLUSIONS

The field experiments proved the possibility of estimating refraction at each station of a double traverse. All the three traverses discussed gave reasonable results with respect to both the values of the coefficients of refraction and the differences between the trigonometric elevations and their check values determined by spirit leveling and three-dimensional triangulation, respectively. However, the results obtained for shorter lines of sight were independent of the choice of the refraction model, and indicate the double traverses have not a major significance for practical applications, except the traverses with longer lines of sight (over 3 km) which are designed in mountain regions without an adjustment to the spirit leveling.

References
Hradilek,L.:1974,Proceedings IAG Symp.Stockholm,Vol.5.
Hradilek,L.:1977,Proceedings IAG Symp.Wageningen,pp.185-190.

DISCUSSION

K. Poder: Thank you very much, professor Hradilek. You have really introduced a lot of new information to the geodetic community. If I may slightly disagree with you, I would say you are happy getting rid of these refraction coefficients, but you must not forget that at the same time you actually get rid of it, you could possible also determine it, and I am quite sure, that the trigonometric network will be very useful to get a better model of the refraction or the refractive index of the air in general. You get a lot of information of its derivatives and this will possibly be useful. But again, I think it is only a matter that instead of just eliminating, you get the information out after all.

L. Hradilek: By mathematical procedures, we can either determine or eliminate coefficients of refraction with the same results in elevations. The elimination of refraction seems to be more convenient with respect to a decrease in computing time. In mountainous regions, the evaluation of refraction, as well as of elevations, is substantially supported by ranging of very inclined distances.

D.G. Currie: I think, that in astronomy we recognize that there is a normal refraction and an anomalous refraction. The normal refraction is what the tables are made to work with. It seems part of what you have just mentioned is connected with trying to make a distinction. Is the data perhaps not so much to determine more coefficients but to evaluate how often the models can be used and of what accuracy, and how much has to be left open because it is out of control of a given class of models. In other words, there is ultimately certainly in our case a variability, that we might never expect to have in the model. And one thing, that I think is important to try in addresses, is how much is the model and how much of it can you not work with.

L. Hradilek: Our refraction models are mostly based on the zero hypothesis assuming a different coefficient of refraction for each observation station and the same refraction for all lines of sight radiating the station (under certain conditions concerning the observation procedure and the height of the observation station above the ground). The zero hypothesis is tested by the actual changes in refraction, which are determined by the measurements of vertical angles. If the test fails, an alternative hypothesis is designed considering the actual changes in refraction for each individual line of sight. Primarily, we are interested in determining elevations and in eliminating the influence of refraction on vertical angles. The value of refraction itself has not so large importance for us. However, when elaborating 7 networks in mountainous and hilly regions, the coefficients of refraction were determined according to our zero hypothesis at all 130 stations; the coefficients of refraction were evaluated within the limits 0.067 - 0.214, the average value was 0.1302. The zero hypothesis failed at several points of another network designed in the plane area of southern Slovakia (with changes in vertical angles attaining 180").

THE ACCURACY OF THE DETERMINATION OF TERRESTRIAL REFRACTION FROM RECIPROCAL ZENITH ANGLES

K. Ramsayer
Geodetic Institute of the University of Stuttgart

To investigate the accuracy of the determination of terrestrial refraction from reciprocal zenith angles and astronomical latitudes and longitudes at both ends of a line a test net with lines from 4 km to 23 km was observed and three dimensionally adjusted. As the measurements of the zenith angles were repeated every hour 40 times in an average the adjusted values were taken as a substitute for the true values. It is shown, that the mean refraction coefficient k, which is changing from k = 0.10 at day up to k = 0.34 at night, and the corresponding refraction angle can be determined very accurately, if both angles are measured simultaneously. Observations with day light are better than observations in the night. For observations with day light the mean difference between the true refraction angle at the observation station and the mean refraction angle of the observed line was smaller than $\pm$ 1" independent of the length of the line. That means that the mean deviation of the true effective refraction coefficient in the observation station and the mean refraction coefficient of the observed line was inverse proportional to the distance.

1. INTRODUCTION

In three dimensional networks and traverses the accuracy of the heights is mainly depending on the accuracy with which the influence of refraction to measured zenith angles can be determined. A well known means to determine this influence is to measure the zenith angles and the directions of the verticals at both ends of a line P_1P_2, Fig. 1. Then we have the following relations

$$z_1 = z_1' + \delta_1 = z_1' + k_1(d/2r_m)\rho,$$
$$z_2 = z_2' + \delta_2 = z_2' + k_2(d/2r_m)\rho,\tag{1}$$

z_1, z_2 = true zenith distances of the line P_1P_2,
z_1', z_2' = measured zenith distances,
δ_1, δ_2 = refraction angles,
k_1, k_2 = refraction coefficients,
r_m = mean radius of the earth,
d = distance between P_1 and P_2,
ρ = 206265".

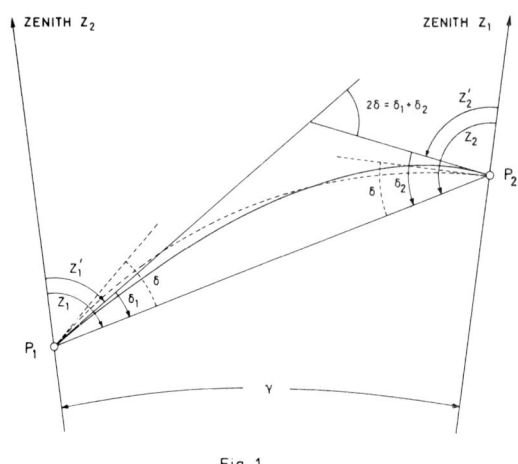

Fig. 1

If we introduce the mean values

$$\delta = (\delta_1 + \delta_2)/2 \quad \text{and} \quad k = (k_1 + k_2)/2 \tag{2}$$

then we get with

$$\Delta\delta = (\delta_1 - \delta_2)/2 \quad \text{and} \quad \Delta k = (k_1 - k_2)/2 \tag{3}$$

$$z_1 = z_1' + \delta + \Delta\delta$$
$$= z_1' + k(d/2r_m)\rho + \Delta k(d/2r_m)\rho,$$
$$z_2 = z_2' + \delta - \Delta\delta$$
$$= z_2' + k(d/2r_m)\rho - \Delta k(d/2r_m)\rho.\tag{4}$$

Between the reciprocal zenith angles we have the relation

$$z_1 + z_2 = z_1' + z_2' + 2\delta$$
$$= z_1' + z_2' + 2k(d/2r_m)\rho = 180° + \gamma. \qquad (5)$$

γ is the angle between the verticals in P_1 and P_2, neglecting the small influence that both verticals are not exactly in the same plane. From (5) we can compute the mean values

$$\delta = (\delta_1 + \delta_2)/2 = \{180° + \gamma - (z_1' + z_2')\}/2, \qquad (6)$$
$$k = (k_1 + k_2)/2 = (\delta/\rho)(2r_m/d). \qquad (7)$$

If we set

$$\delta_1 \simeq \delta_2 \simeq \delta \quad \text{resp.} \quad k_1 \simeq k_2 \simeq k \qquad (8)$$

then we get according to (4) the errors

$$\varepsilon_1 = z_1 - (z_1' + \delta) = +\Delta\delta = +\Delta k(d/2r_m)\rho,$$
$$\varepsilon_2 = z_2 - (z_2' + \delta) = -\Delta\delta = -\Delta k(d/2r_m)\rho. \qquad (9)$$

If we measure both zenith angles simultaneously, we can hope that the influence of refraction to both angles is approximately the same. In this case we can expect that $\Delta\delta$ resp. Δk and hence ε_1 and ε_2 are small.

2. THE TEST-NET

For the investigation of the errors ε and Δk the network shown in Fig. 2 was observed. For each line both zenith angles were measured simultaneously, each by 6 sets with a standard deviation of ± 0.6". These measurements were repeated each hour, partly during the whole day and at different days. The number of repetitions changes from 12 (line 1)

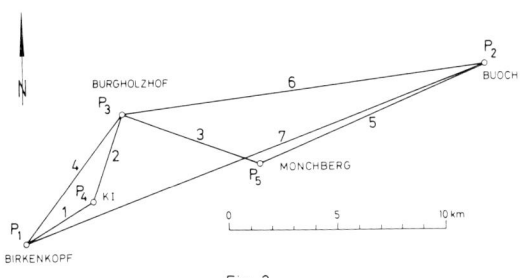

Fig. 2

and 60 (line 7). The mean value per line is 40. Besides the zenith angles the lengths of the lines were measured with Tellurometer CA 1000. Furtheron in all points of the network astronomical latitude and longitude were determined with Zeiss Ni2-Astrolab by one set with 20 stars in an average. The orientation in azimuth was taken from another three dimensional network.

The network was adjusted rigorously three dimensional in an ellipsoidal reference system. The coordinates of point P_1 were given and kept fix. For each line a special refraction coefficient was determined from all zenith angles measured at both ends of the line. The standard deviations of the adjusted zenith angles referred to the adjusted directions of the verticals change from $\pm$ 0.7" (line 1) to $\pm$ 1.4" (line 7). The root mean square is $\pm \overline{1}$". These errors are relatively large, if we consider the great number of zenith angle measurements. They are mainly caused by the moderate accuracy of the astronomical observations and the small redundancy of the network. In the following the adjusted values of the zenith angles were taken as a substitute of the true values.

3. RESULTS

In Fig. 3 to Fig. 6 some results of the investigations are demonstrated. Fig. 3 shows the influence of refraction for

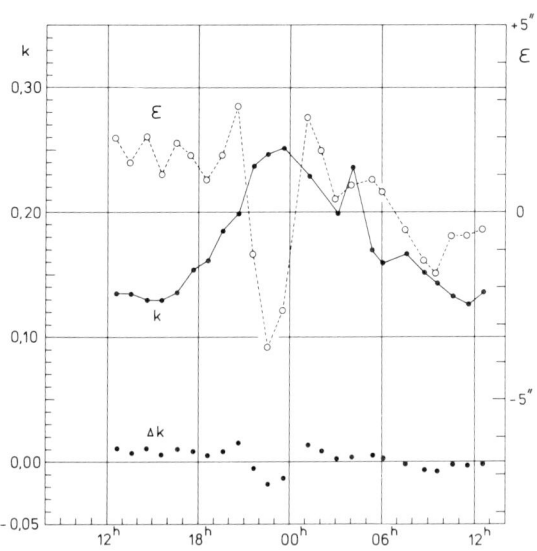

Fig. 3 : Line 5 ($P_5 P_2$) , d = 11.641 km , Aug. 25/26

line 5 from point P_5 to point P_2 between 12 h at 25. August
and 12 h at 26. August. The mean refraction coefficient k is
changing from the standard value 0.13 at mid-day and 0.25
at mid-night. The difference $\Delta k = k_5 - k$ between the true
refraction coefficient k_5 in point P_5 and the mean value k
varies between + 0.015 and - 0.019. The quadratic mean of
Δk is ± 0.009. The error ε resulting from the refraction
error $\overline{\Delta k}$ changes between + 2.8" and - 3.7". Fig. 4 shows
some results of line 6 from P_2 to P_3. We see again that k

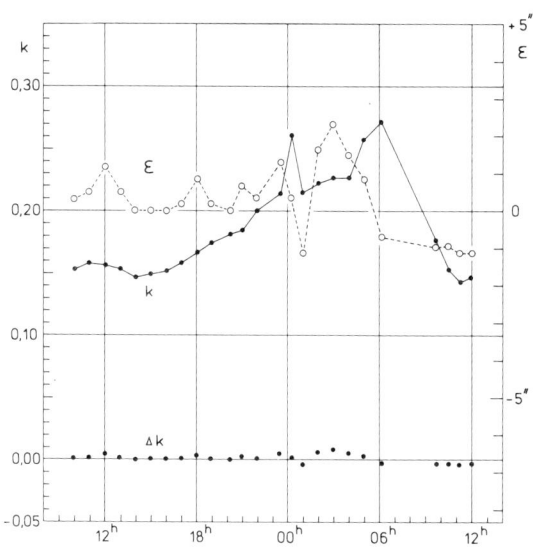

Fig. 4 :   Line 6 (P_2P_3) , d = 16.908 km , Oct. 3/4

has its minimum at mid-day and its maximum at mid-night and
in the early morning. Δk varies between - 0.004 and + 0.008
and ε changes between - 1.2" and + 2.3". Fig. 5 shows the
influence of refraction for the longest line from 10 h in
the morning to 10 h at the following day. Here the change
of k and the errors Δk are relatively small. The errors ε
are small too, although the length of the line is more than
20 km. From Fig. 6 follows again that k is changing very
much and very rapidly during night. Nevertheless Δk does not
exceed 0.022, but by the large distance of the two observation stations the resulting errors ε are enlarged up to 8.2".

Table 1 shows the minimum and maximum of Δk and ε for all
lines seperated for day and night. Further the quadratic
means m_k and m_ε of Δk and ε and the number n of reciprocal
zenith angle measurements are tabulated. We see that for
observations at day m_k is decreasing with distance d from

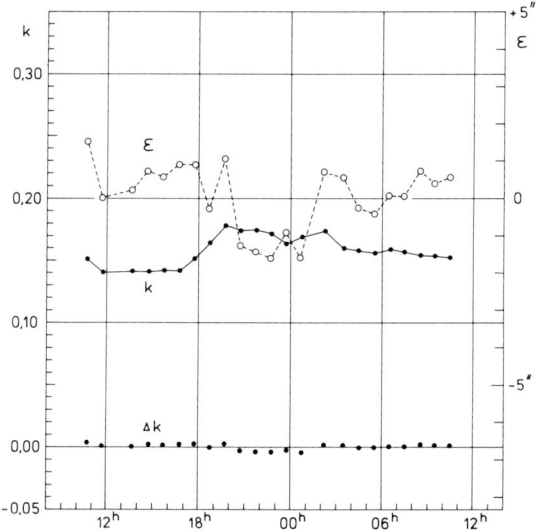

Fig. 5 : Line 7 (P_1P_2), d = 22.848 km, Sept. 12/13

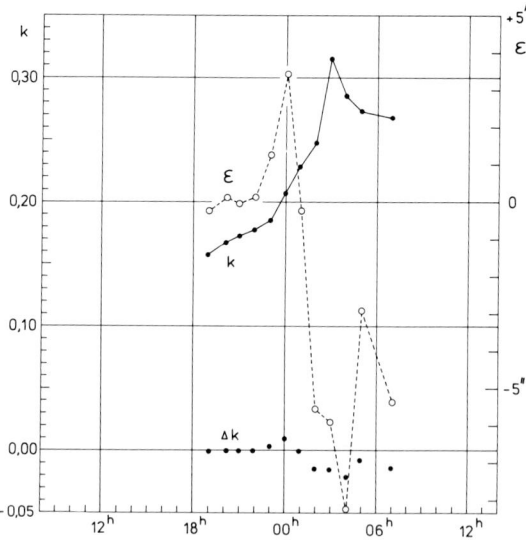

Fig. 6 : Line 7 (P_1P_2), d = 22.848 km, Sept. 16/17

THE ACCURACY OF THE DETERMINATION OF TERRESTRIAL REFRACTION

LINE		d [km]	DAY ($07^h - 19^h$)							NIGHT ($19^h - 07^h$)					
			Δk		$\pm m_k$	ε		$\pm m_\varepsilon$	n	Δk		$\pm m_k$	ε		$\pm m_\varepsilon$	n
NO.	FROM–TO		MIN –	MAX +		MIN –	MAX +			MIN –	MAX +		MIN –	MAX +		
1	$P_1 - P_4$	3.7	–	.026	.018	1.9"	1.6"	1.1"	12	–	.007	.011	1.3"	0.5"	0.8"	5
2	$P_4 - P_3$	4.3	.030	.013	.015	1.2	0.9	1.1	9	.013	.016	.003	2.0	1.7	0.9	26
3	$P_5 - P_3$	6.5	.017	.006	.006	1.3	0.7	0.7	28	.018	.011	.006	0.8	1.4	0.7	13
4	$P_1 - P_3$	7.6	.012	.011	.005	1.6	1.3	0.7	26	.006	.015	.008	3.7	2.9	1.5	24
5	$P_5 - P_2$	11.6	.013	.011	.006	1.9	2.0	1.1	25	.019	.008	.004	2.0	2.3	1.2	24
6	$P_2 - P_3$	16.9	.010	.004	.003	1.7	1.2	0.9	21	.007	.013	.007	8.2	4.9	2.6	35
7	$P_1 - P_2$	22.8	.006	.004	.002	0.4	1.6	0.6	25	.022						

Table 1

+ 0.018 to + 0.002, whilst m_ε differs only slightly from the mean value of + 1". At night m_k and m_ε are somewhat larger than at day with exception of line 2. The decrease of m_k with distance is not so marked as at day time and m_ε is increasing with distance.

Fig. 7 shows m_k and m_ε for day observations depending on

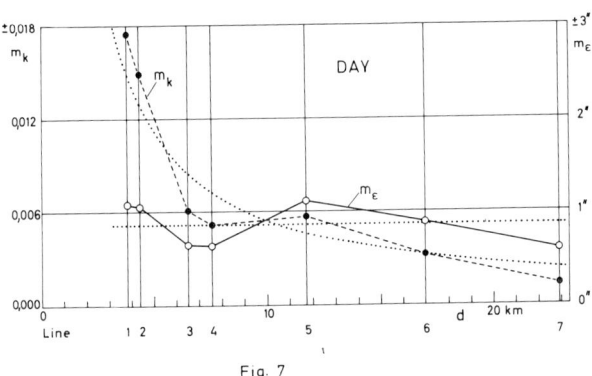

Fig. 7

distance d. In a rough approximation m_ε has the constant value

$$\overline{m}_\varepsilon = \pm 0.9". \tag{10}$$

The corresponding error of the determination of the refraction coefficient is

$$\overline{m}_k = \pm (\overline{m}_\varepsilon/\rho)(2r_m/d) = \pm 0.055 [km]/d[km]. \tag{11}$$

The obvious decrease of m_k with distance may be explained by the fact, that in the test-field the light path runs the more through the free atmosphere the longer the line.

Fig. 8 shows m_k and m_ε for night-observations. Here m_ε increases with distance from + 0.8" up + 2.7", whilst the decrease of m_k with distance is somewhat smaller as for day-observations.

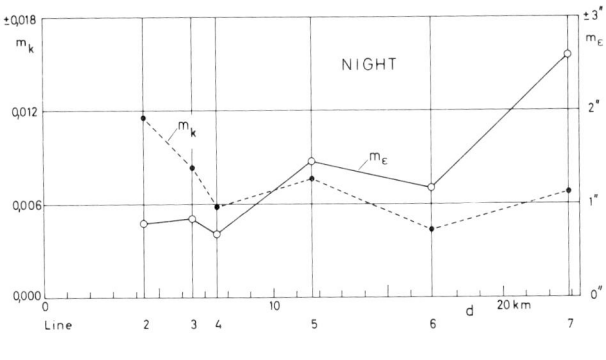

Fig. 8

4. SUMMARY AND CONCLUSIONS

1. The influence of refraction to zenith angles changes very strongly with day time especially during the night. The standard value 0.13 for the refraction coefficient is a rough approximation from 10 h to 15 h. At night k was increasing up to 0.34.

2. The mean refraction coefficient can be determined with good accuracy by measuring the reciprocal zenith angles and the astronomical latitudes and longitudes at both ends of a line.

3. The measurements of the reciprocal zenith angles should be made simultaneously. In this case the differences between the true refraction coefficients at the observation stations and the mean refraction coefficient are small. Observations with day light are better than observations in the night.

4. For observations with day light the standard deviation of a single set of a zenith angle which is reduced for mean refraction was approximately $\pm$ 1" independent of the length of the line.

5. For observations with day light the mean deviation of the true effective refraction coefficient in the observation station and the mean refraction coefficient of the observed line was inverse proportional to the distance.

All together we can say that the determination of refraction by reciprocal zenith angle measurements and astronomical latitude and longitude observations is a surprisingly accurate method.

DISCUSSION

K. Poder: Thank you professor Ramsayer. I must say that apart from professor Hradilek and you a lot of the geodetic community should actually be very ashamed because we have known how to determine zenith distances for more than one hundred years, but obviously nobody has really considered it as seriously as you and professor Hradilek have done. And I think it is of very much interest. We have professionally in Greenland practically only heights by zenith distances, but we have never looked so carefully into the matter, I'm afraid to say. So, I think this is a very interesting paper.

J.A. Hughes: There seems to be an inverse proportion between error and distance. The further you look the better it is. Could you explain that effect to me, a non-geodesist?

K. Ramsayer: You wonder why the deviation of the true refraction coefficient from the mean refraction coefficient of the observed line is inverse proportional to the distance. I think that the reason is, that the longer the distance is the more we come into the free atmosphere. I explain it so. I was also surprised. It is however not always the case. If the lightpath goes very near to the earth's surface you have very uncertain relations.

K. Poder: I can add that this is really going to change the weightning function, which we use for trigonometric levelling. I have assumed that it was independent of the distance. It seems that most observation equations are rather friendly. And the first equation I had was a very unfriendly one, but using your law I think I will get a friendly equation.

NUMERICAL FILTERING OF REFRACTION COEFFICIENTS

H. Kahmen
Geodetic Institute University of Karlsruhe

ABSTRACT

The fluctuations of refraction coefficients can be described by a stochastic process: a two-component process. One component of the process is caused by short-periodic (daily) climatic variations. The other component is influenced by long-periodic (yearly) climatic variations. The central moments of these two components are used to estimate the covariance matrix of time series of refraction coefficients. Different functional and stochastic models are tested in connexion with time averaging of refraction coefficients.

1. INTRODUCTION

For long distances the accuracy of vertical angle measurements depends mainly on the estimation of the parameters of the meteorological field along the ray path. On the basis of atmospheric physics equations can be found which describe the relations between the observations, the model errors and the systematic parameters. In general these equations are named "functional models". The functional models are one supposition for estimating systematic parameters by least-squares adjustment procedures.

Assumptions about the accuracy, the stochastic independence and the correlation of observations are called "stochastic models". The stochastic models are the second supposition for estimating parameters by least-squares adjustment procedures.

The loss of information in connexion with adjustment procedures will be kept small if the functional and stochastic models are estimated as exactly as possible. This paper deals with tests analysing the influence of different functional and stochastic models on the results of adjustment procedures. Though the following tests are based on time series of refraction coefficients the result will be typical for all adjustment procedures using observations influenced by the atmosphere.

2. FUNCTIONAL AND STOCHASTIC MODELS OF TIME SERIES OF REFRACTION COEFFICIENTS

Assumptions concerning the functional and stochastic models of observations can be found best, when a great number of measurements is available. Normally it is difficult to estimate the models directly from geodetic measurements, since for economic reasons only a small number of repeated measurements can be performed. In this paper only those fluctuations of vertical angle measurements will be considered, which are caused by variations of the meteorological field. It is shown (KAHMEN 1977) that these fluctuations can be estimated without direct geodetic measurements using only records of the meteorological field. Long-time records of meteorological parameters are available at many meteorological stations. With these records the physical causes of the fluctuations of the vertical angles can first be found. If the physical causes are well known the fluctuations of the vertical angles can easily be calculated by simple linear transformations. Causes of the fluctuations of vertical angles are the variations of the refraction coefficients.

The figures 2.1 and 2.2 show examples of such time series +). In figure 2.1 we see mean values for one hour of refraction coefficients, plotted daily from January 1962 to December 1963 using observations recorded between 12 a.m. and 1 p.m. . Figure 2.2 shows mean values for one hour of refraction coefficients, plotted every fifth day from January 1962 to April 1971 using observations recorded between 12 a.m. and 1 p.m. .

The figures 2.1 and 2.2 show that the fluctuations of the refraction coefficients can be described by a stochastic process.

+) Further time series have been calculated and will be published.

NUMERICAL FILTERING OF REFRACTION COEFFICIENTS

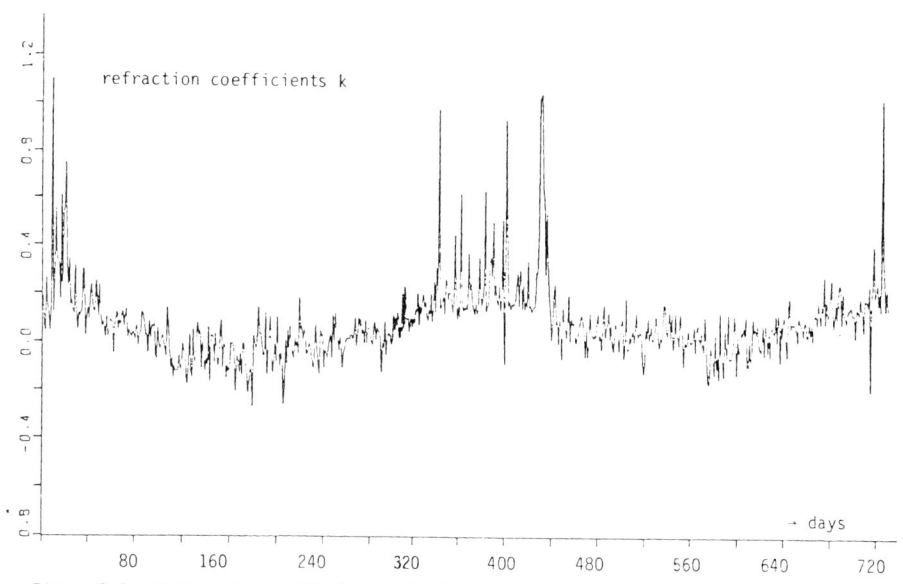

Figure 2.1 Refraction coefficients calculated from 1-1-1962 to 31-12-1963
(time-difference between the single values: 1 day)

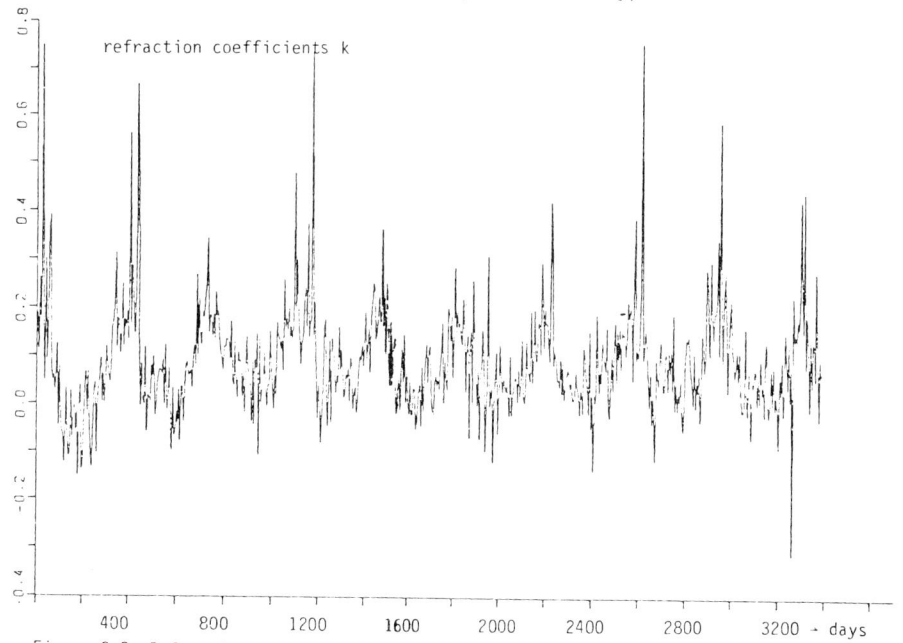

Figure 2.2 Refraction coefficients calculated from 1-1-1962 to 27-4-1971
(time-difference betwen the single values: 5 days)

Figure 2.3 indicates the structure of such a linear stochastic process.

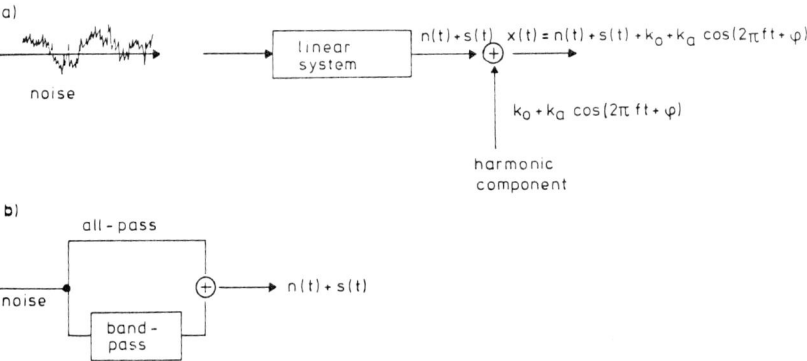

FIGURE 2.3   a) structure of a linear stochastic process
             b) linear system

The linear system is composed of the parallel connection of an all-pass and a band-pass. Consequently the stochastic process, which is superimposed by a trend, can be described by the linear model

$$\mathbf{x(t)} = \mathbf{n(t)} + \mathbf{s(t)} + \mathbf{k_0} + k_a \cos(2\pi ft + \varphi) \qquad (2.1)$$

where

$\mathbf{x(t)} = [x_1, x_2 \ldots]$ and $\mathbf{t} = [t_1, t_2 \ldots]$
$\mathbf{n(t)} = [n_1, n_2 \ldots]$
$\mathbf{s(t)} = [s_1, s_2 \ldots]$
$\mathbf{k_0} \doteq [k_0, k_0 \ldots]$
$\boldsymbol{\varphi} = [\varphi, \varphi \ldots]$

$\mathbf{n(t)}$ and $\mathbf{s(t)}$ describe the non deterministic part of the process and $\mathbf{k_0} + k_a \cos(2\pi f\mathbf{t} + \boldsymbol{\varphi})$ describes the deterministic part of the process.

$\mathbf{n(t)}$ is a short period random process and $\mathbf{s(t)}$ a long period random process. Figure 2.4 for example shows how one signal $x(\mathbf{t})$ is built by additive superposition of the functions $n(\mathbf{t})$, $s(\mathbf{t})$, $k_0+k_a \cos(2\pi f\mathbf{t} + \boldsymbol{\varphi})$.

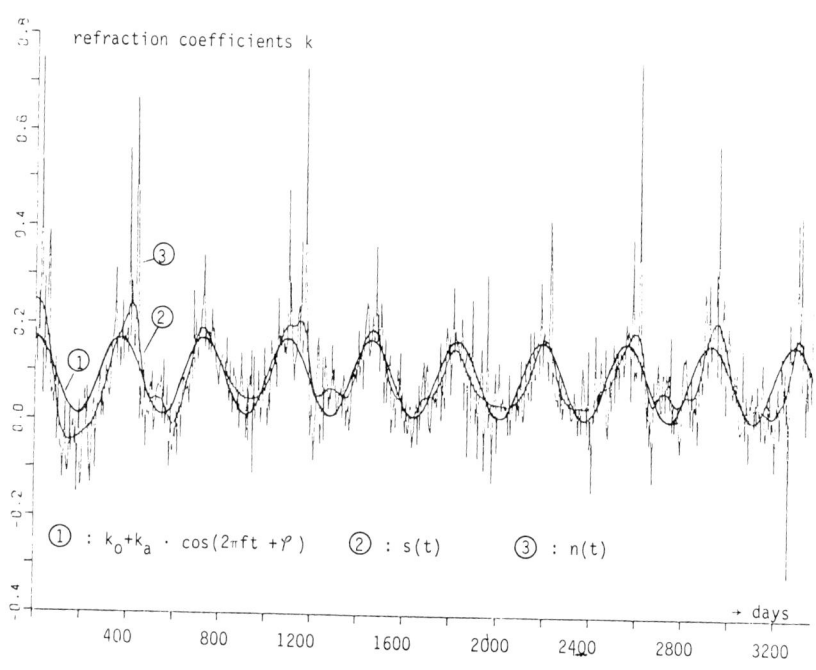

FIGURE 2.4 Superposition of the signals n(**t**), s(**t**), $k_o + k_a \cos(2\pi f\mathbf{t} + \varphi)$

For one signal n(**t**) = **n** we can assume (KAHMEN 1977)

$$E\{\mathbf{n}\} = 0 \tag{2.2}$$

$$E\{\mathbf{n}\mathbf{n}^T\} = \sigma_n^2 \delta(\tau) \tag{2.3}$$

(σ: standard deviation)

where

$$\delta(\tau) = \begin{cases} 1 & \text{for } \tau = 0 \\ 0 & \text{for } \tau \neq 0 \end{cases} \tag{2.4}$$

τ describes the time-difference between single measurements.

$E\{\cdot\}$ is the statistical expectation.

σ_n^2 has different values during the several periods of a year.

One signal $s(t) \equiv \mathbf{s}$ can be approximated by a Gaussian narrow-band noise (KAHMEN 1977)

$$s(t) = s_0(t) \cos [\omega_0 t + \psi(t)] \qquad (2.5)$$

where the probability density of $s_0(t)$ and $\psi(t)$ is

$$p(s_0) = \frac{s_0}{\sigma^2} \exp\left(-\frac{s_0^2}{2\sigma^2}\right), \text{ (Raleigh-Distribution)} \qquad (2.6)$$

$$p(\psi) = \frac{1}{2\pi} [\delta_r(\psi) - \delta_r(\psi - 2\pi)] \qquad (2.7)$$

(δ_r = normed step-function).

Consequently we can assume:

$$E\{\mathbf{s}\} = 0 \qquad (2.8)$$

$$E\{\mathbf{s}\mathbf{s}^T\} = \sigma_s^2 \exp(-a_0^2 \tau^2) \cos \omega_0 \tau = r(\tau) \qquad (2.9)$$

where τ describes the time-difference between single measurements and a_0 is a constant.

3. ADJUSTMENT OF TIME SERIES OF REFRACTION COEFFICIENTS OR REFRACTION ANGLES

The basic equations for the coefficients k of refraction and the angles δ of refraction are (FEARNLEY 1884/85):

$$k = \frac{2}{S} \int_0^S \frac{S-S'}{S} \chi(S') \, dS' \qquad (3.1)$$

$$\delta = \frac{\rho}{R} \int_0^S \frac{S-S'}{S} \chi(S') \, dS' \qquad (3.2)$$

where S is the length of the curved path (the difference between the arc-length and the chord-length is neglected, S' describes the coordinates along the light path, R is the radius of the earth and

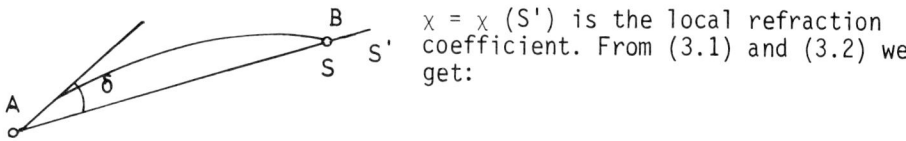

$\chi = \chi(S')$ is the local refraction coefficient. From (3.1) and (3.2) we get:

FIGURE 3.1 Angle of refraction

$$\delta = \frac{S\rho}{2R} \cdot k . \tag{3.3}$$

Approximate refraction free directions can be calculated from time series of vertical angles, if the systematic components of δ are eliminated by subtraction and if the stochastic components of δ are filtered by time averaging procedures. In the following the calculation of refraction free directions will be analysed with different functional and stochastic models. Equ. (3.3) shows that the tests can directly be made with the time series of refraction coefficients shown in figure 2.1 and 2.2 as there is a linear relation between δ and k.

With equ. (2.1) the general model for the vector of the observations $\mathbf{x} \equiv x(\mathbf{t})$ is (Wolf 1977)

$$\mathbf{x} = \mathbf{A}\hat{\mathbf{x}} + \mathbf{B}\mathbf{s} + \mathbf{n} \tag{3.4}$$

where

$\hat{\mathbf{x}}$ ... vector of systematic parameters
$\mathbf{A}$ ... matrix relating $\mathbf{x}$ and $\hat{\mathbf{x}}$
$\mathbf{B}$ ... matrix relating $\mathbf{x}$ and $\mathbf{s}$.

The matrix $\mathbf{C}_{ss} = \text{cov}(\mathbf{s})$ and the matrix $\mathbf{C}_{nn} = \text{cov}(\mathbf{n})$ are given a priori as the vector $\mathbf{n}$ is chracterized by equ. (2.2), (2.3) and the vector $\mathbf{s}$ by equ. (2.8), (2.9). $\mathbf{C}_{ss}$ and $\mathbf{C}_{nn}$ have the form:

$$\mathbf{C}_{ss} = \begin{vmatrix} \sigma_s^2 & r_{12}\sigma_s^2 & \cdots & r_{1u}\sigma_s^2 \\ r_{12}\sigma_s^2 & \sigma_s^2 & \cdots & r_{2u}\sigma_s^2 \\ \vdots & \vdots & \ddots & \vdots \\ r_{u1}\sigma_s^2 & r_{u2}\sigma_s^2 & \cdots & \sigma_s^2 \end{vmatrix} \tag{3.5}$$

$$\mathbf{C}_{nn} = \begin{vmatrix} \sigma_{n1}^2 & 0 & \cdots & 0 \\ 0 & \sigma_{n_2} & \cdots & 0 \\ \vdots & \vdots & & \\ 0 & 0 & \cdots & \sigma_{n_u}^2 \end{vmatrix} \qquad (3.6)$$

For the following calculations we get $\mathbf{B} = \mathbf{I}$ ($\mathbf{I}$: unit matrix) so that we have instead of (3.4) the classical model of collocation

$$\mathbf{x} = \mathbf{A}\hat{\mathbf{x}} + \mathbf{s} + \mathbf{n} . \qquad (3.7)$$

With the side condition

$$\mathbf{s}^T \mathbf{C}_{cc} \mathbf{s} + \mathbf{n}^T \mathbf{C}_{nn} \mathbf{n} = \text{minimum} \qquad (3.8)$$

and condition adjustment with unknowns we get

$$\hat{\mathbf{x}} = (\mathbf{A}^T \bar{\mathbf{C}}_0^{-1} \mathbf{A})^{-1} \mathbf{A}^T \bar{\mathbf{C}}_0^{-1} \mathbf{x} \qquad (3.9)$$

where

$$\bar{\mathbf{C}}_0 = \mathbf{C}_{nn} + \mathbf{C}_{ss} . \qquad (3.10)$$

4. NUMERICAL CALCULATIONS

The observations x are taken from the time series shown in figure 2.1. The estimations of the functional and stochastic models are also partly based on the time series shown in figure 2.1 and 2.2. In order to keep the discussion of the results clean only one parameter will be estimated: the time average of the refraction coefficients. Then the mean square error of the time average is (Wolf 1975):

$$m_{\hat{x}} = m_0 \sqrt{Q_{\hat{x}\hat{x}}} \qquad (3.11)$$

$Q_{\hat{x}\hat{x}}$ is a diagonal element of the matrix $(\mathbf{A}^T \bar{\mathbf{C}}_0^{-1} \mathbf{A})^{-1}$ and

m_0 the mean square error of unit weight of the observations.

The adjustment procedures are based on the following models:

model I (generally used for practical work)
a) k = 0.13
b) all observations have equal σ^2
c) the observations are uncorrelated

model II (assumptions based on the time series shown in figure 2.1 and 2.2)
a) $k = k_0 + k_a \cos(2\pi ft)$
where $k_0 = 0.09$, $k_a = 0.08$, f = 1 oscillation/year,
t = 0 at the beginning of January.
b) $\sigma_n = 0.16$ from November to February
$\sigma_n = 0.08$ from March to April
$\sigma_s = 0.07$ over the whole year
c) the observations are uncorrelated

model III
a) as in model II
b) as in model II
c) r = 0.9 for a time-difference τ up to   1 month
   r = 0   "   "   "   "   of   3 months
   r = 0.8 "   "   "   "   of   6 months
   r = 0   "   "   "   "   of   9 months
   r = 0.7 "   "   "   "   of   12 months

model IV
a) as in model II
b) as in model II
c) r = 0.9 for a time-difference τ up to   1 month
   r = 0   "   "   "   "   of   3 months
   r = 0.4 "   "   "   "   of   6 months
   r = 0   "   "   "   "   of   9 months
   r = 0.3 "   "   "   "   of   12 months

The results of the adjustment procedures are shown in Table 1 and Table 2. For all time-series the time averages $\hat{x}$ (the mean systematic errors of the refraction free directions) and the mean-square errors $m_{\hat{x}}$ become maximum, when model I is used. Comparing the $\hat{x}$ and $m_{\hat{x}}$ of the four different adjustments one can see that the values of $m_{\hat{x}}$ differ much more than those of $\hat{x}$. Analysing the $\hat{x}$ and $m_{\hat{x}}$ when the model II,

TABLE 1

time-series	date	k_i	date	k_i	date	k_i	date	k_i	date	k_i	number of the first... measurements	model I $\hat{x}$	$m_{\hat{x}}$	model II $\hat{x}$	$m_{\hat{x}}$	model III $\hat{x}$	$m_{\hat{x}}$	model IV $\hat{x}$	$m_{\hat{x}}$
1	5-1-62	0.11	5-4-62	0.08	5-7-62	-0.02	5-10-62	-0.06	5-1-63	0.25	7	0.04	0.39	0.03	0.12	0.02	0.16	0.02	0.15
	6-1-62	0.05	6-4-62	-0.06	6-7-62	-0.06	6-10-62	0.11	6-1-63	0.16	9	-0.02	0.36	0.00	0.09	-0.01	0.12	-0.02	0.13
	7-1-62	0.15							7-1-63	0.14	11	-0.04	0.33	-0.01	0.07	-0.01	0.10	-0.01	0.11
	8-1-62	1.10							8-1-63	0.12	16	-0.03	0.27	-0.01	0.03	-0.01	0.08	-0.01	0.08
	9-1-62	0.12							9-1-63	0.17									
2	10-1-62	0.26	10-4-62	-0.03	10-7-62	-0.03	10-10-62	0.03	10-1-63	0.12	7	0.07	0.20	0.05	0.06	0.05	0.07	0.05	0.07
	11-1-62	0.55	11-4-62	0.02	11-7-62	0.03	11-10-62	-0.03	11-1-63	0.14	9	0.05	0.20	0.03	0.05	0.01	0.06	0.00	0.06
	12-1-62	0.34							12-1-63	0.35	11	0.01	0.20	0.00	0.04	0.00	0.05	0.00	0.05
	13-1-62	0.32							13-1-63	0.21	16	0.01	0.17	0.01	0.03	0.00	0.04	0.00	0.04
	14-1-62	0.31							14-1-63	0.16									
3	15-1-62	0.17	15-4-62	-0.06	15-7-62	-0.07	15-10-62	-0.04	15-1-63	0.18	7	0.13	0.25	0.11	0.08	0.10	0.10	0.10	0.10
	16-1-62	0.61	16-4-62	0.15	16-7-62	0.00	16-10-62	-0.02	16-1-63	0.25	9	0.06	0.27	0.06	0.06	0.03	0.09	0.03	0.09
	17-1-62	0.23							17-1-63	0.64	11	0.01	0.28	0.02	0.05	0.02	0.07	0.02	0.08
	18-1-62	0.54							18-1-63	0.13	16	0.03	0.24	0.03	0.05	0.02	0.07	0.02	0.08
	19-1-62	0.55							19-1-63	0.16									
4	20-1-62	0.75	20-4-62	-0.06	20-7-62	-0.10	20-10-62	0.02	20-1-63	0.15	7	0.04	0.29	0.01	0.09	0.00	0.11	0.00	0.11
	21-1-62	0.30	21-4-62	-0.15	21-7-62	-0.04	21-10-62	0.03	21-1-63	0.34	9	-0.02	0.29	-0.02	0.06	-0.05	0.09	-0.05	0.09
	22-1-62	0.23							22-1-63	0.35	11	-0.05	0.26	-0.03	0.05	-0.04	0.07	-0.04	0.08
	23-1-62	0.34							23-1-63	0.27	16	-0.01	0.24	0.01	0.04	-0.01	0.06	-0.02	0.07
	24-1-62	0.15							24-1-63	0.50									
5	25-1-62	0.17	25-4-62	-0.12	25-7-62	0.02	25-10-62	0.11	25-1-63	0.19	7	-0.08	0.14	-0.06	0.04	-0.07	0.04	-0.07	0.04
	26-1-62	0.13	26-4-62	-0.05	26-7-62	-0.09	26-10-62	-0.02	26-1-63	0.24	9	-0.11	0.14	-0.06	0.03	-0.07	0.04	-0.07	0.04
	27-1-62	0.11							27-1-63	0.17	11	-0.12	0.13	-0.06	0.03	-0.06	0.04	-0.06	0.04
	28-1-62	0.31							28-1-63	0.16	16	-0.08	0.12	-0.04	0.02	-0.05	0.03	-0.05	0.03
	29-1-62	0.12							29-1-63	0.23									

TABLE 2

time-series	date	k_i	date	k_i	date	k_i	date	k_i	date	k_i	number of the first ...measurements	model I $\hat{x}$	model I $m_{\hat{x}}$	model II $\hat{x}$	model II $m_{\hat{x}}$	model III $\hat{x}$	model III $m_{\hat{x}}$	model IV $\hat{x}$	model IV $m_{\hat{x}}$
6	1-2-63	0.15	1-5-63	-0.07	1-8-63	0.01	1-10-63	0.12	1-12-63	0.12	7	0.16	0.38	0.06	0.12	0.04	0.14	0.04	0.14
	2-2-63	0.29	2-5-63	-0.12	2-8-63	-0.08	2-10-63	0.08	2-12-63	0.02	9	0.09	0.36	0.02	0.09	-0.01	0.11	-0.02	0.12
	3-2-63	0.29							3-12-63	0.14	11	0.07	0.33	0.02	0.07	0.01	0.09	0.01	0.10
	4-2-63	0.46							4-12-63	0.14	16	0.04	0.27	0.01	0.05	0.00	0.07	-0.01	0.08
	5-2-63	1.02							5-12-63	0.12									
7	6-2-63	1.03	6-5-63	0.11	6-8-63	0.06	6-10-63	0.08	6-12-63	0.19	7	0.41	0.40	0.31	0.13	0.29	0.15	0.29	0.15
	7-2-63	1.05	7-5-63	0.11	7-8-63	0.03	7-10-63	0.19	7-12-63	0.14	9	0.30	0.41	0.21	0.10	0.18	0.13	0.16	0.15
	8-2-63	0.64							8-12-63	0.13	11	0.25	0.39	0.16	0.08	0.15	0.11	0.14	0.13
	9-2-63	0.59							9-12-63	0.05	16	0.17	0.34	0.13	0.06	0.12	0.10	0.12	0.11
	10-2-63	0.28							10-12-63	0.18									
8	11-2-63	0.54	11-5-63	0.02	11-8-63	-0.03	11-10-63	0.26	11-12-63	0.17	7	0.05	0.20	0.03	0.05	0.07	0.07	0.03	0.07
	12-2-63	0.32	12-5-63	0.03	12-8-63	0.03	12-10-63	0.14	12-12-63	0.19	9	0.01	0.19	0.01	0.04	0.01	0.06	0.00	0.06
	13-2-63	0.19							13-12-63	0.15	11	0.02	0.17	0.03	0.03	0.03	0.05	0.03	0.06
	14-2-63	0.12							14-12-63	0.24	16	0.03	0.14	0.03	0.02	0.03	0.04	0.03	0.04
	15-2-63	0.01							15-12-63	0.19									
9	16-2-63	0.08	16-5-63	-0.02	16-8-63	0.04	16-10-63	0.10	16-12-63	0.22	7	-0.03	0.09	-0.02	0.02	-0.02	0.03	-0.02	0.03
	17-2-63	0.20	17-5-63	0.02	17-8-63	-0.03	17-10-63	0.21	17-12-63	0.23	9	-0.04	0.09	-0.02	0.02	-0.02	0.02	-0.03	0.03
	18-2-63	0.24							18-12-63	0.40	11	-0.04	0.09	-0.01	0.02	-0.01	0.03	-0.01	0.03
	19-2-63	0.08							19-12-63	0.19	16	-0.01	0.12	-0.01	0.02	-0.01	0.03	0.01	0.03
	20-2-63	0.09							20-12-63	0.21									
10	21-2-63	0.00	21-5-63	0.16	21-8-63	0.07	21-10-63	0.21	21-12-63	0.27	7	-0.06	0.08	-0.01	0.04	0.00	0.05	0.00	0.05
	22-2-63	0.10	22-5-63	0.05	22-8-63	0.04	22-10-63	0.12	22-12-63	0.24	9	-0.07	0.07	0.00	0.03	0.01	0.04	0.02	0.04
	23-2-63	0.06							23-12-63	0.53	11	-0.05	0.08	0.03	0.02	0.03	0.04	0.02	0.04
	24-2-63	0.06							24-12-63	0.21	16	0.02	0.14	0.03	0.02	0.03	0.04	0.03	0.04
	25-2-63	0.15							25-12-63	0.19									

model III and model IV are used, we can notice that there is only a small difference between the $\hat{x}$ while there is a greater difference between the $m_{\hat{x}}$. Sometimes the $m_{\hat{x}}$ calculated with model IV are three times greater than the $m_{\hat{x}}$ calculated with model II. This difference increases with the number of the observations. There is no significant difference between the results of the adjustment based on model III and model IV. Consequently only a rough approximation of the factor $\exp(-a_0^2 \tau^2)$ of (2.9) is needed. The values of $m_{\hat{x}}$ calculated with model III and IV are nearly always larger than the values of $\hat{x}$. We will not find that as often when considering the values calculated with model II. Therefore the results we get with model III and IV appear more realistic.

5. OPTIMUM ARRANGEMENT FOR THE TIME OF MEASUREMENTS

The vector

$$z = (A^T \bar{C}_0^{-1} A)^{-1} A^T \bar{C}_0^{-1}$$

of equ. (3.9) can be used to find an optimum arrangement of the time of the measurements, as its components characterize, how much the single measurements affect the time average. Table 3 for example shows the normed vector z_0 for the first 7, 9, 11 and 16 measurements of the time series 1 ... 5 of table 1, when model III is used.

TABLE 3

number of the observations	z_{0_7}	z_{0_9}	$z_{0_{11}}$	$z_{0_{16}}$
1	0.090	0.050	0.041	0.033
2	0.090	0.050	0.041	0.033
3	0.090	0.050	0.041	0.033
4	0.090	0.050	0.041	0.033
5	0.090	0.050	0.041	0.033
6	0.275	0.207	0.129	0.117
7	0.275	0.207	0.129	0.117
8		0.167	0.138	0.102
9		0.167	0.138	0.102
10			0.129	0.117
11			0.129	0.117
12				0.033
13				0.033
14				0.033
15				0.033
16				0.033

For example the first column z_{07} shows that the first five Observations are less significant in the final result than the last two observations.

6. CONCLUSIONS

The foregoing considerations are model studies. As the results of the adjustment procedures are not derived form direct geodetic measurements they cannot generally be compared with those calculated with direct measurements. The model studies however can help us to estimate suitable functional and stochastic models for adjustment procedures, to interpret the results and finally to find an optimum arrangement for the time of the measurements.

LITERATURE:

FEARNLEY, C.: Zur Theorie der terrestrischen Refraktion.
Forhandlinger: Videnskabs-Selskabet; Christiane 1884, Christiane 1885.

KAHMEN, H.: Some Considerations on the Stochastic Behaviour of the Angles of Refraction and of the Refraction Indices, concerning Laser- and Microwave Distance Measurements. Int. Symp. on EDM and the Influence of Atm. Refraction. Wageningen, Netherlands, 1977.

WOLF, H.: Ausgleichungsrechnung.
Ferd. Dümmler's Verlag, Bonn 1975.

WOLF, H.: Die Sonderfälle der diskreten Kollokation.
Österr. Zeitschrift für Vermessungswesen und Photogrammetrie, 3/4/1977.

DISCUSSION

L. Hradilek: Can your method be used for a smaller number of observations, let us say 5 repetitions of vertical angle measurements within one day?

H. Kahmen: Yes, the arrangement of the measurements is included in my model. I have taken this measurement together to an alternative time equation. The recordings of the meteorological parameters were done between 12 a.m. and 1 p.m. and there were recordings every ten minutes. I time-averaged these values and the single refraction coefficients sought for were time-averages of one hour. So I think they include several measurements during one day.

B. Garfinkel: Can you give us a definition of the coefficient of refraction?

VERTICAL REFRACTION ANGLE DERIVED FROM THE VARIANCE OF THE ANGLE-OF-ARRIVAL FLUCTUATIONS

F.K. Brunner
University of New South Wales,
Kensington, 2033, Australia.

ABSTRACT. A theory is developed for evaluating the vertical refraction angle from the variance of the angle-of-arrival fluctuations, assuming a horizontally homogeneous turbulent atmospheric surface layer. The vertical refraction angle is mainly a function of the vertical temperature gradient, and the variance of the angle-of-arrival is related to the temperature structure parameter C_T^2. However, surface layer similarity theory states that both the mean vertical temperature gradient and C_T^2 are functions of the same scaling temperature T_* and a thermal stability parameter. This therefore provides an indirect method of determining the vertical refraction angle from a measurement of the variance of the angle-of-arrival and an estimate of the thermal stability parameter. Advantages of this method over other techniques of evaluating vertical refraction are discussed.

1. INTRODUCTION

The precise determination of the atmospheric effect on vertical angle measurements must still be considered a primary research topic in geodesy. An improvement of precision in determining the vertical refraction angle would greatly benefit many geodetic operations.

Considering the present state-of-the-art in determination of the vertical refraction angle, two different approaches can be distinguished (Prilepin, 1974): the meteorological and the instrumental solution. The meteorological solution is based either on the selection of favourable observation times when refraction effect prediction is more reliable, or on design of a realistic atmospheric model for which meteorological parameters may be determined from measurements such as temperature gradients or heat fluxes (Brunner, 1978). Much of the present understanding of the nature of atmospheric refraction must be attributed to the meteorological approach. It seems unlikely, however, that it will generally yield an accuracy of 0.5" for the vertical refraction angle.

The underlying principle of the instrumental solution is the dispersion effect of light waves propagating through the atmosphere. Several technical solutions have been proposed (Prilepin, 1974), and a few have resulted in actual prototypes (Tengström, 1978; Glissmann, 1976; Williams, 1978). It is not unrealistic to predict that these instruments will yield an accuracy of 0.5" for the vertical refraction angle in the near future. However, test measurements have shown that atmospheric turbulence causes considerable problems in measuring the small dispersion angle, and precise measurements are only possible during favourable observation times.

The theory of a new approach to the determination of the vertical refraction will be presented here. In principle, it utilizes exactly that effect of the turbulent medium on light wave propagation, which has caused difficulty for the instrumental solution. A single vertical angle observation of a remote target through the telescope of a theodolite can be considered as the sum of the mean value and the momentary deviation from this mean value caused by turbulence in the atmosphere. For all derivations in this paper the ergodic hypothesis is invoked, replacing ensemble averages by time averages (denoted by overbars). The angle-of-arrival fluctuations are defined as the fluctuations of the normal on to the arriving wave front at the telescope (Lawrence and Strohbehn, 1970). The variance of the vertical component of the angle-of-arrival fluctuations is denoted by σ_α^2.

It has been shown (e.g., Brunner, 1978) that the mean angle of refraction is related to the mean vertical refractive index gradient, and (Appendix) that the variance of the angle-of-arrival fluctuations is related to the refractive index structure parameter C_n^2, characterising the structure of the atmospheric turbulence along the line of sight. Atmospheric surface layer theory (Appendix) states that statistics of the mean and the turbulent flow fields are functions of scaling and a stability parameter. For dry air it follows that the mean gradient and the structure parameter of the refractive index can both be expressed as functions of the scaling temperature T_* and an atmospheric stability parameter. Eliminating T_* in these expressions, it is possible to calculate the mean angle of refraction from a measurement of the variance of the angle-of-arrival and an estimate of the atmospheric stability parameter.

The remainder of this paper explains the principal features of the theory, treating the simple case of a horizontal line of sight in a horizontally homogeneous turbulent medium. Humidity effects have also been neglected in the present derivations. The possible extension of this theory to more realistic cases (inclined line of sight, general topography, humidity effects for water crossings) is briefly discussed. In anticipation of this extension the theory is based on the height-independent scaling temperature T_*, rather than on the mean temperature gradient which is a function of height.

2. THEORY

2.1 Mean angle of refraction

The curvature of an optical ray is related to the vertical gradient of refractivity ($\partial N/\partial z$). The time average of the refraction angle, $\bar{\delta}$, observed at A for a horizontal line of sight from A to B, can then be derived as (Brunner, 1978)

$$\bar{\delta} = -10^{-6} \int_0^S \frac{\partial \bar{N}}{\partial z} \left(1 - \frac{x}{S}\right) dx \qquad (1)$$

where $(\partial \bar{N}/\partial z)$ is the mean vertical refractivity gradient, S is the path length, and x is an integration variable, see Figure 1. The second term of the integral of equation (1) represents a weighting function for the refractivity gradients.

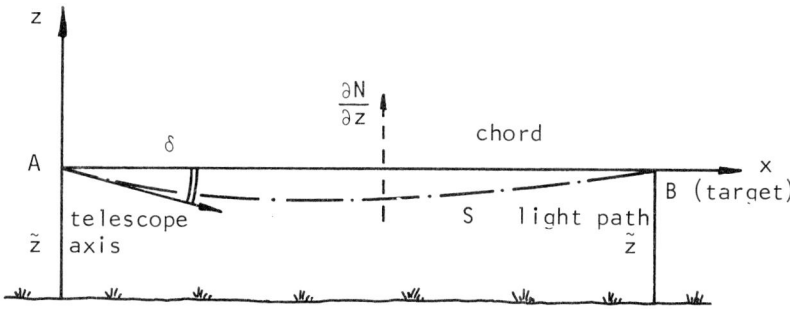

Figure 1. Geometry of the observation set-up.

Using the formula of Barrel and Sears, equation (A12), the gradient of refractivity can be derived as

$$\frac{\partial N}{\partial z} = \frac{N}{T} \left(\frac{T}{P} \frac{\partial P}{\partial z} - \frac{\partial T}{\partial z} \right) \qquad (2)$$

where T is the temperature in degrees Kelvin, $\partial P/\partial z$ is the vertical pressure gradient, and $\partial T/\partial z$ is the vertical temperature gradient. The small effect of the water vapour pressure gradient has been neglected in equation (2). Substituting the hydrostatic equation

$$\frac{\partial P}{\partial z} = -\frac{g\,P}{R\,T} \qquad (3)$$

and the relationship (sufficiently accurate in the present context) for the potential temperature gradient

$$\frac{\partial \Theta}{\partial z} = \frac{\partial T}{\partial z} + \Gamma \qquad (4)$$

in equation (2) yields

$$\frac{\partial N}{\partial z} = -\frac{N}{T}\left(\frac{g}{R} - \Gamma + \frac{\partial \Theta}{\partial z}\right) \qquad (5)$$

where Γ is the adiabatic lapse rate (0.0098 K m^{-1}), and g/R is the ratio of gravity to the gas constant (0.0342 K m^{-1}). Transition of equation (5) to the average values of the involved parameters, and substitution of the flux-profile relationship (A5) for the mean potential temperature gradient $(\overline{\partial \Theta / \partial z})$ yields

$$\frac{\partial \overline{N}}{\partial z} = -\frac{\overline{N}}{\overline{T}}\left(0.0244 + \frac{T_*}{k\tilde{z}}\phi_h\right) \qquad (6)$$

where T_* is the scaling temperature, k is the von Kármán constant, $\tilde{z}$ is the path height above the ground (see Figure 1), and ϕ_h is the flux-profile function.

For a light ray parallel to the ground in a horizontally homogeneous surface layer, all the terms in equation (6) are constant. Therefore the integration of equation (6) according to equation (1) yields the final result

$$\overline{\delta} = 10^{-6}\frac{\overline{NS}}{2\overline{T}}\left(0.0244 + \frac{T_*}{k\tilde{z}}\phi_h\right) \qquad (7)$$

In this equation for $\overline{\delta}$, the first part represents the effect of the atmosphere for neutral conditions (Appendix), and the second part, through $T_*\phi_h$ accounts for deviations from neutral conditions, the diabatic effect. The sign of this second term is determined by the sign of the sensible heat flux in equation (A2) for the scaling temperature T_*, which will be negative for unstable (clear day) conditions, and positive for stable (clear night) conditions.

2.2 Variance of the angle-of-arrival

Much of the theory of wave propagation in a turbulent atmosphere has been given by Tatarskii (1971). The papers by Lawrence and Strohbehn (1970) and de Wolf (1974) also have been found very useful. When an electromagnetic wave propagates in a turbulent medium, it experiences random fluctuations of amplitude, intensity, phase and angle-of-arrival due to the refractive index fluctuations.

It can be shown that the variance of the angle-of-arrival, σ_α^2, is a function of the phase structure function, $D_\phi(b)$, of the electromagnetic wave propagation

$$\sigma_\alpha^2 = \frac{D_\phi(b)}{\kappa^2 b^2} \tag{8}$$

where b is the interferometer separation and κ is the wave number $\kappa = 2\pi/\lambda$, with λ being the wave length. Equation (8) is derived under the assumption that the geometrical optics approximation is valid (Tatarskii, 1971).

The phase structure function has been derived by Tatarskii (1971) for a spherical wave propagating through a distance S in a homogeneous and locally isotropic turbulent medium with

$$D_\phi(b) = 1.09\ \kappa^2 C_n^2\ S\ b^{5/3} \tag{9}$$

where C_n^2 is the refractive index structure parameter. Equation (9) is applicable when

$$l_0 \ll (\lambda S)^{1/2} \ll L_0 \tag{10}$$

where $(\lambda S)^{1/2}$ is the radius of the first Fresnel zone, l_0 is the inner scale of the turbulence (in the order of a few millimetres), and L_0 is the outer scale of the turbulence (about twice the path height above the ground).

The angle-of-arrival variance for a telescope can be obtained from the above equations, if the interferometer separation b is replaced by the diameter D of the receiving objective

$$\sigma_\alpha^2 = 1.09\ C_n^2\ S\ D^{-1/3} \tag{11}$$

and is valid for $(\lambda S)^{1/2} \ll 2D$ (Lawrence and Strohbehn, 1970). In the Appendix the relationship between C_n^2 and the temperature structure parameter C_T^2 is given, and subsequently an expression is derived for C_T^2 as a function of the scaling temperature T_*, the stability parameter (z/L) and the height above the ground $\tilde{z}$. Substituting (A13) and (A17) into equation (11) yields

$$\sigma_\alpha^2 = 1.09 \cdot 10^{-12}\ S\ D^{-1/3}\ (N/T)^2\ \tilde{z}^{-2/3}\ T_*^2\ f(z/L) \tag{12}$$

where N is the refractivity of air, T is the temperature in degrees Kelvin, $\tilde{z}$ is the height of the optical path above the ground, T_* is the scaling temperature, and $f(z/L)$ is a function given by equation (A18). Equation (12) represents an expression for the magnitude of T_* only.

2.3 Method

It has been shown in the previous derivations that the mean angle of refraction $\bar{\delta}$ and the variance of the angle-of-arrival σ_α^2 are both functions of $|T_*|$ which can therefore be eliminated in equation (7) using (12). In order to retain the sign of T_* in equation (7), which equals also the sign of the stability parameter, sign (z/L) is incorporated in the final equation for $\bar{\delta}$

$$\bar{\delta} = 1.22 \cdot 10^{-8} S\bar{N}/\bar{T} + \text{sign}(z/L) \; 0.479 \; D^{1/6} \tilde{z}^{-2/3} S^{\frac{1}{2}} \sigma_\alpha p(z/L) \quad (13)$$

where the stability function $p(z/L)$ is given as

$$\begin{aligned} p(z/L) &= \phi_h [k^2 f(z/L)]^{-\frac{1}{2}} \\ &= \phi_h^{\frac{1}{2}} (\phi_m - z/L)^{1/6} \end{aligned} \quad (14)$$

In equation (13) the first term once again accounts for adiabatic conditions in the atmosphere, and the second term represents the diabatic correction. The discontinuity of this second term at $z/L = 0$ is caused by the change of the sign of z/L. This will not be very critical, however, as for neutral conditions the value of the whole second term in (13) tends towards zero. The stability function $p(z/L)$ should be evaluated for $\tilde{z}$, using the formulae for ϕ_h and ϕ_m given in the Appendix. The form of the stability function $p(z/L)$ versus z/L is shown in Figure 2.

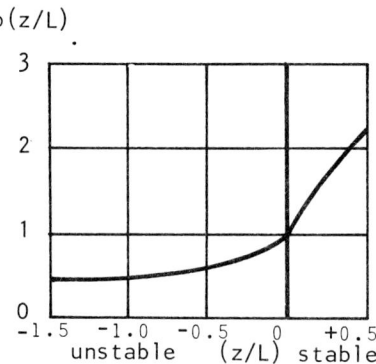

Figure 2. Stability function $p(z/L)$ versus z/L.

The theory developed above expresses the mean angle of refraction $\bar{\delta}$ as a function of the length of the line of sight, average value of refractivity and temperature, the telescope diameter, height above the ground of the line of sight, the standard deviation of the angle-of-arrival fluctuations σ_α, and a stability function $p(z/L)$. If the standard deviation of $\bar{\delta}$ should not exceed ±0.5", then an error analysis of equation (13) shows that the determination of those parameters is uncritical with the exception of σ_α and $p(z/L)$. σ_α should be determined with a relative precision of 10 to 20% which is certainly not an impossible task, even for visual observations through the theodolite telescope. Excellent reviews of the experimental determination of σ_α^2 have been given by Lee (1969), Lawrence and Strohbehn (1970), and Tatarskii (1971). For the practical evaluation of $p(z/L)$, the value of the Obukhov length L can be calculated using equation (A3) when wind speed and sensible heat flux are measured or estimated from empirical formula (Webb, 1964; Brunner and Fraser, 1977). Figure 2 indicates that for daylight observations, when z/L is negative, the determination of $p(z/L)$ is not too critical. For night observations, the stability function $p(z/L)$ is not extended in Figure 2 beyond $z/L = +0.5$, because the atmosphere shows generally insufficient thermal fluctuations during strong stability conditions (Okamoto and Webb, 1970), and consequently the proposed method will not be applicable for such conditions.

3. DISCUSSION

A new approach for determining the mean angle of refraction, $\bar{\delta}$, has been developed. For dry air $\bar{\delta}$ is mainly a function of the mean temperature gradient which is generated by the turbulent processes in the atmosphere. The mean temperature gradient can be obtained from a measurement of the variance of the angle-of-arrival, σ_α^2, using the temperature structure parameter C_T^2. The theory is developed for a horizontal line of sight parallel to the ground. Horizontally homogeneous turbulence is assumed for the derivations.

An obvious advantage of this method is that the effects of the turbulent medium on wave propagation which have been found adverse to other techniques are utilised here to advantage. Several pointings through the telescope of a theodolite are usually carried out to obtain a representative value for a vertical angle. During this time period the variance of the angle-of-arrival can be evaluated using the same telescope and along the same line of sight. It is beyond the scope of the present paper to give conclusive recommendations about the measuring technique of σ_α^2. However, without employing additional instruments, σ_α^2 could be inferred from the blurring of a target (Wesely and Derzko, 1975) or from the spread of the image dancing (Kukkamäki, 1950), estimated by visual observations through the telescope. Thus the method will not require new instrumental developments.

In the meteorological solution of the refraction problem point measurements generally are used for some atmospheric parameters. Representative values of these parameters require long averaging times. These averaging times are drastically reduced when path averaged values can be used, such as C_T^2 derived from σ_α^2 measurements (Wyngaard and Clifford, 1978), illustrating a further advantage of the method developed here.

The method presented here has not as yet been tested in field experiments. However, results of two independent experiments may be considered as preliminary verification of the method. Vertical refraction was successfully determined from measured sensible heat fluxes (Brunner, 1978), and sensible heat fluxes were determined from image blurring (Wesely, 1976). Further proof of the second step may be seen in the measurements by Coulman (1966).

The theory which has been intentionally derived for telescope observations, could easily be recast for laser beam propagation, using the appropriate beam equations. For laser beam propagation the variance of the log-amplitude, the phase-angle or the vertical displacement could be utilized for the determination of C_T^2. Appropriate corrections for the humidity effects of light wave propagation, significant for water crossings, can be incorporated in the present theory without great difficulties. Special attention must be given to the weighting functions in the integrals for the refraction angle and the variance of the angle-of-arrival, when the extension of this theory is considered for a more realistic line of sight with general topography. These additional considerations will be treated by the author in the future.

ACKNOWLEDGEMENTS

The writer is indebted to R.B. Forrest, J.R. Pollard and E.K. Webb for criticism that has led to clarification of the content of this paper, and to Mrs. S. Lennon for typing the manuscript.

APPENDIX: MICROMETEOROLOGICAL BACKGROUND

The surface layer of a horizontally homogeneous atmospheric boundary layer can be effectively described by a few ensemble-average statistical properties. For a comprehensive treatment, reference should be made to Priestley (1959), Lumley and Panofsky (1964), Webb (1964, 1965), Busch (1973), Businger (1973) and Wyngaard (1973).

The turbulent structure of the atmospheric surface layer may be expressed by simple similarity scaling. Neglecting the humidity effects in the atmosphere in the present context, basically three scaling parameters are adopted, defined as the friction velocity u_*, the scaling temperature T_*, and the Obukhov length L:

$$u_* = (\tau/\rho)^{1/2} \tag{A1}$$

$$T_* = - H(\rho c_p u_*)^{-1} \tag{A2}$$

$$L = u_*^2 T(k g T_*)^{-1} \tag{A3}$$

where τ is the shearing stress (downward flux of horizontal momentum), H is the sensible heat flux, ρ is the air density, c_p is the specific heat of air at constant pressure, T is the air temperature, g is the acceleration due to gravity, and k is the von Kármán constant. The Obukhov length L is used to form a dimensionless atmospheric stability parameter z/L, where z is the height above the ground. The atmospheric conditions characterised by negative values of z/L are called unstable. Positive values of z/L indicate stable conditions, and for z/L equal to zero or nearly zero, neutral conditions prevail.

In the atmospheric surface layer (a few tens of metres thick) fluxes of momentum and heat are considered essentially constant with height. Applying scaling to the vertical gradients of mean horizontal windspeed, $\overline{u}$, and mean potential temperature, $\overline{\theta}$, the following flux-profile relationships are obtained:

$$\frac{\partial \overline{u}}{\partial z} = \frac{u_*}{kz} \phi_m \tag{A4}$$

$$\frac{\partial \overline{\theta}}{\partial z} = \frac{T_*}{kz} \phi_h \tag{A5}$$

where the profile shape functions ϕ_m and ϕ_h account for the stability effect, and are functions of z/L. The forms of ϕ_m and ϕ_h have recently been reviewed by Dyer (1974), with the following results:

Unstable conditions ($z/L < 0$):

$$\phi_m = (1 - 16 \, z/L)^{-\frac{1}{4}} \tag{A6}$$

$$\phi_h = \phi_m^2 \tag{A7}$$

Stable conditions ($z/L > 0$):

$$\phi_m = \phi_h = 1 + 5 \, z/L \tag{A8}$$

For neutral conditions, where z/L approaches zero, both ϕ_m and ϕ_h tend to go to unity. For the numerical values in the above equations the value of the von Kármán constant has been assumed to be $k = 0.4$.

The determination of z/L, u_* and H from meteorological measurements has been discussed in great detail by Webb (1965) and the evaluation of these parameters in connection with refraction studies has been reported recently by the author (Brunner, 1978; Brunner and Fraser, 1977).

The structure function $D(r)$ is defined as the mean square difference of the values of a variable at distance r apart. If the Kolmogorov law is applicable, then

$$D(r) = C^2 \, r^{2/3} \tag{A9}$$

where C^2 is the structure parameter. If the variable is a scalar, the structure parameter C^2 is given (e.g. Panofsky, 1968) by

$$C^2 = a \, \varepsilon^{-1/3} \, \chi \tag{A10}$$

where a is a constant, ε is the rate of dissipation of turbulent energy, and χ is the rate of dissipation of the fluctuations of the scalar.

The refractive index of air, n, is often conveniently described by the refractivity, N,

$$N = (n - 1) \, 10^6 \tag{A11}$$

For the wavelength $\lambda = 0.56 \, \mu m$ the refractivity of air is given with sufficient accuracy as (Barrel and Sears, 1939)

$$N = 79 \frac{P}{T} - 11 \frac{e}{T} \tag{A12}$$

where P is the total air pressure in mb, T is the temperature in degrees Kelvin, and e is the water vapour pressure in mb. Using (A12) the refractive index structure parameter C_n^2 of dry air can be related to the temperature structure parameter C_T^2 (Bouricius and Clifford, 1970)

$$C_n^2 = 10^{-12} \, (N/T)^2 \, C_T^2 \tag{A13}$$

According to Panofsky (1968) the rate of dissipation of turbulent energy ε for dry air can be expressed as the sum of mechanical and thermal production rates for turbulent energy, and ε may then be expressed as (Busch, 1973)

$$\varepsilon = \frac{u_*^3}{kz} (\phi_m - z/L) \tag{A14}$$

where all parameters used here have been explained previously. The rate of dissipation of temperature fluctuations χ can be expressed as (Lumley and Panofsky, 1964)

$$\chi = K_h \left(\frac{\partial \overline{\theta}}{\partial z} \right)^2 \tag{A15}$$

where K_h is the temperature exchange coefficient ($K_h = k \, u_* \, z/\phi_h$), and ($\partial \overline{\theta}/\partial z$) is the mean temperature gradient. Substitution of the flux-profile relationship yields

$$\chi = \frac{u_* T_*^2}{kz} \phi_h \tag{A16}$$

The numerical value for a still is the subject of conjecture (Panofsky, 1968; Wesely and Alcaraz, 1973; Wyngaard et al., 1971), but the value

3.2 is used here. Accordingly, substituting (A14) and (A16) into equation (A10) yields for C_T^2

$$C_T^2 = T_*^2 \, z^{-2/3} \, f(z/L) \tag{A17}$$

where

$$f(z/L) = 3.2 \, k^{-2/3} \, \phi_h (\phi_m - z/L)^{-1/3} \tag{A18}$$

REFERENCES

Barrel, H. and Sears, J.E.: 1939, *Phil. Trans. Roy. Soc. London* A-238, pp. 1-64.
Bouricius, G.M.B. and Clifford, S.F.: 1970, *J.Opt. Soc. Am.* 60, pp. 1484-1489.
Brunner, F.K.: 1978, in P. Richardus (ed.) *Proc. Int.Symp. EDM and Influence Atmos.Refraction*, Wageningen, Publ. Netherlands Geodetic Commission, pp. 245-255.
Brunner, F.K. and Fraser, C.S.: 1977, *Unisurv* G27 (Univ. NSW, Sydney), pp. 3-26.
Busch, N.E.: 1973, in D.A. Haugen (ed.), *Workshop on Micrometeorology*, Am. Met. Soc., Boston, pp. 1-65.
Businger, J.A.: 1973, in D.A. Haugen (ed.), *Workshop on Micrometeorology*, Am. Met. Soc., Boston, pp. 67-100.
Coulman, C.E.: 1966, *J. Opt. Soc. Am.* 56, pp. 1232-1238.
de Wolf, D.A.: 1974, *Proc. IEEE* 62, pp. 1523-1529.
Dyer, A.J.: 1974, *Boundary-Layer Meteorol.* 7, pp. 363-372.
Glissmann, T.: 1976, *Zur Bestimmung des Refraktionswinkels über die Dispersion des Lichtes mittels positionsempfindlicher Photodioden*, Wiss. Arb. Geod. Photogramm. Kartogr. T.U. Hannover, Band 62, Hannover.
Kukkamäki, T.J.: 1950, *Geofisica pura et applicata* 18, pp. 120-127.
Lawrence, R.A. and Strohbehn, J.W.: 1970, *Proc. IEEE* 58, pp.1523-1545.
Lee, R.W.: 1969, *Radio Science* 4, pp. 1211-1223.
Lumley, J.L. and Panofsky, H.A.: 1964, *The Structure of Atmospheric Turbulence*, Interscience-Wiley, New York.
Okamoto, M. and Webb, E.K.: 1970, *Quart. J. Roy. Meteorol. Soc.* 96, pp. 591-600.
Panofsky, H.A.: 1968, *J. Geophys. Res.* 73, pp. 6047-6049.
Prilepin, M.T.: 1974, Elimination of angular refraction by means of multiple wavelength methods, *Proc. Int. Symp. Terrestrial EDM and Atmos. Effects on Angular Measurements*, Stockholm, 1974, Vol. 5.
Priestley, C.H.B.: 1959, *Turbulent Transfer in the Lower Atmosphere*, Univ. of Chicago Press, Chicago.
Tatarskii, V.I.: 1971, *The Effects of the Turbulent Atmosphere on Wave Propagation*, (Translated from the Russian), NTIS, Springfield, Va.
Tengström, E.: 1978, in P. Richardus (ed.), *Proc. Int. Symp. EDM and Influence Atmos. Refraction*, Wageningen, Publ. Netherlands Geodetic Commission, pp. 101-124.
Wesely, M.L.: 1976, *J. Appl. Meteorol.* 15, pp. 1177-1188.

Wesely, M.L. and Alcaraz, E.C.: 1973, *J. Geophys. Res.* 78, pp.6224-6232.
Wesely, M.L. and Derzko, Z.I.: 1975, *Applied Optics* 14, pp. 847-853.
Williams, D.C.: 1978, in P. Richardus (ed.), *Proc. Int. Symp. EDM and Influence Atmos. Refraction*, Wageningen, Publ. Netherlands Geodetic Commission, pp.163-170.
Webb, E.K.: 1964, *Applied Optics* 3, pp. 1329-1335.
Webb, E.K.: 1965, *Meteorol. Monographs* 6(28), pp. 27-58.
Wyngaard, J.C.: 1973, in D.A. Haugen (ed.), *Workshop on Micrometeorology*, Am. Met. Soc., Boston, pp. 101-149.
Wyngaard, J.C. and Clifford, S.J.: 1978, Momentum, Heat and Moisture Fluxes from Structure Parameters, *J. Atmos. Sci.* 35, (in press).
Wyngaard, J.C., Izumi, Y. and Collins, Jr., S.A.: 1971, *J. Opt. Soc. Am.* 61, pp. 1646-1650.

DISCUSSION

J. Milewski: I think it is a very interesting paper and promising idea, but practically I suppose that we can use this method only in that case if the ratio between the systematic influence, dependent on the atmospheric parameters and the random error of observation is rather large. Such conditions are typical for big turbulence of unstable status of atmosphere. If, however, we had a very stable condition, we have usually a very great value of absolute refraction index, but its variation is then very small. Under such a status of atmosphere the ratio between the variations of refraction and random errors of observation will be rather small, creating unconvenient situations for the use of Brunner's method.

P.V. Angus-Leppan: Dr Brunner is proposing the method for unstable conditions, where there is visible shimmer, and it is easy to estimate the variations of δ with some precision from simple telescope observations. Under stable conditions you may have to use some method other than estimation, because there are large but slow movements of the image.

D.G. Currie: One question I might have on that, is in some work of propagation of laser beams over horizontal paths. There seem to be other parameters that come in to make a significant variation, e.g. the way in which the turbulence varies above the ground over very smooth fields depends quite a bit on the wind. And I think the scaling height, or the height at which you get significant changes in the structure, does depend on the wind velocity, and therefore I suspect that the wind velocity will have a strong influence on the magnitude as well as the frequency of the variation.

P.V. Angus-Leppan: In region II under unstable conditions it is independent of wind. But under other conditions wind certainly is a factor. I think that is taken into account in the parameter T_*.

K. Poder: May I add that the people at the University of Hannover have made some experiments with laser propagation under wind and turbulence.

As far as I am recalling they are published in the proceedings from the Wageningen symposium.

D.G. Currie: Is that true? Thank you.

RESULTS FROM AN ABSOLUTE TEST OF THE NPL DISPERSOMETER OVER 4 KM

D. C. Williams
National Physical Laboratory
Teddington, Middlesex, United Kingdom

ABSTRACT

The NPL instrument to correct for angular refraction in geodesy measures the small dispersion between images of red (633 nm) and blue (442 nm) laser sources formed by a telescope. It uses a rotating chopper disc in association with an optical compensator plate and a phase null meter. The instrument and some early field tests have been described in detail elsewhere (Williams, D.C., 1978).

Absolute tests of the prototype have recently been made over a 4 km suburban range at an average height of 30 m. The apparent elevation of the red laser was observed with a T3 theodolite and corrected for refraction, the laser and theodolite being referred to local bench marks with a known height difference determined by spirit levelling.

On an occasion in late May 1978, the effects of turbulence were particularly small. The refraction-corrected theodolite readings then agreed with the expected value to within one sexagesimal second. Other results obtained during the preceding two months indicate that increasing turbulence causes the dispersion angle to appear too large, the greatest error observed being about 3 seconds of refraction. It is thought that most of the turbulence was due to effluent from chimneys, and that the apparent error may be due to an instrumental effect which has not yet been identified.

The performance of the instrument has also been evaluated on a 20 km range near Uppsala during the present Symposium, by kind invitation of Prof. E. Tengström. The principal difference from the 4 km range was that the intensity scintillations of the red and blue signals were no longer correlated when the refraction was large. Independent compensation of red and blue signals for this effect should enable good results to be obtained over this distance.

The author wishes to acknowledge the help of S-G. Mårtensson and S. Eklund (University of Uppsala) and A.J. Griffiths (NPL) in performing the tests.

REFERENCES

Williams, D.C.: 1978, "First field tests of an angular dual wavelength instrument", Proc. of the International Symposium on Electromagnetic Distance Measurement and the Influence of Atmospheric Refraction, Netherlands Geodetic Commission, pp. 163-170.

EXPERIENCES FROM IDM MEASUREMENTS AT THE TEST BASE OF THE GEODETIC INSTITUTE OF UPPSALA UNIVERSITY

S-G. Mårtensson
Geodetic Institute
Uppsala, Sweden

ABSTRACT

IDM measurements carried out in Uppsala sometimes show violent changes in refraction over short periods of time.

The explanation for these changes might be found if the dynamics of atmospheric turbulence is taken into consideration.

It is too early to make scientific conclusions from the material collected so far, but observations show the way this particular problem can be handled in the future.

The paper presented, reflects some of the turbulence effects achieved by the IDM instrument constructed by prof. E. Tengström.

INTRODUCTION

From measurements carried out on the Björklinge - Flogsta refraction base in Uppsala, it seems as if the refraction contains of two (for IDM detectable) essential spectral parts. One part with a high frequency and the other one with a low frequency.

The high frequency spectrum is caused by the atmospheric turbulence and the low frequency spectrum corresponds to the refraction obtained, for instance by vertical angle measurements (or atmospheric models), (fig. 1).

This is, of course, nothing new. It is a wellknown phenomenon and has been mentioned in the literature by several authors, among others by G. Dietze (Dietze, G.: 1957), who tried to explain the turbulence spectrum by convection theories only. Today we know, from micrometeorologists, more about turbulence in the atmosphere, e.g. that convection is not the only contributor but probably responsible for the dominating part when dealing with vertical refraction.

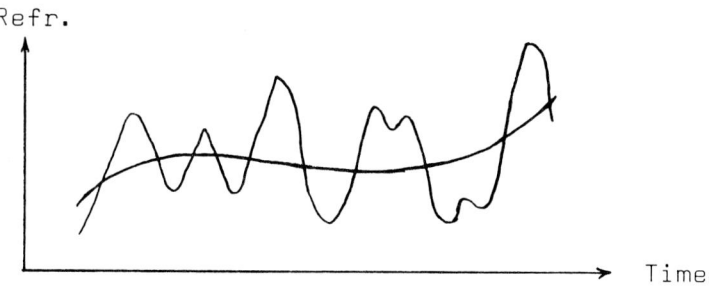

Fig. 1 High and low frequency spectrum of refraction.

The new and interesting thing is that parts of the high frequency spectrum have been detected by IDM measurements in Uppsala, and the contents of this paper deal with certain properties of that spectrum.

GENERAL

The detectings of the atmospheric turbulence became reality when the light sources were exchanged from mercury lamps to lasers. The intensive light of the lasers decreased the exposure times from a question of minutes to a question of seconds and even to parts of a second.

When the registrations were in the "minute area", the high frequency spectrum averaged out to the low frequency spectrum, a fact which is documented in the Niinisalo measurements in 1971 (Tengström, E.: 1974), where the IDM measurements fit very well to the refraction computed from an atmospherical model.

In the "second area", which is the operating area of today, the high frequency spectrum dominates. Measurements carried out in this exposure time area indicate sometimes violent amplitudes of the spectrum. The greatest amplitude detected so far is 1140 cc (on 20 km) in 20 minutes of time. This is an extreme value, normally detected amplitudes seem to be about 150 - 250 cc.

Some measurements indicating this will be given in the next part.

MEASUREMENTS

Fig. 2 shows observations carried out on a very turbulent evening. The theodolite readings (by a Wild T3) show a change in refraction from 565 cc to 870 cc during 1 hour 40 minutes. IDM observations during the first part of the evening indicates even greater amplitudes, a change from 123 cc to 1263 cc during 20 minutes of time! (Refraction obtained from a standard atmospheric model gives a refraction value of 140 cc on the 20 km base). All IDM observations were performed with He-Ne red (6328 Å) and Ar blue (4880 Å) lasers and with 1/2 sec. exposure time.

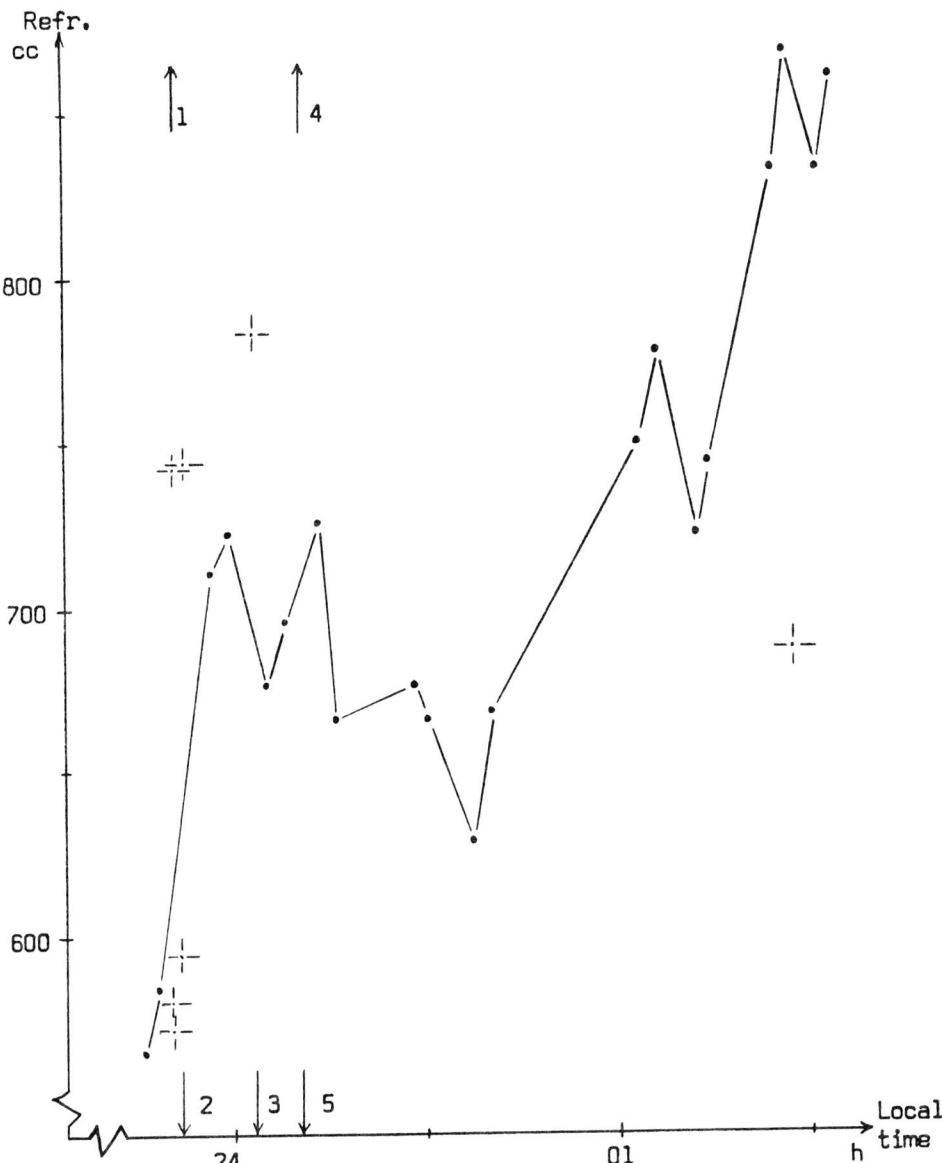

Fig. 2 Refraction computed from • vertical angle measurements and -¦- IDM measurements, 1976-08-16.

Additional IDM measurements that do not fit the range of fig. 2 (arrows in the figure indicating the time):
1) 903 cc, 950 cc, 965 cc
2) 431 cc, 209 cc, 123 cc
3) 427 cc
4) 1211 cc, 1263 cc
5) 231 cc

Fig. 3 shows observations performed under much more favourable atmospheric conditions (at least from a geodesists point of view). The vertical angle readings were very stable (within 10-20 cc) and the refraction computed from those readings does not differ too much from the "standard refraction value".

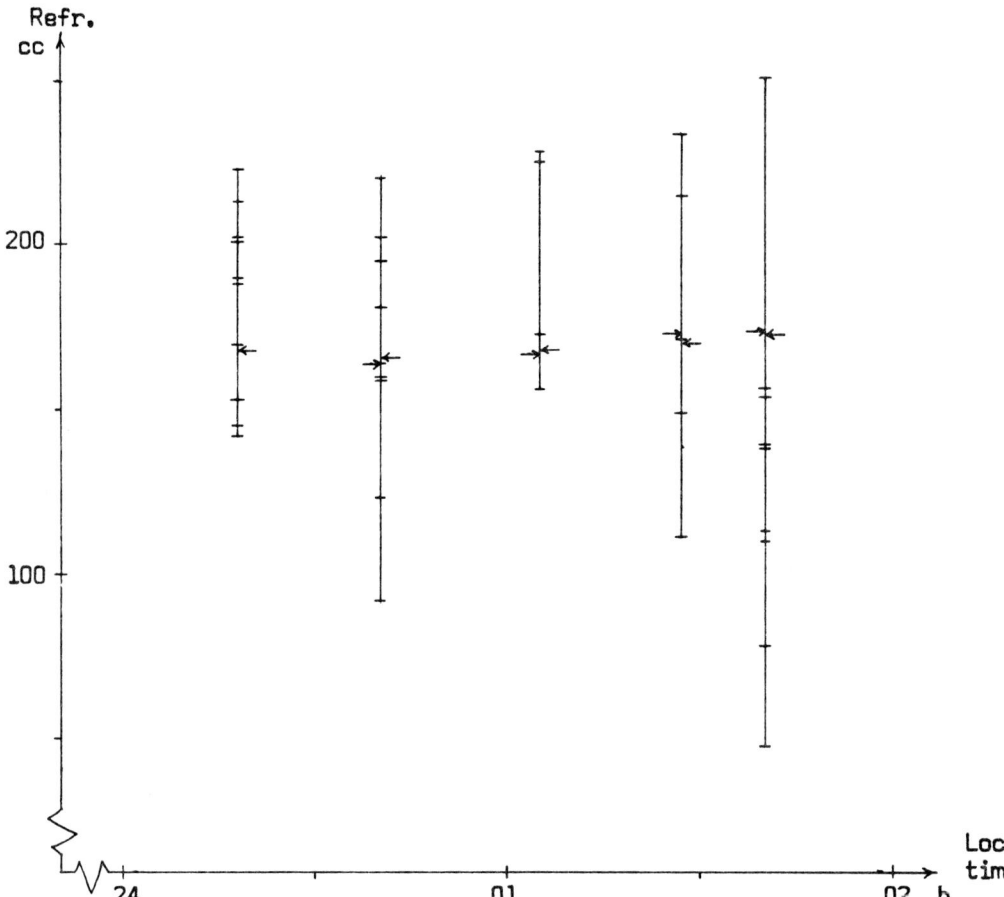

Fig. 3 Refraction computed from ← mean of four vertical angle measurements, † IDM measurements (1/2 sec. exp.time) and → IDM measurements (5 sec. exp.time), 1977-06-21.

EXPERIENCES FROM IDM MEASUREMENTS

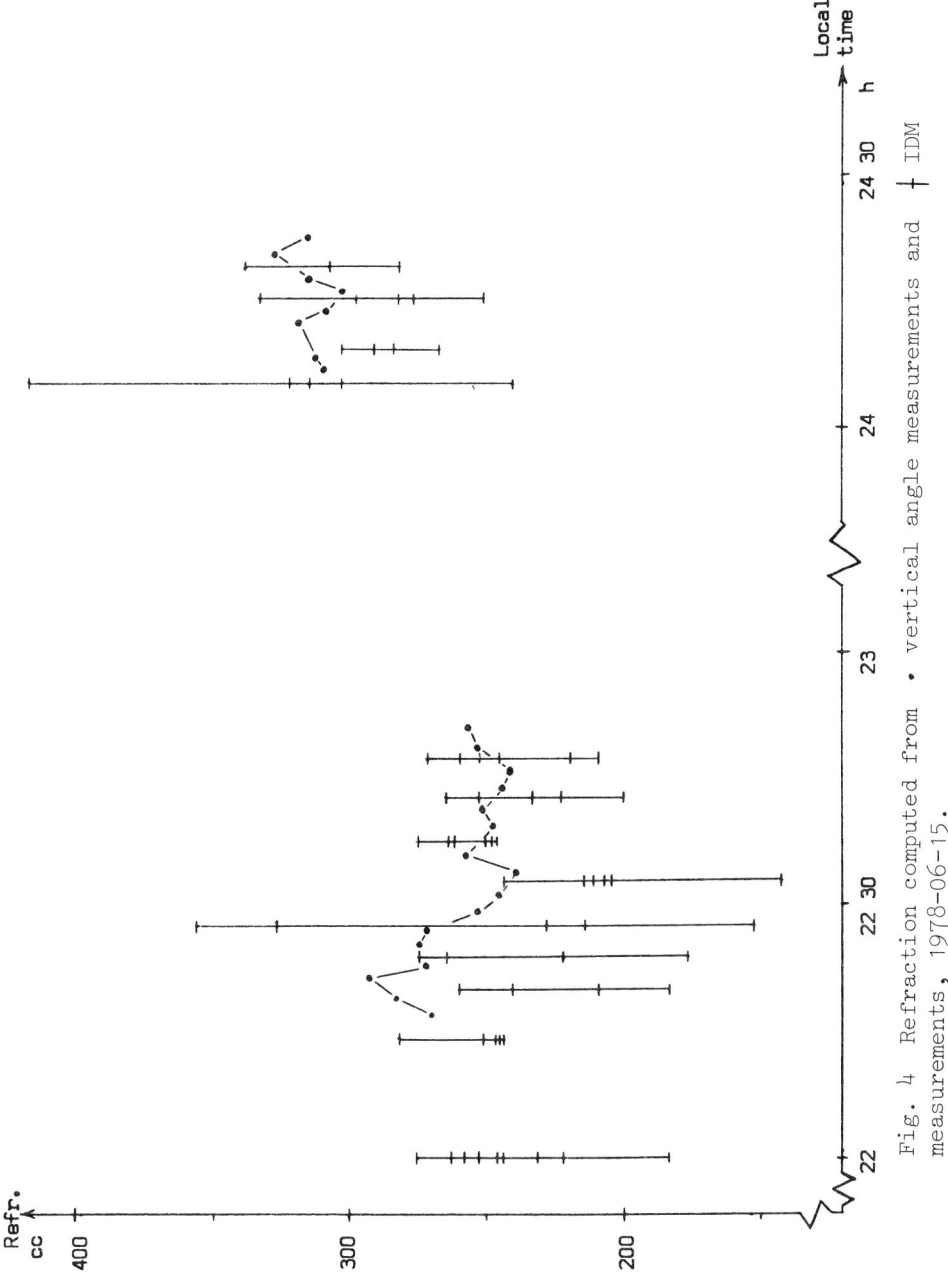

Fig. 4 Refraction computed from • vertical angle measurements and ┼ IDM measurements, 1978-06-15.

Although the IDM observations show quite different values, they differ from 48 cc to 250 cc during a time period less than 30 sec. The IDM observations were performed with He-Ne red and He-Cd UV (3250 Å) lasers and the exposure times were mainly 1/2 sec. except for a few observations which had an exposure time of 5 sec. These observations are very interesting, because it looks as if the 5 sec. exposure times were enough to give a good integration of the high frequency spectrum. It can be seen from figure 3 that the 5 sec. exposures fit very well to the values achieved by vertical angle readings, in fact better than 3 cc.

Fig. 4 shows observations performed in the middle of June 1978. The interesting thing is, that this particular evening an integration time as long as 20 sec. was not enough to give an agreement with the vertical angle readings. The time interval between two succesive exposures was 5 sec. He-Ne red and He-Cd lasers were used.

DISCUSSION

The exposure times in the "parts-of-a-second-to-several-seconds" area seem to be too short. For geodetic refraction studies it is necessary to bring them back into the "minute area" again.

This can be done in several ways, some of which may probably, by the present method, decrease the accuracy in obtaining the dispersion. To maintain the required accuracy on long time exposure observations and to avoid subjective measurements, densitometer readings in combination with Fourier analysis will be used (see paper presented at this symposium by J. Milewski).

The possibility that the light rays travel through different turbulent media when the lasers are separated, as in our case by one meter, should not be excluded. However, preliminary studies of this problem do not indicate any apparent differences.

Acknowledgements

The author of this paper is indebted to:
- Prof. E. Tengström and Dr J. Milewski for general remarks,
- S. Eklund for assistance in the observations,
- M. O'Shaughnessy for checking of the English text,
- I. Ohlsson for typing the manuscript.

REFERENCES

Dietze, G.: 1957, Einführung in die Optik der Atmosphäre, Leipzig.

Tengström, E.: 1974, "Report on the results of the IDM experiments at

Uppsala 1970 and at the Finnish base of Niinisalo 1971", Proc. of the symposium EDM and Refraction, Stockholm, Sweden.

Tengström, E.: 1977, "Some absolute tests of the results of IDM measurements in the field with a description of formulas used in the tests", Proc. of IAG symposium, Wageningen, The Netherlands.

DISCUSSION

D.G. Currie: I think in both this case and Williams' presentation there may be part of the problem on the instability, maybe correctable with an instrument, which measures both the angle and the dispersion at the same time. It may very well be that a large part of the motion or jumping around that you are seeing, is actually a change in dispersion and a change in refraction, so that an instrument which was tracking this would find that when refraction increased the dispersion also increased. That is one thing at least on the astronomical side that seems like it may be necessary, if we have to deal with trying to remove the high frequency components.

S-G. Mårtensson: Yes, I think you are right. It is necessary to examine if and when there is a correlation between dispersion and refraction. Both Williams and I have seen, that if we have a big refraction we also have a poor correlation between the red and the blue images. But if we have a small refraction, there seems to be a much better correlation. I think this is important, and I think it must be examined when and where we have correlation between dispersion and refraction.

E. Tengström: We have to look at two types of averaging - we have to try spatial averaging and we have to try time averaging. Both must be investigated, and we are going to do that.

POSSIBILITIES OF INCREASING THE ACCURACY IN THE DETERMINATION OF
REFRACTIONAL ANGLES WITH TENGSTRÖM'S IDM

J. Milewski
Higher School of Engineering,
Koszalin, Poland

ABSTRACT

The main purpose of the IDM research of prof. Tengström was to confirm
the possibility of determining instrumentally the refraction angle in
the field even over long distances.

An appreciable increase of the accuracy could only be achieved by
further improvement of the instrument. Even with the first laboratory
model it seems possible to achieve an increase of the accuracy in the
final results. This paper discusses the above possibilities and proposes some methods to achieve this improvement in the accuracy.

1. INTRODUCTION

Experience from investigations in the field indicates that the angular
refraction is varying with amounts, which can rarely be calculated from
theoretical formulas, containing measured meteorological parameters,
valid along the ray at the instant of observation.

Even if the theoretical model atmosphere is correct, which probably will
certainly never be the case, the discrete meteorological measurements
along the path above a non homogeneous surface will not be able to give
information for an accurate calculation of the curvature integral. In
practice it is also extremely difficult to collect data which all correspond to the epoch of observation.

Formulas containing various meteorological parameters have been elaborated by several scientists (Angus-Leppan, P.V., 1971, Brunner, F.K. 1977,
Brunner, F.K. & Fraser, C.S., 1977, Rinner, K., 1977, Saastamoinen, J.,
1974) and their achievements are of great importance.

But a severe difficulty when applying such formulas is the more or less
stochastic behaviour of grad n for certain terrain conditions (topographical features, vegetation, variation of wind and cloud cover etc.),

especially near the ground.

From what is said above one must conclude, that a direct determination of the whole atmospherical integral at the epoch of observation is necessary in order to master the problem of eliminating the refractional influences both in EDM measurements and, above all, in angular measurements.

The n-integral at EDM and the $\frac{dn}{dh}$ - integral at vertical angle measurements can be evaluated by multi-wave approaches.

The success in EDM has recently been demonstrated by Hugget and Slater (Hugget, G.R., Slater, L.E., 1977), whose terrameter, using He-Ne and He-Cd lasers, are soon available on the market by Terra Technology, Redmond, Washington (Hugget, G.R., 1978).

It seems also, that the two-colour systems (Glissman, T., 1977, Prilepin, M.T., 1974, Tengström, E., 1967, Tengström, E., 1977, Williams, D.C., 1977) promise to solve the refraction angle problem with high accuracy in the nearby future. Prilepin (Prilepin, M.T., 1975) has also suggested the use of a single-wave system. Perhaps Tengström's IDM-system can also be used under favourable atmospheric conditions with a small stellar interferometer set up, which will increase the accuracy.

Of all above mentioned methods I have had the possibility to learn exactly in practice the achievements of the IDM method. On this base I think that this method have certain possibilities to increase the accuracy and reliability of its result. The proposal for such possibilities are the essence of this paper.

2. ANALYSIS OF CURRENT ACHIEVEMENT OF THE REFRACTION ANGLE USING IDM

To have out of account the whole description of practical realization of setting of the refraction angle by the dispersion given detailed earlier (Tengström, E., 1967) we only would analyse the possible sources of errors and the total error resulting by them.

From the principle of IDM measurement the vertical angle of refraction is calculated

$$\alpha = K_{12}^{(1)} \delta_{12} \tag{1}$$

where

$$K_{12}^{(1)} = \frac{N_{01}}{N_{02} - N_{01}} \tag{2}$$

is the dispersion coefficient dependent only on the refractive index n_{oi}

$$N_{oi} = n_{oi} - 1 \tag{3}$$

and δ - the dispersion angle

$$\delta_{12} = \alpha_2 - \alpha_1 \tag{4}$$

In prof. Tengström's IDM measurements, laser sources of monochromatic light are used, He-Ne (red 0.6328 μm) and He-Cd (UV 0.3250 μm). They should be in the same horisontal plane and in a straight line perpendicular to the direction of wave propagation.

As a result of dispersion and refraction in atmosphere a difference δ appears between the refraction angles of the two light beams. It is the angle between the planes of the R and UV wave fronts. After interference in the 6 slit grating and focusing by the Cassegrain camera (f = 6260 mm) the interference fringes produced are photographed and then δ is calculated as below

$$\delta_{RUV} = \frac{z_{RUV}}{f} = \frac{\lambda_R \, z_{RUV}}{l_R \, d} \tag{5}$$

where λ_R the length of the red beam

z_{RUV} the distance between the central fringes of red and UV (fig. 1) on the photograph

l_R the distance between succeeding fringes of red

d the slit distance

The z and l are measured on precise comparator. Let us discuss the accuracy of this method. From (1) we have

$$\left(\frac{m_\alpha}{\alpha}\right) = \sqrt{\left(\frac{m_K}{K}\right)^2 + \left(\frac{m_\delta}{\delta}\right)^2} \tag{6}$$

From the Edlen's formula, for these laser sources (dry air at T = 273.16°K, p = 760 Torr, contain of 0.03% CO_2)

$$N_{0_{UV}} = 304.24 \times 10^{-6} \pm (<4 \times 10^{-8})$$

$$N_{0_R} = 291.76 \times 10^{-6} \pm (<4 \times 10^{-8})$$

whence $\Delta N = N_{0_{UV}} - N_{0_R} = 1248 \times 10^{-8} \pm (<6 \times 15^{-8})$

$$K = K_{RUV}^{(R)} = 23.38$$

and
$$\frac{m_K}{K} = \sqrt{\left(\frac{m_N}{N}\right)^2 + \left(\frac{m_{\Delta N}}{\Delta N}\right)^2} \simeq 5 \times 10^{-3} \qquad (7)$$

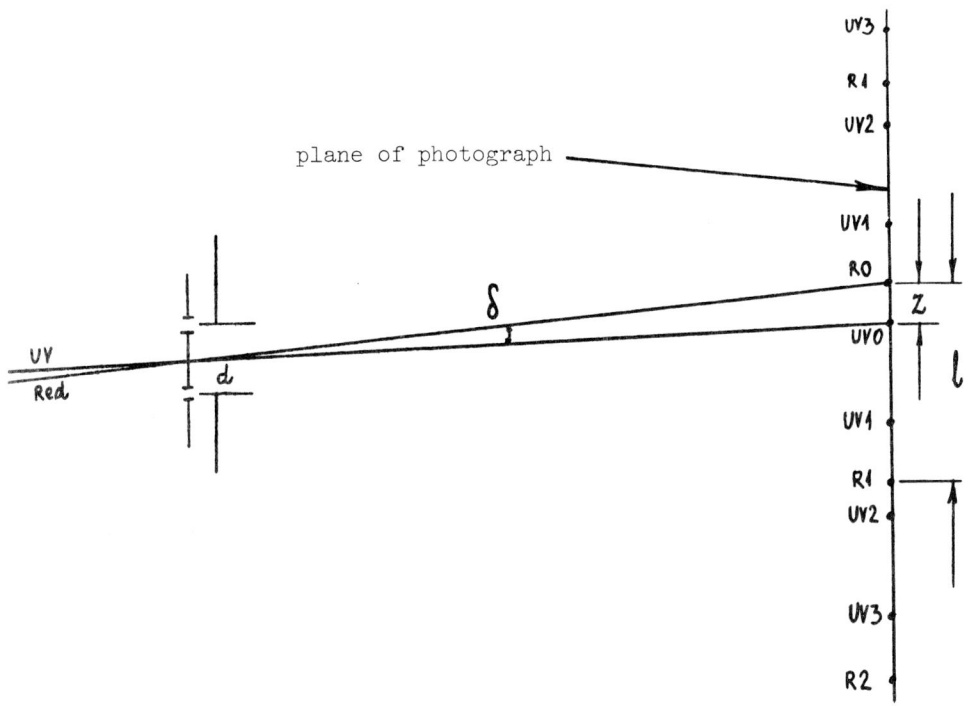

Fig. 1

Nowadays, neglecting the influence of the humidity of the air (Tengström, E., 1974), this value limits the relative accuracy of the refraction angle determination. The magnitude corresponds with an absolute accuracy of about 0.03^{cc}/km with average meteorological conditions.

From (5) we get

$$\left(\frac{m_\delta}{\delta}\right)^2 = \left(\frac{m_\lambda}{\lambda}\right)^2 + \left(\frac{m_d}{d}\right)^2 + \left(\frac{m_z}{z}\right)^2 + \left(\frac{m_l}{l}\right)^2 \qquad (8)$$

where $\lambda = \lambda_R = 0.6328$ μm

with the spectral width of about ± 1.5 GHz

Since $\lambda = \dfrac{c}{f}$

and $\dfrac{m_c}{c} \simeq 4 \times 10^{-9}$ $\dfrac{m_f}{f} = 3 \times 10^{-6}$

hence $\left(\dfrac{m_\lambda}{\lambda}\right)^2 = 9 \times 10^{-12}$ \hfill (9)

In the current IDM measurements d = 5 mm is obtained directly with an accuracy of about 0.05 mm, whence

$$\left(\dfrac{m_d}{d}\right)^2 \simeq 1 \times 10^{-4} \qquad (10)$$

Estimation of the influences of obtaining of the z and l errors requires more penetrating consideration.

These distances are measured on 35 mm film. In this case the following sources of partial errors are possible:

a. The change of z and l result from a non perpendicular situated photosensitive plane relative to the main optical axis of Cassegrain camera (fig. 2). If we describe this deviation by means of its component angle, θ lengthwise to the fringe lines and φ perpendicular to these lines, then;

a.1. The θ deviation causes a change in the distances z and l to $z' = z \sec\theta$ and $l' = l \sec\theta$; these changes do not introduce any errors since in δ exists only the quotient

$$\dfrac{z'}{l'} = \dfrac{z}{l}$$

a.2. The φ deviation causes a change in the scales between the lines of red and UV fringes. Assuming C for the scale along the red fringes, the scale along the UV fringes is

$$C' = C(1 + \dfrac{t}{f}\sin\phi)$$

where t is the distance between the red and UV fringe lines ≃ 200 μm, and f is the focal length of camera = 6260 mm.

φ can be established approximately from the required conditions of perpendicular of the optical axis in 35 mm cameras of about $1°$ whence

$$\left(\frac{C1}{C}\right)_{max} = 1 + \frac{0.2}{6260} \sin 1° \simeq 1 + 6 \times 10^{-7} < 1 + 1 \times 10^{-6}$$

Because $z_{max} < 100$ μm, then the maximum error caused by this change of scale gives a value of $z_{max} < 10^{-4}$ μm which can be neglected.

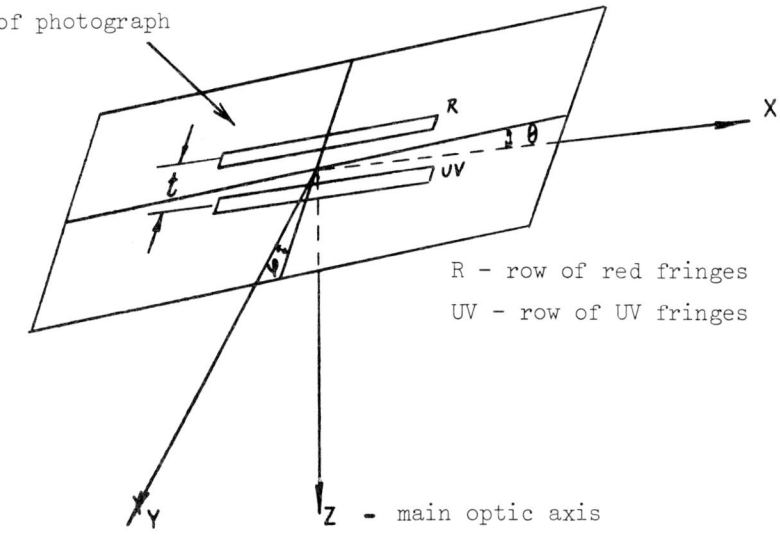

Fig. 2

b. The deviation θ (fig. 3) of the direction of line of fringes according to the direction of the motion of comparator carriage. In this case we obtain instead of z and l, $z' = l \sec \theta$ and $l' = l \sec \theta$ and it results as the case described in a.1.

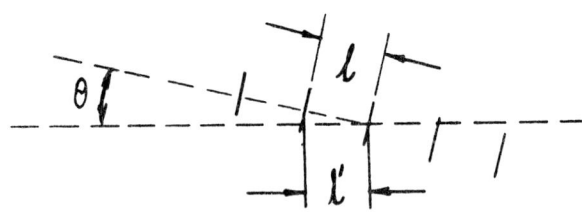

Fig. 3

c.   The error of positioning the stadia hairs of the comparator on the image of the fringe which are more or less sharp.

It is the main most decisive source of error, especially in obtaining z, which is measured between the central fringes of two different spectra-images. Each of them is situated in two different lines separated by distance of about 200 µm.

As result of the unequal film sensitivy and coefficient of absorption of the atmosphere to the red and UV the shape, the extent and the intensity of the red and UV fringes are rather different.

Now the accuracy of obtaining z and l by one observer is estimated at about 1-2 µm (internal). In spite of this internal accuracy there can occur divergences of about a few µm between the results of two different observers. For further consideration we assume $D \simeq 20$ km, $l = 800$ µm, $z \simeq 67$ µm and

$$m_z = m_l = 2 \text{ µm} \tag{11}$$

whence $\left(\dfrac{m_z}{z}\right)^2 \simeq 9 \times 10^{-4}$ and $\left(\dfrac{m_l}{l}\right)^2 \simeq 6.25 \times 10^{-6}$ (12)

Let us consider one more possible factor of error in z as result of non exactly adjusting the two laser sources with each other. Some deviations of the required equal level (fig. 4a) and from the plane perpendicular to the direction to IDM (fig. 4b) are possible. The deviations appear when $h \neq 0$ or $p \neq 0$.

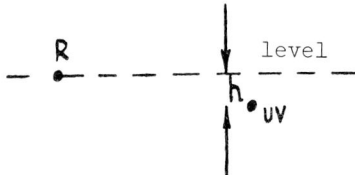

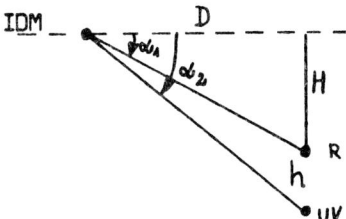

Fig. 4a

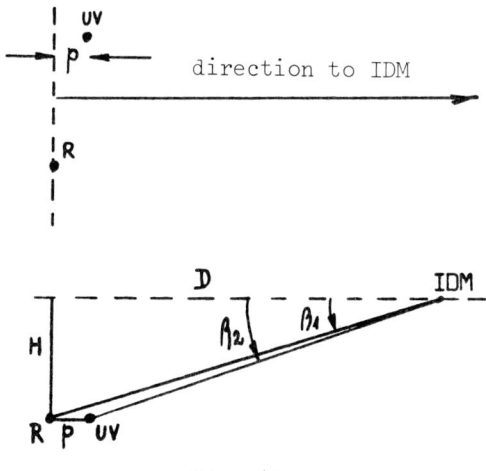

Fig. 4b

In this instance we obtain

$$tg\alpha_1 = \frac{H}{D} \ , \ tg\alpha_2 = \frac{H+h}{D} \ , \ tg\beta_1 = \frac{H}{D+p} \ , \ tg\beta_2 = \frac{H}{D}$$

and after simple transformation, expansion in series and assuming that $h \ll D$, $p \ll D$, $H+h \simeq H$, $D+p \simeq D$, we obtain

$$\Delta\delta_1^{cc} = \alpha_2 - \alpha_1 \simeq \frac{h}{D}(1 - 2\frac{H}{D})\rho^{cc}$$

$$\Delta\delta_2^{cc} = \beta_2 - \beta_1 \simeq \frac{pH}{D^2}(1 - 2\frac{H}{D})\rho^{cc}$$

(13)

and the resulting systematic error in δ is

$$S^{cc} = \Delta\delta_1 + \Delta\delta_2 = (h - \frac{pH}{D})(1 - 2\frac{H}{D})\frac{\rho^{cc}}{D}$$

(14)

To ensure the error in α less than 1.0^{cc} one must require an accuracy in the positioning of the laser sources (by $D \simeq 20$ km, $H \simeq 0.06$ km)

$$h < 1.25 \text{ mm} \ , \ p < 0.4 \text{ m}$$

(15)

It is very easy to preserve these requirements and in the following the influence of S from (14) can be neglected.

But it seems that it is not to neglect a possible difference in z as a result of the possible differences of atmosphere medium in non space-identical light path of the red and UV beams.

After taking into account the above consideration and neglecting the influence of humidity (Tengström, E., 1974, Tengström, E., 1977) we obtain from (6) and (8)

$$\frac{m_\alpha}{\alpha} = \sqrt{A + B} \qquad (16)$$

where

$$A = \left(\frac{m_K}{K}\right)^2 + \left(\frac{m_\lambda}{\lambda}\right)^2 + \left(\frac{m_d}{d}\right)^2 \simeq (5 \times 10^{-3})^2 + (3 \times 10^{-6})^2 +$$

$$+ (1 \times 10^{-2})^2 \simeq (1.1 \times 10^{-2})^2 \qquad (17)$$

$$B = \left(\frac{m_z}{z}\right)^2 + \left(\frac{m_l}{l}\right)^2 \simeq (3 \times 10^{-2})^2 + (2.5 \times 10^{-3})^2 \simeq$$

$$\simeq (3.0 \times 10^{-2})^2 \qquad (18)$$

A is the part of relative error of absolute value of refraction angle. It depends only on the accuracy of Edlen's formula and the IDM's construction and adjustment. It does not change during repeated measurements of the same α or different α angles by this same instrument. It does not influence the variations in the refraction angle. But it introduces a systematic error into absolute value of the refraction angle.

Because the A-value is limited by a magnitude of

$$\frac{m_K}{K} \simeq 5 \times 10^{-3} \qquad \text{(Edlen's formula limitation)}$$

we can consider that:

- the highest possible theoretical accuracy of the IDM is limited to approximately $\pm 5 \times 10^{-3} \alpha$,
- it would be useful to increase the accuracy of the term $\frac{m_d}{d}$ so that the term will be in a magnitude of about $1 - 2 \times 10^{-3}$.

B is the part of random relative error in absolute value of refraction angle.

The accuracy of z is of importance for this part of error. It is clear that increasing of the accuracy of z has the crucial significance and decides the accuracy of the whole method.

3. PROPOSAL OF ACCURACY INCREASING OF THE MAIN FACTOR OF IDM METHOD

The discussed analysis showed that the main factor of accuracy is the accuracy of obtaining z. It would also be advisable to increase the accuracy of obtaining of the value d.

A. The method that have been used up to the present to obtain z and l described above, is characterized by many negative aspects like:

A.1. The necessity of the displacement of the red and UV laser sources and their inconvenient adjustment.

A.2. The measurement of distance between fringes which are not lying in one straight line.

A.3. Pointings at fringe shape images that does not guarantee the best obtainable positions.

A.4. Relatively large deviation between values of z as determined by two different observers.

The reasons mentioned in A.3. and A.4. have the greatest weight in the budget error. They have been caused on the one hand by the unsharpness of the fringe shape produced by short time variation of refraction, vibration of IDM and light sources, imperfect optics and film, and on the other hand by imperfect and subjectivity between different observers. An additional factor is the difference between the images of the red and UV fringes being caused by the fact that the photographic film has different sensitivity to the red and to the UV spectrum.

Apart from the above mentioned reasons the best definition of the fringe could be obtained by the objectively positioning for maximum opacity. Such a position could be determined by the measurement of the fringe image's optical density.

The equality I_2 of the energy produced in each part of its spectrum by one energetic constant light beam crossing through partially transparent medium, is dependent on its blackness. The extent E of blackness is dependent of quantity of received photoenergy I_1. The relation between the above mentioned values are not directly or inversely proportional but they are given by monotonic functions,

$$E = f(I_1) \quad \text{increasing function}$$

$$I_2 = g(E) \quad \text{decreasing function}$$

hence $\quad (I_2)_{min} = F(I_{1_{max}}) \quad , \quad (I_2)_{max} = F(I_{1_{min}})$

where $F(x) = g(f(x))$

The I_1 is the effecting stream by exposure. Its mean average position estimates E_{max} on the film. Therefore the received $(I_2)_{min}$ by photometric scanning points to the right definition of fringe.

This value can be realized using the microphodensitometer in two ways, one by plotting a graphical curve, or two by directly obtaining the position of maximum by comparison with the standard curve. One proposed method of improvement is to have all the red and UV fringes in one straight line. This excludes the source of error mentioned in A.1 and A.2.

Let us consider the proposed device. After plotting a photodensity curve which is the sum of red and UV diffraction curves deviated from each other by the value z (fig. 5).

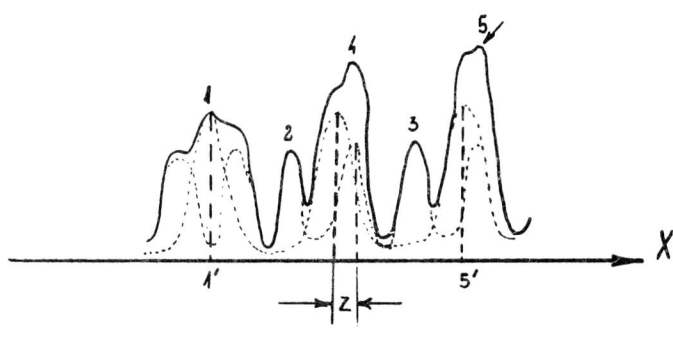

Fig. 5

Then the z value can be obtained

$$z = \frac{x_3 + x_2}{2} - \frac{x_5 + x_1}{2} \tag{19}$$

and
$$l_r = \frac{x_5 - x_1}{2} \tag{20}$$

or
$$l_r = \frac{\lambda_r}{\lambda_{UV}} l_{UV} = \frac{\lambda_r}{\lambda_{UV}} \frac{x_3 - x_2}{2} \tag{21}$$

where for current used sources of light $\frac{\lambda_r}{\lambda_{UV}} = 1.94708$, that means

$l_r \approx 2l_{UV}$.

Therefore, by $z \ll l_{UV}$ the difficulty appears in obtaining the right position of the red fringes (fig. 5 points N = 1',5'), because every red maximum is lying near each second maximum of UV fringes, shifted only by means of z.

It seems to be possible to solve the question in two different ways.

One is to increase the resolving power of IDM grating to about $\frac{1}{10}$ z, that means for the red and UV laser a value of about

$$\Delta\lambda = \frac{1}{10} (0.6328 - 0.325) \simeq 0.031 \text{ µm} \tag{22}$$

According to (Born, M., Wolf, E., 1959) the resolving power of a grating is defined by the ratio

$$R = \frac{\lambda}{\Delta\lambda} = mN \tag{23}$$

where   m   the order of fringe image

            N   the number of grating slits.

Substituting in (23) for $\lambda = 0.6328$ µm, $\Delta\lambda = 0.031$ µm, m = 1 (the principal maximum) we obtain

$$N \simeq 20 \tag{24}$$

which proposes to use grating of non less than 20 slits, but to achieve comparable light power on film surface it must keep hold the same sum of transparent surface of grating.

It is clear that the above mentioned proposal depends only on theoretical considerations and the practical investigations in this question are of great importance.

A second way one can try is analyzing the image of the film density curve of the photometric poorly devided red fringe maxima. From such an image each second maximum (fig. 5 points 2 and 3) is apparently divided, which gives an accurate position of the other UV maximum position.

The relative ratio of the maximum magnitude of red can be obtained from the red fringes alone. In the same way one can also obtain the ratio of UV alone.

Now one has:

B.1.   The ratios $I_1:I_2:I_3:I_4:I_5$ for maximum of red fringes alone.

B.2. The means of I_1', I_2', ... I_{11}' for the UV fringe maximums calculated from the relative ratio of UV fringes only and from the absolute magnitudes of selected well divided UV fringes and red together.

B.3. The mean $f(x)$ of sum density function of red and UV, where x is the argument of position (fig. 5).

By simple Fourier's analysis we obtain

C.1. The mean of UV density function $A(x)$.

C.2. The mean of red density function $B(x) = f(x) - A(x)$.

C.3. The maximum position x_i of function $B(x)$

$$J_i^{(r)} = \max(B(x_i))$$

The results calculated in such a manner can be objectively controlled by:

D.1. $\quad J_r = \frac{1}{2}(x_c - x_{c'}) = \text{const.}$ \hfill (25)

where x_c and $x_{c'}$ are the suitable fringe maximum position lying symmetrical to the central fringe.

D.2. The ratios of $J_i^{(r)}$ should be equal to those obtained experimentally.

Then z from formula (19) is given by

$$z = J_{UV} - J_r \hfill (26)$$

It can be assumed that the proposed procedure should give the z mean with a higher accuracy. But even if the accuracy would only be comparable with that from the comparator reading the absolute value of z would be much more objective and independent of the observer.

B. The d value was obtained by direct measurement of slits. In the formula (5) there acts some actual value which effects the interference. Therefore it is proposed to obtain the value of d from observations of the laser source interference image produced by the IDM grating. When we use for measurement the micrometer of the Wild T4 theodolite we can determine the angle θ with an accuracy of about 0.5^{cc}.

Because $\quad d = \dfrac{n\lambda}{\sin\theta}$ \hfill (27)

where n is the order of interference fringes, the relative accuracy of d when $d \simeq 5$ mm and $n = 3$ is

$$\frac{m_d}{d} \simeq 2 \times 10^{-3} \hfill (28)$$

Such a measurement can be realized in the laboratory under controlled conditions.

A further step to increase the possibility of an absolute test of IDM measurements is to correlate fully the observed vertical angle with the determinated dispersion. In this case it seems to be advantageous to make simultaneous exposures of the fringes of IDM and of the image of the red laser source at the focal plane of the theodolite.

In this way one observes both the actual time of observation and average time of the derived phenomena thus excluding the instrumental sources of error connected with the direct measurement of the vertical angle. Of course, it is clear that such a change involves some increase of the following evaluations for achieving the final result of the measured vertical angle.

4. CONCLUSIONS

1. The above mentioned proposals should increase the accuracy of the results of the present IDM measurement.

2. The increase of the accuracy in the present results can facilitate the explanation of the most important causes of problems in the IDM method and would assist in its further improvement.

ACKNOWLEDGEMENTS

The author is indebted to:
Professor E. Tengström and S-G. Mårtensson for the opportunity to investigate the IDM results and for their criticism, advise and profitable discussions that have led to clarification of the content of this paper,
Miss I. Ohlsson for the help in typing.

REFERENCES

Angus-Leppan, P.V.: 1971, "Meteorological Physics Applied to the calculation of Refraction Corrections". Conf. of Commonwealth Survey Officers, Cambridge.

Angus-Leppan, P.V., Webb, E.K.: 1971, "Turbulent Heat Transfer and Atmospheric Refraction", General Ass. of IUGG, Moscow.

Born, M., Wolf, E.: 1959, "Principles of Optics", Pergamon Press, London.

Brunner, F.K.: 1977, "Experimental Determination of the Coefficients of Refraction from Heat Flux Measurements", Proc. of IAG Symposium, Wageningen, The Netherlands.

Brunner, F.K., Fraser, C.S.: 1977, "An Atmospheric Turbulent Transfer Model for EDM Reduction", Proc. of IAG Symposium, Wageningen, The Netherlands.

Glissman, T.: 1977, "A Coincidence Method for Refraction Eliminating Angle Measurement", Proc. of IAG Symposium, Wageningen, The Netherlands.

Huggett, G.R.: 1978, Letter to prof. Tengström of July 12.

Huggett, G.R., Slater, L.E.: 1977, "Recent Advances in Multiwavelength Distance Measurement", Proc. of IAG Symposium, Wageningen, The Netherlands.

Prilepin, M.T.: 1974, "Elimination of Angular Refraction by means of Multiple-wavelength Method", IAG Symposium, Stockholm, Sweden.

Prilepin, M.T.: 1975, "Interferometer for Determination of Geodetic Refraction", XVI General Ass. of IUGG, Grenoble, France.

Rinner, K.: 1977, "Meteorological Correction of Laser and Microwave Distances", Proc. of IAG Symposium, Wageningen, The Netherlands.

Saastamoinen, J.: 1974, "Theory and Calculation of Refraction Effects on Directions from Meteorological Information", IAG Symposium, Stockholm, Sweden.

Tengström, E.: 1967, "Elimination of Refraction at Vertical Angle Measurements using Laser of Different Wavelengths", Proc. of the Int. Symposium on Figure of the Earth and Refraction, Vienna, Austria.

Tengström, E.: 1974, Appendix to Dr Tengström's letter to Dr Brein, Jan. 10, 1969, concerning the influence of humidity on refraction determinations with the dual wavelength method, Proc. of IAG Symposium, Stockholm, Sweden.

Tengström, E.: 1977, "Some Absolute Tests of the Results of IDM Measurements in the Field with a Description of Formulas used in the Tests", Proc. of IAG Symposium, Wageningen, The Netherlands.

Walsh, J.W.T.: 1953, "Photometry", 2nd Edition, London.

Williams, D.C.: 1977, "First Field Tests of an Angular Dual Wavelength Instrument", Proc. of IAG Symposium, Wageningen, The Netherlands.

DISCUSSION

D.G. Currie: I wonder, if your humidity problem could perhaps be amenable to the techniques that we look forward to use, and that is: there are absorption bands in both the 6000 and the 8000 Ångström area which are accessable with silicon diods, and in our case where it is not gradients that we are interested in but actual water vapor content, we would sample the entire beam and switch between the in the band and the out of the band. And as long one does not operate in the saturated band, that can give a rather straight-forward determination of the water vapor in that column. In your case, where you have two separate columns, and you are asking for the gradient between these, it seems that you might also be able perhaps to switch between these two columns, giving the water vapor gradient in the same direction, which you are interested in, giving that number for a humidity correction on your vertical angle.

J. Milewski: I think, that the main part of the refraction angle dependent on humidity can be calculated by the meteorological data. But I suppose it must be investigated which parameters explain the occasional great uncorrelation between dispersion and refraction when using widely separated spectral lines. It is therefore very important also to study the humidity factor. I believe it would be possible by investigating simultaneously the fluctuations of microwave beams.

E. Tengström: We have studied the humidity question rather carefully since the time as de Munck pointed out to me the importance of the vertical water vapor gradient. We investigated wet models using Laplace formula, and found that during inversions with great positive temperature gradients the humidity corrections can take negative values which can not be neglected. Meteorological experiences here show, however, that the humidity gradient can reach values more than 20 times those predicted from the temperature gradient by Laplace, so that already at an inversion of $+2^\circ C/100$ m and a humidity gradient of -30 mm/km, the correction amounts to $-8"$, the total refraction being as high as $130"$. Great refraction values, which occur during inversions in the evening some hours before midnight, seem to correspond to high negative humidity gradients, at least in our experiments, and it is therefore - during such conditions - necessary to have relevant gradient observations done, so that the humidity integral can be computed with sufficient accuracy. We have started a cooperation with the Meteorological Institute here, and our common plan is to use balloons evenly spread out along the line of sight of our test base for assembling meteorological informations at short intervals of time, to be transmitted from the sensors in the balloon sondes and recorded at the observation site. From the vertical gradient distribution of humidity, pressure and temperature obtained, we may calculate the humidity correction under various conditions of inversion with high accuracy. In the future field work we have to avoid measuring during such inversions, which correspond to high positive values of the vertical gradient of temperature. A rough observation of this gradient before the measurement can help us to decide if we shall observe or not. Under normal conditions ($\sim -1°C/100$ m) with medium horizontal windspeed, the humidity gradient has been found to be small and the humidity correction is negligable or can be derived from rough data. To calculate the total refraction angle by means of merely meteorological data seems to be impossible to realize even in the future, because we shall never in practise be able to achieve simultaneity with the angle observations and a sensor spacing along the line of sight, which satisfy our needs for deriving refraction with an accuracy aimed at for the moment (that is to within $0".2 \equiv$ mean error of today's star catalogues). What the turbulence effects concerns, we have to choose integration times of dispersion measurements equal to the angle observation times. And these integration times have to be evaluated according to the specific use of the results. With lasers it will - to my opinion - also be necessary to focus a wider part of the incoming red and blue wavefronts, which have different irregularities in shape, and then by means of a small concave collimator mirror direct the integrated plane fronts into the small aperture of the observing instrument. I believe that this spatial integration (see my previous remark under Mårtensson's paper) will reduce the occasional bad correlation between dispersion and refraction, now observed with our short laser exposures. I don't think the definition of the fringes will be essentially deteriorated, because we will have the same situation as in the case of ordinary light of various wavelengths, the wavefronts of which are very regular at great distances.

J. Dommanget: I would like to inform you about some research I made on the effect of water vapor pressure in the atmosphere on the image quality observed with our 45 cm aperture Cooke-Zeiss refractor. At Brussels, the Royal Observatory and the Royal Meteorological Institute are located at the same spot. So this was a good opportunity for observing image quality (Danjon's estimator scale) - refractor diameter shut down to 25 cm - when radiosounding were made. A statistical study showed a good correlation between water vapor pressure and image quality when in the same time shearing (important wind gradient) between two contiguous layers was observed. This seems to support the idea just considered, of the importance of water vapor pressure also into the general problem of atmospheric refraction. (See: Communication de l'Observatoire Royal de Belgique, no 231, 1964).

E. Tengström: Mr Kahmen and Mr Brunner said, that there is a clear correlation between the magnitude of the refraction angle and the fluctuations caused by the turbulence. Then your statement would mean that the refraction is strongly correlated to the humidity pressure itself, which I find a little strange.

J. Milewski: I think it is rather clear, because there is a very close correlation between the gradient of temperature and the gradient of humidity. One may riskily admit that it should also be a close correlation between the humidity pressure and the turbulence.

APPLICATION OF REFRACTIONAL EFFECTS IN GEODETIC FRAMEWORK
OF IRAQ BY CONTINUOUS TRILATERATION

R. Pażus
Polservice-PPG Survey Office; Baghdad, Iraq

1. INTRODUCTION

Terrestrial refraction effect has been of considerable interest to
many geodesists for many decades and there is a rich litterature on
the subject, of long standing. Due to its local character this
problem is still topical. Geodetic determination of how far terrestrial
refraction influences geodetic observations is, therefore, very often
studied.

"POLSERVICE", the Polish Foreign Trade Enterprise, was charged with
a task to establish the network, the general contractor being the State
Enterprise for Survey and Cartography, Warsaw, Poland. Works were
started in 1974 and the completion of the network is to take place in
the middle of 1979. As a result, among others, geodetic network was
established, of average distance between points 15 kilometers (the
shortest side about 8 km, the longest about 35 km). The structure of
the network is as shown on Figure 1, that is all the distances between
neighbouring points are measured (there is no deviation from this
principle on the total area of Iraq) and, additionally, an angle
between two, most clearly seen, directions is measured on each point
(deviation from this principle is possible).

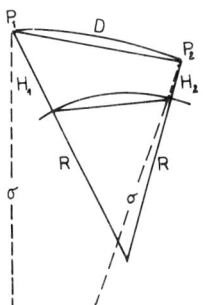

Figure 1. The structure of the network.

So far over 8000 sides were measured by AGA model 8 geodimeters, which shows how large is the scope of the works.

2. OPTICAL EFFECTS

During initial observations many interesting remarks about the refraction effects were made. In the evening receding of the horizontal line was being observed and objects on the horizon had shapes unnaturally elongated. This phenomenon occurs during the whole night and early in the morning - later on the profile of objects diminishes and they conceal under the horizon. It happened that light of desert settlements situated at a distance of over 100 km was being observed at night.

A phenomenon of "picture division" is also interesting; the point is that during a very short time in the morning and in the evening, a picture being at a distance of several kilometers is seen double. At the beginning, in the evening a target picture elongates; later on it divides into two and, for example the target is seen distinctly at a higher level, and in less sharp outline, at a lower level. After several minutes this phenomenon disappears. However, very often during the day-time two horizon lines separated by a layer giving the optical illusion of water surface is being observed.

3. COEFFICIENT OF REFRACTION AND ITS EFFECT ON LENGTH OF MEASURED SIDE.

The ratio of the radius of curvature of the ellipsoid (R) to the radius of curvature of the wave path (ς) in a specified point is defined as the accurate coefficient of refraction (see Figure 2):

$$K = \frac{R}{\varsigma} \qquad (1)$$

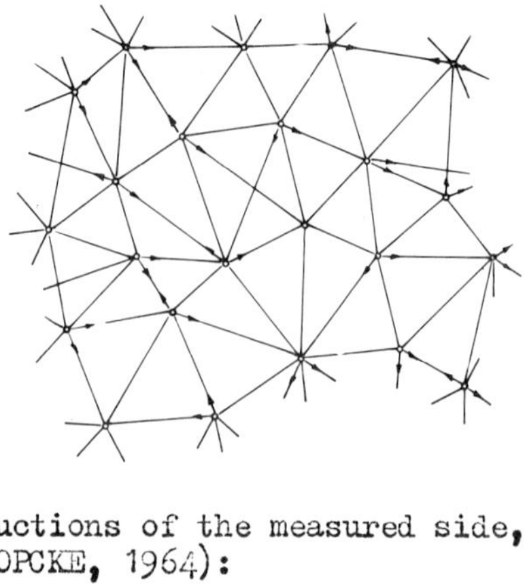

Figure 2.

From several reductions of the measured side, two depend on coefficient of refraction (HOPCKE, 1964):

- correction for path curvature, i.e. the reduction to chord distance (P_1P_2):

$$k_1 = K^2 \cdot \frac{D^3}{24 R^2} \qquad (2)$$

- correction for variation in the velocity for propagation along the line:

$$k_2 = - (K - K^2) \cdot \frac{D^3}{12 R^2} \qquad (3)$$

These corrections can be combined:

$$\text{corr.} = - (2K - K^2) \cdot \frac{D^3}{24 R^2} \qquad (4)$$

Quantities of reduction corrections according to formula (4) for the range of lengths measured in Iraq geodetic network (R = 6 370 km) is indicated on the following table:

K	10 km	15 km	20 km	30 km	35 km
− 0.5	+0.001	+0.004	+0.010		
0.0	0.000	0.000	0.000	0.000	0.000
+ 0.5	−0.001	−0.003	−0.006	−0.021	−0.033
+ 1.0	−0.001	−0.003	−0.008	−0.028	−0.044
+ 1.5	−0.001	−0.003	−0.006	−0.021	−0.033
+ 2.0	0.000	0.000	0.000	0.000	0.000

It results from the above that at distance measured up to 15 kilometers there is always accuracy of $2 \cdot 10^{-7}$ (regardless the accuracy of K); for longer distances, however, not more than 35 kilometers, the error of specifying K quantity, equal to 0.5 does not cause that the reduction error is more than $1 \cdot 10^{-6}$.

Before each measurement of length, simultaneous zenith distance measurements from the two terminals were carried out, enabling accurate introduction of the said reductions. The observations were carried out for laser beam – therefore the coefficient of refraction was specified for helium-neon (HeNe) gas laser beam. For some areas simultaneous zenith distance measurements were repeated after the distance measurement.

4. TERRESTRIAL REFRACTION IN IRAQ

Investigations on terrestrial refraction carried out in Iraq were limited to determinations of refraction coefficient in ground atmospheric layer of 2 - 25 meters. The basic examination depended on 24-hour determination of quantities and changes of refraction coefficient for the South and West Desert area. Eight determinations were made in different places and different seasons of the year.

Figure 3 indicates 3 chosen diagrams of daily changes of refraction coefficient, made on the Iraq Desert area. Due to the homogeneity of basement soil and invariability of weather, the daily changes of refraction coefficient in the middle of the day are very regular during the yearly period.

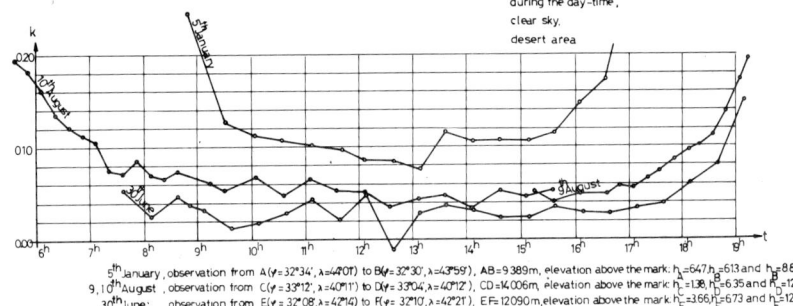

Figure 3. Daily refraction coefficient from non-reciprocal vertical observation when the height difference was known.

The shortest day of the year (22nd December) is characterized by refraction coefficient $K = 0.10$; the longest day of the year (21st June) $K = 0.02$. Individual determinations (average from two targets) are marked with circles on the diagram. It was found that there was a constant difference between the determinations or the higher and lower targets equal to about $K = 0.03$. The determination was made while aiming at the opaque beacons.

Figure 4 specifies three chosen diagrams of 24-hour changes of refraction coefficient. These determinations showed that during the night the refraction coefficient would be most probably $K > 0.7$. The determinations were made while aiming at lamps, heliotropes and opaque beacons.

Observations as shown in Figure 3 and Figure 4 were made by the method of vertical angle measurement at the known difference of target height and length. While calculating the refraction coefficient, influence of deviation of the vertical was disregarded; its influence on quantity of the coefficient is approximately ± 0.01.

Figure 5 illustrates all determinations of the refraction coefficient calculated for the specified area during the yearly period. Reciprocal observations were made each time for a different side of the specified area (hence 315 determinations were made). These observations were carried out before the measurement by geodimeter.

They were made at the same time, with accuracy ± 1 minute (when the fixed signal through the radio-telephones was given).

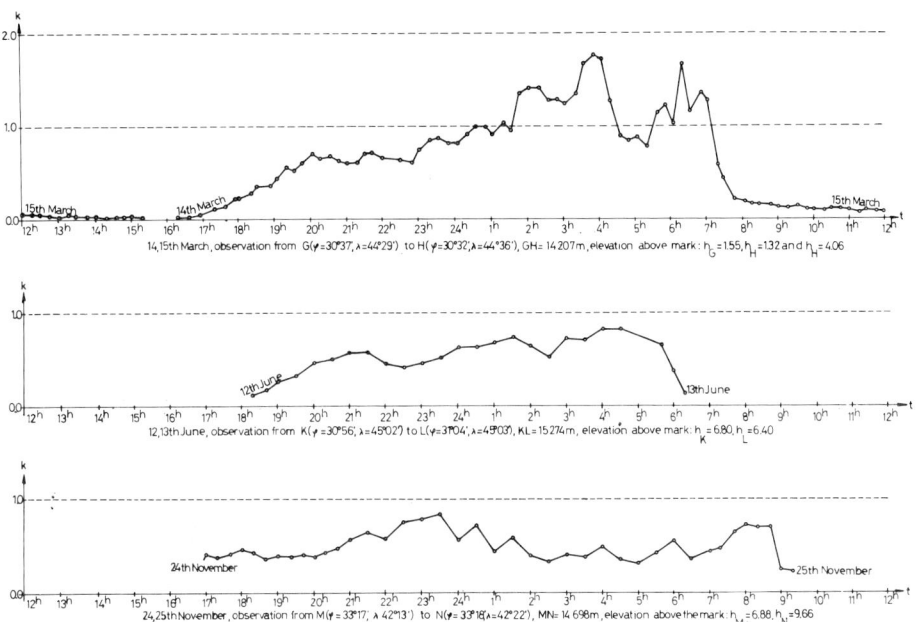

Figure 4. 24-hour changes of refraction coefficient.

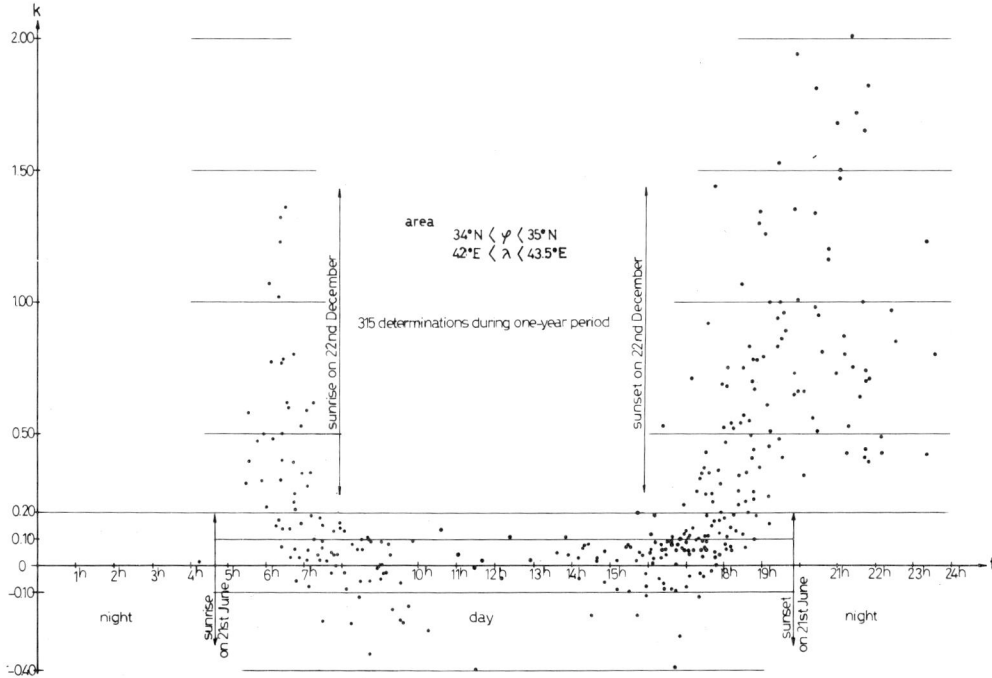

Figure 5. Refraction coefficient from reciprocal zenith angles.

Laser beam emitted by a geodimeter or its reflection in a reflector was the target. The diagram includes all the observations, which correctness was controlled by trigonometric levelling (the sum of the height differences in the three sides of a triangle). The refraction coefficient from the said determinations concerns helium-neon (HeNe) gas laser beam.

Practical importance of dispersion of light for the used measurements was not ascertained.

5. CONCLUSION

Results of research on terrestrial refraction in atmospheric layer up to 25 meters, reviewed in this paper, had considerable influence on establishment of geodetic network in Iraq.

Owing to high coefficients of refraction it was possible to carry out observations from ground stations (apart from obvious exceptions) hence limiting the necessity to use towers and masts. These had considerable importance for the organization of works in the establishment of such a big network. During a ground reconnaissance, geodimeter lines attainable during the day and those requiring uplifting of the target above the horizon by night terrestrial refraction were determined.

This paper reviews only the influence of terrestrial refraction on the desert area. However, it should be pointed out that while measuring the Marshy area in the Southern region of Iraq, more advantageous results with regard to terrestrial refraction, were attained (higher coefficient of refraction during the night).

Preliminary adjustment for about 40% of the area confirmed high accuracy of measurements - the acquired accuracy was 1 : 650 000 at each point of the network.

Formulas and accuracy of determinations were not included in this paper in order to make it perspicuous. They are well-known by geodesists dealing with this problem.

This article reviews the problem of how the phenomenon of refraction, being generally considered as troublesome for geodesists, occured to be very useful and it was utilized in the establishment of geodetic control in Iraq.

REFERENCES:

HOPCKE, W.: 1964, Uber die Bahnkrümmung elektromagnetischer Wellen und ihren Einfluss auf die Streckenmessungen, Zeitschrift für Vermessungswesen, Vol.6, pp.183-200.

DISCUSSION

T.J. Kukkamäki: I was very pleased to see that in Iraq Mr Pazus has got the same kind of results that we have got here in northern latitudes, when we measured a 900 km long traverse through Finland. We got exactly the same results for k. We used towers, but the towers were only in order to rise the instruments above the treecrowns. From the refractional points of view, the treecrowns are comparable to the ground in Iraq, where the landscape was open. During the daytime k was surprisingly consistent. Trigonometrical height determinations gave good results for these sides with an average length of 30 km and with height differences of ±50 m our results agree with the Iraq results, which means that these results seem to be global.

J. Milewski: In spite of the fact that Iraq represents rather a very special area of atmospheric conditions?

T.J. Kukkamäki: Yes, our results agree well with the Pazus' ones. Have other colleagues any comments to this?

K. Poder: If you are assuming the model of the atmosphere used for deriving the conventional formulae for refractional effects in distance measurements, then you think it is fine, and you have k as the ratio of the curvature of the earth through the curvature of the lightbeam, and you get the curvature correction which roughly and more practically is written in a way which satisfies surveyors' needs, and which also includes some humidity reductions. Now, assume you have a small disturbance near the terminal point where you observe the zenith distances, then you may start by observing that, and your observation indicates that the target is there, due to a small local disturbance, but the major part of the path goes mostly and regularly in free air. In that case the use of k, which also can be put into the formulae, from zenith distances, will give you a very wrong correction. In that case it might be better to take just the classical value of 0.12 or 0.13. If you have that large variation it is a sign that your model is not good enough, and it may be better just to take a more theoretical correction.

J. Milewski: You are right, Mr Poder, as to the distance correction. Theoretically it could arise a disagreement between the real k as average of the whole distance and that obtained from the vertical angles. But we can control how the Hoepke's formula works by the right Edlén's formula for reduction od distance measurements using meteorological parameters. They installed some control stable base points, and at the base points they measured all the time, every day, all the parameters for obtaining the gradient of temperature and of pressure. And the corrections derived from Hoepke's and from Edlén's formulae agreed with each other.

K. Poder: If you use Hoepke's formula you are actually missing a term. Hoepke's formula needs a correction also for the second order derivative of the vertical gradient. This is needed in very sloping lines, and if you use his formula for very sloping lines you will get into troubles. This is important in the case of sloping lines.

T.J. Kukkamäki: So, now you have heard one pessimist and two optimists. Any more pessimists here? If not, we may continue our programme.

COMPUTING PARALLACTIC REFRACTION FOR STELLAR TRIANGULATION

Juhani Kakkuri and Ossi Ojanen

Finnish Geodetic Institute, Helsinki

1. INTRODUCTION

The parallactic refraction correction should be applied to the photographic observations of the light signals carried either by artificial satellites or by meteorological balloons. The correction to the right ascension and declination are made with the equations

$$\left.\begin{array}{l}\Delta\alpha = \dfrac{1}{15} \dfrac{\cos\phi \, \sinh}{\cos\delta \, \sin z} \Delta\zeta \\ \Delta\delta = \dfrac{\sin\phi \, \cos\delta - \cos\phi \, \sin\delta \, \cosh}{\sin z} \Delta\zeta\end{array}\right\} \quad (1.1)$$

where ϕ is the latitude of the observation station, z the zenith distance, α the right ascension, δ the declination and h the hour angle of the light signal. Parallactic refraction is denoted by $\Delta\zeta$.

Numerous formulae have been derived for computing the parallactic refraction term $\Delta\zeta$. The form of the formula depends on the purpose for which the parallactic refraction is needed. In the case of distant light signals, such as artificial satellites or natural moon, the formula for $\Delta\zeta$ is very simple. In the case of balloon observations the formula for computing the term $\Delta\zeta$ is more complicated and depends on the atmospheric model used. Oterma (1960) and others have derived formulae for both cases. Her formula in the case of a distant light signal is

$$\Delta\zeta = 0.0012568 \, \frac{1-s}{s} \, \zeta_\infty \quad (1.2)$$

and in the case of balloon-borne light signals

$$\Delta\zeta = (\frac{p}{760} \, \eta + \frac{t}{10} \, \kappa) \, \zeta_\infty \quad (1.3)$$

where the whole astronomic refraction

$$\zeta_\infty = \frac{1}{10} \tan z_0' \, (601\overset{''}{.}7052 - 66\overset{''}{.}968 c + 20\overset{''}{.}971 c^2 \\ - 10\overset{''}{.}704 c^3 + 7\overset{''}{.}655 c^4 - 7\overset{''}{.}091 \frac{c^5}{1-c}) \quad (1.4)$$

applies to standard conditions at which the temperature is 0 °C and the air pressure 760 mmHg, and the parameter c has the form:

$$c = (\frac{1}{10} \sec z_0')^2 \tag{1.5}$$

The air pressure p is taken in mmHg and the air temperature in $^{\circ}$C and both are measured at the station. The terms

$$\left. \begin{array}{l} \eta = \dfrac{\Delta \zeta_0}{\zeta_\infty} \\ \kappa = \dfrac{\Delta \zeta_{10} - \Delta \zeta_0}{\zeta_\infty} \end{array} \right\} \tag{1.6}$$

where $\Delta \zeta_{10}$ and $\Delta \zeta_0$ are values of the parallactic refraction corresponding to the temperatures 10 $^{\circ}$C and 0 $^{\circ}$C; they are given in tables compiled by Oterma (1960) for given s and apparent zenith distance z_0'. The parameter s depends on the height H of the light signal from the ground level according to

$$s = \frac{H}{r_0 + H} \tag{1.7}$$

where r_0 is the mean curvature radius of the earth (6368.8 km used). Oterma's tables are based on the assumption that the quantity $\mu-1$, in which μ is the index of refraction, is directly proportional to the air density ρ and that the vertical temperature decreases linearly from t_0 on the ground to t_1 of the tropopause and then remains constant to relatively great heights. Oterma states that

$$t = t_0 - ks \quad (k = \text{constant}), \quad 0 \leq s \leq s_1,$$
$$t = t_1, \quad s \geq s_1,$$

where $t_0 = 0$ $^{\circ}$C, $t_1 = -50.687$ $^{\circ}$C and $s_1 = 0.0014$ corresponding to H = 8.929 km for the tropopause, and the air pressure on the ground is taken to be 760 mmHg.

The astronomical refraction formula (1.4) is given in a series in powers of the secant of the apparent zenith distance. Such a series can be used only up to a certain zenith distance. In Oterma's formula the limit is around 85°. The value of the formula (1.4) in this case differs insignificantly from the numerical solution of the original refraction integral formula used by Oterma (1960).

Investigations in which the curvature of the light ray is of principal concern suggest that formula (1.3) should be used instead of formula (1.2) if the height of the light signal is below 100 kilometres.

2. APPLICATION OF OTERMA'S THEORY TO FINNISH STELLAR TRIANGULATION

The above refraction model prepared by Oterma has been applied to Finnish Stellar Triangulation. In this work balloon-borne flash lights were used, and because the balloon always remains below a height of 40 km formula (1.3) should be used. In the first computation (Kakkuri 1973) the ratios η and κ as functions of given s and z_0' were taken from Oterma's tables and nomographs and the astronomic refraction calculated from formula (1.4). An electronic computer program was later prepared for computing the parameters η and κ with the help of numerical integration. The refraction integral has been written in the form

$$\zeta = \int_0^\omega \frac{\beta \, d\omega}{1-\beta\omega} \frac{\sin z_0'}{\sqrt{\cos^2 z_0' + \psi}} \tag{2.1}$$

for obtaining rapid converging. When following Oterma's model the variables and constants in formula (2.1) are:

$$\left. \begin{array}{l} \mu_0 - 1 = (\bar{\mu}_0 - 1) \dfrac{p_0}{\bar{p}_0(1+\alpha t_0)} \\[6pt] \beta = \dfrac{\mu_0 - 1}{\mu_0} \\[6pt] \omega = 1 - (1 - Bs)^{(A/B)-1}, \text{ when } 0 \leq s \leq s_1 \\[6pt] \omega = 1 - (1 - \omega_1) e^{A_1 s_1} e^{-A_1 s}, \text{ when } s \geq s_1 \\[6pt] \psi = \left(\dfrac{1-\beta\omega}{1-s} \right)^2 - 1 \end{array} \right\} \tag{2.2}$$

where $\bar{\mu}_0$ is the index of refraction on the ground at standard conditions $\bar{t}_0 = 0\ ^\circ\text{C}$ and $\bar{p}_0 = 760$ mmHg and

$$\left. \begin{array}{l} A = \dfrac{796.8}{1+\alpha t_0}, \\[6pt] A_1 = \dfrac{796.8}{1+\alpha t_1}, \\[6pt] B = \dfrac{\alpha k}{1+\alpha t_0}, \\[6pt] \alpha k = \dfrac{\alpha t_0 - \alpha t_1}{s_1}, \\[6pt] \alpha = 0.003668, \\[6pt] s_1 = 0.0014. \end{array} \right\} \tag{2.3}$$

The above parameters can naturally be defined according to other atmospheric models.

The parallactic refraction $\Delta\zeta$ can be calculated from the true zenith distance z and the height parameter s using the following algorithm:

$$z_0' = z - \zeta_\infty \tag{2.4}$$

where ζ_∞ is obtained from formula (2.1) integrating between the limits 0 and 1 or from formula (1.4).

Using some additional notations we have

$$z' = z_0' + \zeta \tag{2.5}$$

where ζ is the integral (2.1) in which the upper integration limit ω comes from formulae (2.2) as a function of s,

$$\sin i = \frac{1-s}{1-\beta\omega} \sin z_o \quad (2.6)$$

$$\frac{H}{r} \sin z' = \sin i - (1-s)\sin z' \quad (2.7)$$

and

$$\sin(z' - \bar{z}) = \frac{H}{r} \sin z' \frac{\sin \bar{z}}{\sin(z'-i)} \quad (2.8)$$

where $z' - \bar{z}$ is obtained by iteration.

Finally we have the parallactic refraction

$$\Delta \zeta = \zeta_\infty - \zeta + z' - \bar{z} \quad (2.9)$$

The results obtained are encouraging. One test is as follows:

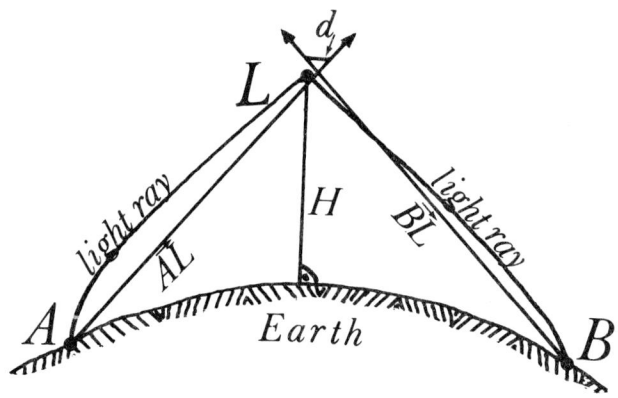

Fig. 1. A and B are the stations, H is the height of the signal L, $\vec{AL}$ and $\vec{BL}$ are uncorrected direction vectors and d is shortest segment between them.

A light signal was photographed from stations A and B, both of which were geographically known. In order to make the parallactic refraction correction to the direction vectors $\vec{AL}$ and $\vec{BL}$ determined from the background stars, the height of the light signal and hence the parameter s should be known (Fig. 1). This is done using an iterative process. In the first iteration step the height of the light signal is approximated with that of the shortest segment (d) between the uncorrected direction vectors $\vec{AL}$ and $\vec{BL}$. This height approximation is used for computing the parallactic refraction correction, which is then applied to the uncorrected direction vectors. The corrected direction vectors obtained redetermine the height, and a new value is then calculated for the parallactic refraction using recalculated height, etc. The iteration is continued until the shortest segment becomes constant. This is generally obtained after three repetitions. If the geographical stations' coordinates are correctly known and if there are no systematic errors in timing, or in modelling the parallactic refraction, etc., the segment d between the direction vectors, which have been finally corrected, should be zero in the limits of random errors. The accuracy of the parallactic refraction correction can then be estimated.

The analyses show that parallactic refraction can be calculated using Oterma's theory with an accuracy of about ± 2 sec of arc for zenith distances from $70°$ to $75°$. The accuracy of this order is also obtained for zenith distances from $75°$ to $80°$, but from $80°$ to $85°$ the segment seems to become greater and correlated to the zenith distance, which may be an indication of unsatisfactory modelling of the atmosphere.

3. REFERENCES

Kakkuri, Juhani (1973), Stellar Triangulation with Balloon-Borne Beacons. Ann. Acad. Sci. Fenn. A III, 113. Helsinki.

Oterma, Liisi (1960), Computing the Refraction for the Väisälä Astronomical Method of Triangulation. Ast. Opt. Inst. Turku Univ. 20. Turku.

DISCUSSION

T.J. Kukkamäki: Thank you and prof. Oterma for this presentation. I hope that nobody has anything against it, neither prof. Oterma herself.

DISCUSSION ABOUT COMPUTING PARALLACTIC REFRACTION IN MODEL ATMOSPHERES

Chairman: T.J. Kukkamäki

J. Saastamoinen: (introduction to the discussion)
The problem with parallactic refraction for points within the atmosphere is that - in my opinion - there is no model atmosphere that can be used for calculating it. In astronomical refraction, the thing is different. We measure the refractive index at the ground level and we know that at the top of the atmosphere it is equal to 1. But in parallactic refraction, when the flash is inside the atmosphere, we should have meteorological information at that point, too. Otherwise we can compute the parallactic refraction only on the basis of our particular model, which may be not good. It may well be good for astronomical refraction but not for parallactic refraction. There is also the effect of uncompensated isopycnic tilts. If we have a level surface and constant temperature, everything is fine, as the isopycnic layers are horizontal . But going from South to North, the temperature gradient will cause a meridional tilt of the isopycnics at the ground level. At a height of approximately 8 km this tilt has become zero, and above that it takes the opposite direction. Now in astronomical refraction the combined effect of these tilts is largely cancelled out. But when we deal with parallactic refraction we miss the upper compensating layers, and there will be quite a sensible correction for tilt, which is difficult to take into account in atmospheric models.

B. Garfinkel: Regarding your statement, that the models are good for astronomical refraction but not for parallactic refraction, I am inclined to question the statement. If the model is good for infinity, it should be good for a distance which is less than that.

J. Saastamoinen: I stated that if the object is inside the atmosphere, the model is not good, because we don't use the upper part of the atmosphere, that compensates for the lower part.

B. Garfinkel: But you use the upper part if you have astronomical refraction for zenith distances large enough where the profiles make a difference.

J. Saastamoinen: But we know the refractive index only at the ground level and that it is 1 at the top level.

B. Garfinkel: But you also need the intermediate values if you have large zenith distances.

J. Saastamoinen: We don't usually know the values in between.

B. Garfinkel: But you assume, that you know what you need to be able to calculate the astronomical refraction. And the same information could be used to calculate the parallactic refraction just as well.

J. Saastamoinen: No, it can not. You see, in parallactic refraction the main term depends on the difference in barometric pressure between the two points, and this term may change quite a lot without any change in astronomical refraction.

B. Garfinkel: Well, I don't quite agree.

G. Teleki: I support completely Dr Saastamoinen's conclusions. It is a realistic conception. Your conception is idealistic. From Dr Kakkuri's investigations it is visible, that at higher zenith distances there are some differences from the theory. But it is a result of the tilting of the realistic atmosphere.

J. Saastamoinen: You see, the ground pressure at the sea level is quite constant. It only changes about 5% at the most. If you take the vertical differences of pressure, they change much more because the air is free to go up and down.

B. Garfinkel: But when you calculate astronomical refraction for large zenith distances and have no accurate data from the intermediate layers how can you get a reliable result? You have to make use of all information about the structure of the atmosphere.

J. Saastamoinen: We don't have at all the information we need in the case of parallactic refraction. But for astronomical refraction we have reasonable information.

B. Garfinkel: For large zenith distances?

J. Saastamoinen: Not for large.

B. Garfinkel: This is just the point. If you work with small zenith distances then you have a different situation.

J. Saastamoinen: Yes, but for parallactic refraction we don't have the information even for small zenith distances.

J.A. Hughes: Regardless of the model atmosphere which is used, be it good or bad, it can be applied to any kind of refraction. Parallactic refraction however, involves geometric considerations which depend upon where the object and observer are. One may have one model for parallactic refraction and one for astronomical refraction, but the essential difference is a geometric one, it seems to me.

J. Saastamoinen: Well, the thing is that for astronomical refraction, if the model is bad, the result is not that bad because we measure through the whole atmosphere whose refractive index at the upper boundary is necessarily 1. And we also know from ground pressure the total weight of the column. But if we stop somewhere inside at a point we don't know anything about, we need some observations made at that point.

J.A. Hughes: I agree that in this case a correct model is more important.

D.G. Currie: I think that one relationship that Dr Garfinkel was mentioning, if you have a model which has purely horizontal layers, then the only information you obtain for the distribution of layers are, at least in the astronomical language, very extreme zenith distances where there is not a lot of data. With the parallactic refraction you are within these layers. Neglecting the tilting of the layers, you can have a good number for the astronomical, and a weak determination of that distribution of layers will cause problems if you are asking what happens when you are half way down to the layer.

J.A. Hughes: We don't have to discuss parallactic refraction in terms of the atmospheric structure only. The moon has a parallactic refraction which can amount to a couple of arc seconds from geometry alone.

J. Saastamoinen: In that case, the question is extremely simple.

J.A. Hughes: Yes, of course, as you say, with objects within the atmosphere you have an especially difficult case.

B. Garfinkel: The polytropic model that I assumed in my theory provides information for the calculation of the astronomical refraction, as well as parallactic refraction. The question of whether or not the model is sufficiently accurate can be settled only by observational checks of the theory. I recommend that such checks be carried out.

J. Saastamoinen: The main correction term in astronomical refraction depends only on the ground pressure and the angle.

EDM PANEL ON INSTRUMENTS AND ATMOSPHERICAL CORRECTIONS

I. Brook (chairman)
D.G. Currie, A.H. Dodson, K. Poder

I. Brook: The aim of this session is to discuss EDM instruments which do not require the normal type of refraction corrections. The Georan is such an instrument and there are, of course, others, often still at the prototype stage, which also utilize two-colour techniques. I think we can even stretch us to including the Mekometer ME 3000 in this instrument category even though it is a single-colour instrument. We in Sweden use Mekometer equipment; and only have a theoretical knowledge of two-colour systems. Very few instruments seem to advance beyond the prototype stage and I think it is a question of nonprogress in many ways. We have read about many interesting developments such as dispersion studies done by Prilepin, Bender-Owen, Wood and Thompson. In the early 70's the National Bureau of Standards in Colorado produced a prototype two-colour instrument. The problem at that time was the size of the components and difficulties associated with suitable blue light sources. The equipment was not field equipment. At the 1974 Stockholm symposium a paper was presented by Hugget and Slater, where they described the equipment they were working on. I have been told that their instrument the Terrameter will in fact be produced commercially. We had two papers prepared by Bradsell and Shipley on the Georan 1 and the Georan 2, and were told that the instruments would be on the market within a few months after the Stockholm symposium, but as far as I know the instrument is not yet available. At the Land Survey of Sweden we have a ME 3000, and my experience of the instrument is not wholly positive. We have seen progress in developing better light sources, which can be used in practical field equipment. But we have seen also a change in the economic climate and escalating costs. The cost of a two-colour instrument today is of the order of a quarter of a million dollars. I think we should discuss to what extent we require and can afford two-colour equipment as well as the problems associated with calibrating and checking it. Dr Currie, what are your impressions of developments in this field?

D.G. Currie: From recent discussions it would appear, that instruments are being produced for the U S government. It is clear that the size and type of the market will determine prices and what manufacturers would aim for, as far as requirements are concerned. My impression was that it was not clearly defined, if the potential market was for ten units or one hundred units over the next few years. And that is something that will clearly affect prices, and it would seem to me that these kind of discussions will be valuable for potential manufacturers. I am not familiar with the details of what Hugget is planning, but that kind of information will be a very large factor in what would appear on the market. Another aspect is the question of when one needs one tenth of a millimeter and where one is going to be able to make use of such data.

I. Brook: I think Dr Dodson could comment on practical applications and requirements in this field. He has worked on deformation research. The question is do we need one millimeter, half a millimeter or a hundredth of a millimeter?

A.H. Dodson: On the shorter ranges instruments there is first of all the practical question of whether the two-colour instruments can solve the refraction problem because of the small amount of dispersion that is available. I think we have to look at what sort of accuracy we are aiming at. At the moment we can obtain 1-3 millimeter accuracy fairly easily in small engineering networks. This involves, however, a lot of theodolite work, and if we get more accurate EDM instruments we can cut out a lot of angular observations. But I think the problem of refraction, as far as engineering size work is concerned, is mainly what we have been discussing during the previous few days, that is, the vertical refraction problem. I do not think elimination of refraction in EDM instruments is so important to us, unless people are looking for higher accuracies than we at the moment think are necessary in the horizontal plane. The Mekometer we know was launched as being a 1 or 2 ppm instrument. My experience and Brook's experiences are that it is not that accurate. You can get differential measurements which are certainly that accurate, but the absolute accuracy is not as good. Do we need some further development along the lines of the Mekometer in order to produce the 1 ppm instrument without going to two colours? I think two-colour is out of the question for the short range instruments. But there is another aspect to this discussion. What sort of accuracy are we looking for at this shorter range? And should we be looking to improve the instruments or the techniques for eliminating the errors mathematically? Should we be looking in the latter direction rather than trying to improve instruments further, which undoubtedly is a very expensive business these days?

K. Poder: Up to now we have been speaking about the short ranges. If we consider long lines in the range from 40 to 80 km, then undoubtedly an 0.1 ppm instrument, which is technically feasible, with two colours should give you an accuracy of 4 to 8 mm. But then, of course, the geodetic large-scale community would have to deal with variation in time of the coordinated stations. This is one negative aspect. Then you have the curvature problems. This term will in most cases reach a magnitude of up to 0.5 m when you get up to these ranges. This means that the curvature, and here you will call it a large scale curvature, will possibly be the limiting factor for two-colour instruments on long ranges where it is very tempting to use them to get rid of the primary effect of the refraction. You are then back to the classical problem in all distance measurements, namely your metric unit will curve. You may hope to correct it slightly statistically, but the fact remains that for long ranges the main problem will be one of curvature effects. If you go to shorter ranges - here I am speaking as a geodesist interested in that scale of magnitude - then you could make networks with elements of 2 to 10 km. Most experiments, and also theory, indicates that there will be no need for two-colour instruments. The already well established techniques will certainly give you an accuracy of at least 1 ppm. There is one point more, namely the problem of the index error. Most instruments have troubles in the region from 1 mm to 1-2 cm, and I am putting the provocative question: what are you going to do with this index error? I was told that it was only a matter of proper technology, and I would, therefore, like some of the two-colour instrument makers just to join the teams making one-colour instruments and have them produce a more stable and smaller index correction.

T.J. Kukkamäki: · I think that with Mekometers we get somewhat better accuracy than 1 ppm at distances of between some hundred meters and one kilometer. Even without two colours it is not difficult to determine the effect of the refraction when the lightbeam is going close to the ground. These kinds of measurements are, of course, very important for engineering and geophysical work. But then for these longer distances between 40 and 70 km it is not so easy to put the thermometers in the beam path. But with the two-colour method it should be easier to get an accuracy of 1 ppm or even better. And as regards curvature - my theory is not too strong - but I think it should be possible to determine from two-colour results the curvature with sufficient accuracy to compute the necessary corrections. Then one might ask why we need this high accuracies on these 40 km distances. We need such accuracies very urgently in areas of crustal movement.

P.V. Angus-Leppan: To me the Mekometer seems to be a slightly mixed up instrument. It has extremely high specifications and I think that the accuracy, specified, can be achieved, but only if one uses it as a normal instrument, that is applying an atmospheric correction. On the long range instruments, my feeling is that we will not have problems with curvature. In fact, the rather elegant equation set up by Moritz will solve that, even with a fairly crude model, even if the curvature is changing along the path. The model is obviously not accurate enough to give you the first velocity correction, but second velocity in the curvature correction can be worked out to sufficient accuracy with atmospheric models. What I think we could do constructively, and perhaps this conference could think about is: if we could draw up a series of specifications for an instrument, so that everybody would be satisfied with one instrument. Then we wouldn't get the various manufacturers competing with each other, each producing a small number of expensive units.

L. Hradilek: I agree with professor Kukkamäki. The use of accurate instrumentations for the determinations of the earth's crustal movements is of very great importance. Such movements are of the magnitude of 5 mm a year and having such a precise ranging instrument we won't need to measure vertical angles for estimating the movement of mountain peaks. When the distances are inclined about 20-30 degrees, the determination of elevation differences - from the geometrical standpoint - is better than by other methods. For this purpose it is very important to have this instrumentation.

J. Milewski: I think that the two-colour instrument represents the future only for measurement of distances up to about 50 km, because at these lengths a very accurate reduction of the optical path is correlated with the accuracy of the length. This is very difficult because of the generally limited knowledge of the geometrical path, as we can only model the real path from the averaging coefficients. If these reductions can be made then we can make accurate measurements with two-colour equipment. For studies of crustal movements over lines of 20-30 km the two-colour instruments will probably prove to be of great importance.

K. Poder: The only long range two-colour measurements I have heard of, where we really get out to 80 km, were made on Hawaii many years ago. And the blue laser was of the size of a middle size field haubitser. So technically it is very difficult for a two-colour instrument to reach such distances. There is another purely theoretical problem, namely when you have a long range you get a separation between the red and the blue, which means what you are aiming at you do not really get, because the two waves will propagate in different atmospheric layers. Speaking of measurements of short lines, you can either measure a long line as the sum of small elements, or you can try to observe it directly and observe the refractive index along the line. So what you do if you break it up, is to get the index distribution along the broken line. For the extra effort involved you can spread your observations over a larger time and get a better randomizing. At the Helsinki symposium, I was very happy that we supported the short range instrument (2-3 km) concept and wanted them to be improved. And I still think this is the correct approach.

T.J. Kukkamäki: I cannot agree. It is correct to measure these sections, which are individually very accurate. But there is a problem. How can we project these individual sections to the chord? We need very accurate break angles, which are hard to obtain. For instance, when we determined the 900 km long traverse through Finland, the individual sections were not difficult to determine with Geodimeters. The main problem was to determine the angles.

I. Brook: May I also comment, when Froome produced the Mekometer he was talking about 3 ppm. We have had the use of four Mekometers and we were looking for an accuracy of 1 ppm, but ran into problems. The first instrument had a 14 ppm frequency error. The second one had a 10 ppm

frequency error. The third one did not work properly. And the fourth
one we are testing now. One must, of course, accept frequency drift,
as frequency checks must be part of an EDM routine. What worries us is
frequency instability. I have been in touch with Kern and explained
that we have difficulties because we cannot get an agreement with our
measured interferometer lines. According to Kern, the inconsistency
must depend on the interferometer. So we have borrowed a new interferometer and are checking the indoor calibration baseline again. But
I think, if you speak to Alan Dodson, who has worked very much with
Mekometers, he will tell you that he is of the same impression as I am.
I spoke to one of Kern's applications engineers when we had a one day
symposium on the ME 3000, and his reaction was that we are not really
sensible people, we geodesists, when we talk about absolute measurements, because it is not possible to measure an absolute distance with
any instrument. With the Mekometer one should not at all discuss
measuring absolute distances, you should only talk about measuring
differences. So Kern themselves do not appear wholly to share your
opinion about an absolute accuracy of 1 ppm.

T.J. Kukkamäki: When we purchased our Mekometer we calibrated it very
carefully against our calibration line. The length of that calibration
line cannot be absolute, not at all, but its accuracy of 0.1 ppm is
enough for the calibration of the Mekometer.

I. Brook: We have, in fact, calibrated the Mekometer on a 140 m baseline, and on a 50 m interferometrically measured baseline. We checked
periodical errors. I am not quite sure how you at the Geodetic Institute
check the modulation frequency. The instrument works with a very high
frequency and to check it one needs relatively complicated electronic
equipments. I would also like to comment on the calibration of two-
colour instruments. How shall we calibrate these instruments? If we
shall come down to 0.1 mm, we must have standards which are a power of
ten better than the actual instrument.

T.J. Kukkamäki: We had standard frequencies and calibrated with those.
But that was only the partial calibration. The total calibration was
against the calibration line. And we have a calibration line of half
a kilometer with the accuracy ±0.05 mm, that is 50 µm accuracy.

K. Poder: I would like to comment on the earlier discussion. The projection effect is such that for a typical line you will lose a maximum of about 10% of the accuracy on the projection. Normally, the loss
will be only a few percent.

A.H. Dodson: I would like to come back to the Mekometer for a while.
I agree with both professor Kukkamäki and Ian Brook. We found with
certainly more than one Mekometer, that the variations of the instruments are quite large and at times quite alarming. With full calibration: electronic calibration, baseline calibration, periodic calibration, and meteorological data along the line, we can get 1 ppm accuracy
over short lines. But it is not an ideal instrument, if you are looking

for one to do away with refraction. The problems of calibration are also quite considerable and to get a calibration better than 0.5 ppm is certainly not easy. Frequency variations can be quite large and we found day to day variations. So you have to make the calibration immediately before the observations and then immediately afterwards. And there is another factor. I think that the cavity is the main problem with the Mekometer. Certainly it could have been placed in a 5 mm engineering instrument, which would have done away with the need for refraction corrections; but to put it in a 0.1 mm resolution instrument was, I think, unwise.

D.G. Currie: Concerning the remarks by Kukkamäki and Poder: Since there are going to be at least one and probably several of the two-colour devices, would it be worth suggesting a study of data from them, which could permit a quantative answer to most of the discussion, which to a certain extent have been intuitive. I am sure instrument constructors would be interested in such requirements as Angus-Leppan has said earlier.

H. Kahmen: In Karlsruhe we have used Mekometers for several years, and I think we have had the same experiences as professor Kukkamäki. We measured many engineering networks of about 6 or 7 hundred meters and when we determined the coordinates, the mean square errors were always less than 1 mm.

A.H. Dodson: We have also had similar experiences, but your values are a measure of the internal consistency of the instrument. We found a very good internal consistency over short periods, but the absolute accuracy - unless we carefully calibrate frequencies - can be much worse. Certainly a calibration on an accurate baseline will show if we get an accuracy of 1 ppm or not. But the absolute accuracy is much more difficult to determine. Professor Kukkamäki's baseline will give him an absolute measure against the Väisälä comparator, and he is getting a very good agreement there, but both in Brook's and my own case it does not agree with the laser interferometer, that is, different laser interferometers. So somewhere there is something wrong. Perhaps in Germany and Finland you get better Mekometers than those available to us.

T.J. Kukkamäki: Maybe you have not been careful enough when using your instruments?

A.H. Dodson: It is quite possible.

D.G. Currie: Did you say that you used your Mekometer to measure your baseline, or did you use your baseline to calibrate the frequency of the Mekometer?

T.J. Kukkamäki: We used our baseline to calibrate the Mekometer. At first we calibrated our Mekometer for periodical errors and the frequencies, but we considered this only as a partial calibration. We need some total calibration so that we can practically use our instrument. And when we got, with these different distances, consistent results, then we were satisfied. The accuracy was better than 1 ppm.

D.G. Currie: But are you not then doing with your absolute baselines the same as what they did with their electronic facilities? Are you not calibrating your baseline on your half kilometer and carrying that to the other network? If you adjust the measure, according to what your baseline was, are you not doing the same as they did by adjusting the Mekometer to what their laboratory frequency standard said?

T.J. Kukkamäki: Yes, yes.

J. Kakkuri: May I add a little to this contribution by professor Kukkamäki. This Mekometer was studied on the Nummela standard baseline. The length of the baseline is 864 m, and it is accurate to 1 to 15 millions. On the baseline there are shorter lines also, 24 m, 64 m, 216 m and so on, and those distances were measured with the Mekometer and the differences between Mekometer determinations and the real baseline distances were compared with each other. And the agreement was better than 1 to 2 millions and in some cases 1 to 4 millions.

D.G. Currie: So you say you did not need to calibrate the Mekometer?

H. Kahmen: In Karlsruhe we repeated the measurements during several periods of the year, and the differences between the absolute and the Mekometer results were not greater than the mean square errors of the coordinates. In Karlsruhe we have done much work in connection with calibrating the frequencies. We have developed a special instrument for calibrating the frequencies. But I must agree with you, that high quality coordinates can only be achieved after very accurate calibration of the frequencies, immediately before and immediately after the measuring periods.

I. Brook: The major part of the discussions have dealt with the Mekometer, due, I suppose, to the fact that it is an instrument that we have seen or used. But few of us have had an opportunity to see even a prototype two-colour equipment, although we are conversant with the theory of the construction. I would like to echo what Currie said: it would be very interesting if an evaluation of Hugget's or similar equipment could be carried out. And I think that in Europe we are very willing to assist in these studies. We know, that in Finland there are very high quality baselines of varying lengths. We must have some accepted reference standard if we are going to evaluate the equipment. I am sure that professor Kukkamäki is willing to make baseline facilities in Finland available for these evaluations. Is that not correct?

T.J. Kukkamäki: Before midsummer we had a meeting in Helsinki with the U S Defense Mapping Agency and it was agreed that they should come with their two-colour instrument to check it on the Nummela baseline and on the 22 km long baseline at Niinisalo. Whether it will be the Hugget instrument I am not sure.

SUMMARY OF THE DISCUSSION

1) Two-colour or three frequency EDM devices may be important at medium distances (up to 50 km) for several purposes, where high absolute accuracies are required. But they should be handy to use in the field and easy to calibrate.

2) Short range instruments for absolute measurements of the Mekometer type, are extremely valuable, but such instruments need further improvements as regards facilities for frequency control.

3) Instability in and the size of the index error of existing EDM equipments should be decreased as far as possible.

4) The alternatives a) sum of small legs, or b) direct total distance should be investigated as regards ultimate accuracy achievable.

LEVELLING REFRACTION RESEARCH, ITS PRESENT STATE AND FUTURE POSSIBILITIES

T.J.Kukkamäki
Finnish Geodetic Institute, Helsinki

When I had to start the Second Levelling of Finland, 43 years ago, after the first observing summer I made one important and happy finding. The vibration of the image in the levelling instrument made the levelling difficult in middle summer while in spring and autumn this trouble was not so bad. So I proposed that levelling work must be interrupted for the six weeks from the midsummer to the beginning of August and that was accepted. The fact that just these weeks are the best vacation time in Finland made our colleagues in the institute jealous, but it was not our fault, and we were happy.

Then I made one other finding which turned out to be rather stupid. The vibration of the rod image in levelling instrument's view was bad at noon but was the better the farther from the noon the observation was made. Even when the sun was below the horizon this image seemed to be clear and unmovable. So I proposed that we should construct fitting instrumentation for observations in dark period, i.e. some devices for lighting the rod scale and the levelling instrument itself. Before starting the practical measures in that direction, we made a lot of research into the microclimatic and also optical conditions in the first three meters of atmosphere above the ground, and I think, we got rather clear picture of the conditions there. As these details seem to be unclear to many scientists, even to those who have published about the levelling refraction, I should like tell something about our experience on the matter.

We knew from the numerous meteorological publications the general variation of the temperature in the three lowest meters of air. In night, because of Austrahlung, the ground is colder than the air just above it. Thus the temperature of the air increase with the height, at least up to three meters; the temperature gradient is positive. Soon after the sunrise the Einstrahlung surpasses the Austrahlung, and the temperature of the air is decreasing with the height; the vertical temperature gradient is negative. The absolute value of the gradient increases with the morning hours up to 13-15 hours in the afternoon. Since that the absolute value of the gradient decreases and reaches zero about at the sunset. In course of the time when the sun is below the horizon the positive gradient increases lowly until the rising sun turns the vertical temperature gradient to zero and to the negative values.

The absolute values of the vertical gradient are the greater the clearer the sky is on day or night.

According to the information given in the microclimatological text-books with the negative gradient the warmer, i.e. lighter, air near the ground is rising turbulently. It means that air bubbles, some meters large and differing about $1^{\circ}C$ from the surrounding air, are rising in the lowest air layers. This upwards moving in bubbles makes the vibration of the rod scale and one can compute with some theoretical assumptions that the vibration increases with the 1.5 power of the sight length and is for a sight lenght of 50 meter in average conditions about 1 mm. The amplitude depends strongly on the vertical gradient.

In order to verify this reasoning, we made observations series with sight lengths 25, 50, 75, 100, 125 and 150 m around the whole 24 hours. We got the exponent 1.68 instead of the theoretical 1.5 and the amplitude was 1.2 mm for 75 m sight length. The frequency of the vibration was for small gradient one in second and with great gradient 5 - 10 in a second.

Our observations agreed the theoretical values satisfactorily. The vibration allows relatively safe pointing to the center of the vibrating scale image unless the vibration is not too large. Strong vibration force the observer to shorten the sight length and further to stop the observation. It was proved that our decision to keep summer vacation of levelling observers just in the finest summer weather period was scientifically grounded.

But now, how it is with the night observations. In the night the vertical gradient is positive, i.e. the colder and heavier air layers are under the lighter ones and no turbulence appears. Under the open sky, however, a wind of some decimeters in second always appears. This causes a slow swaying of the air layers in periods on tens of seconds or even minutes. The image of rod scale under these conditions is moving up and down in the same slow periodes. The image looks sharp and beautiful and not moving, but when you make a new pointing after tens of seconds or minutes the reading might deviate from the earlier considerably.

In order to get some idea of the amount this movement we continued the above mentioned vibration observations in night time also. We found that the amplitude of the swaying increases with the second power of the sight length. The amplitude was with a temperature gradient of $+0^{\circ}.2/1m$ in average 1 mm for 75 m sight length.

So this result showed that the levelling in night, when the vertical gradient always is positive, is not possible with an accuracy requested for precise levelling.

The vibration is symmetrical in vertical and horizontal directions. It is strongly depending on the gradient and consequently on cloudiness. That cannot cause any systematical error. That only limits observing possibilities and makes the accidental error larger. Normal sight length in precise levelling is 40, 50 or 60 m. I cannot understand that e.g. in USA even 150 m sight length was accepted still in the instructions of 1929. When vibration makes observing difficult the sight length is to be shortened. When even 30 m is not short enough the observation is to be interrupted. I think that this understanding about the vibration is accepted generally.

The understanding of the slow swaying is not so selfevident, but fortunately our colleagues have not been so stupid as I was in the beginning of my work that had tried to observe when the gradient is positive.

Naturally the gradient is greater nearer to the ground. So the horizontal sight in levelling bends more in uphill sight where it comes nearer to the ground than in the downhill sight. In order to get some picture of this phenomenon and possibly to derive some quantitative correction we studied the text-books and all possible other sources. We found names like Lallemand, Hugershoff, Kohlmüller, de Graaf-Hunter, Cole and Bomford. All these and many others had made researches into the levelling refraction. They based their computation on different functions for temperature as

$$t = a + b \cdot \log(z+c)$$
$$t = a + b \cdot z^2$$
$$t = a + b \cdot z + c \cdot z^2$$

Most of these scientists were satisfied with qualitive results. Rather few direct observations were carried out about the temperature variation with the height in different hours of day and night.

Further we have noticed that in some cases rather strange apprehensions came in view. For instance one serious scientist said in discussion with me 40 years ago, that light is bending in morning upwards as the temperature is rising and in afternoon it is bending downwards as the temperature is going down. In an article of Bulletin géodésique 20 years ago it is said that soon after the noon when the turbulent mixing reaches its maximum the temperature gradient is zero and no bending of light beam appears.

In the state of this kind of uncertainty we decided to make our own observations of the temperature in air up to three meters and simultaneously to try to observe any possible deviations in levelled height differences.

At first we observed with thermocouple the temperature at 1/3, 1 and 3 meter, but we found soon that the second derivative of the temperature is not possible to determine from single set of temperature readings. One needs long recordings. Since that we have observed only one temperature difference between 0.5 and 2.5 meter. The temperature function

$$t = a + b \cdot z^c$$

is used. The linear factor in it is computed from the direct observation but the exponent c is computed from the long temperature recordings made by Best in South England. I derived the exponent c for different hours, different months and different latitudes. This is, however, too sophisticated, and my colleague Professor Hytönen has shown that one can use the average value $c = -0.1$ in all circumstances without causing any significant error.

In order to see the refraction effect in levelling as clearly as possible we thought it is better to use an extraordinary sight length e.g. 100 meter as then the effect should be four times larger than with 50 meter, and the effect is not covered by the other observing errors. So we levelled a slope 2 km long with 18 meter elevation difference in all hours of day and night, with an exception of period from 22 to 03. From the graph we see that the observed deviations agree the values computed from the simultaneous gradient observations beautifully. The deviations reached values over 10 mm with 100 meter sight length and 4 mm with the regular 50 meter sight length.

As the gradient observations and the used levelling refraction formulas gave such a consistent result, we concluded that though other errors cover the refraction effect at a single instrument station, it must give a result which has at least a correct direction on longer observation series as at one bench mark interval.

As further the observing of the temperature gradient with a resistance thermometer was easy causing no waisting of time or needing no additional personnel, since 1937 we have observed the gradient at each instrument station and computed levelling refraction correction for each instrument station and applied this correction in the reduction of the whole Second Levelling of Finland. In average this correction is 0.06 mm per 1m height difference, i.e. we get a hill, 1000 m high, because of systematic levelling refraction 60 mm too low.

Here I have described our efforts to determine and to correct the levelling refraction in Finland. In the years after the Second World War there have been several researches into the levelling refraction and I should like to mention here the names Brocks, Reissmann, Hase, Balasz, Entin, Strusinsky and Kneissl, who all have studied this interesting question. Yesterday we heard already important reports on theoretical or practical results and today shall hear more.

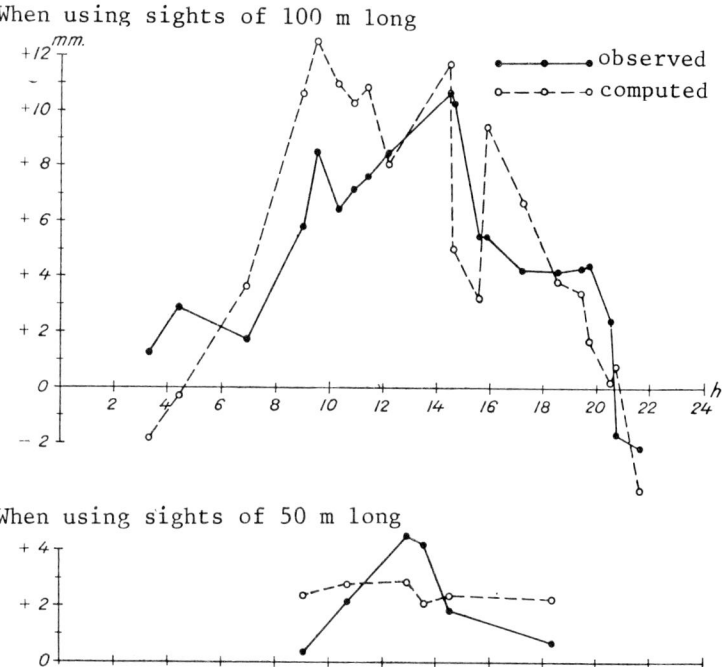

Fig. 1. Observed and computed values of the levelling refraction on the test line of 1.958 km long.

The purpose of this presentation of mine is to suggest some you will study the levelling refraction still more and more, as I think that this phenomenon is not yet unravelled satisfactorily. You may use classical methods but it would be still better to use electronics and electronic computors which have yielded so many magical solutions nowadays.

DISCUSSION

B. Garfinkel: I have two questions to ask. I was under the impression that in the Kukkamäki formula for the temperature profile, there are two important parameters, called b and c. c is the exponent. And furthermore the exponent was expected to be affected by the time of observation, particularly the time of the day, and also the season of the year. Now today you said that one can choose c as a fixed constant, something like -0.1. Now how do you reconsile these two opinions?

T.J. Kukkamäki: It is so: the first opinion was the old opinion of a young man, then this second one is the new opinion of an old man. I think that the last one is wiser than the first one.

B. Garfinkel: I am very pleased, it is now much simpler to make the calculations. I have, however, a second question. Suppose you want to measure these parameters b and c in the field at a particular time of observation. Could you use three temperature measurements at different heights, obtaining two differences and then from those differences calculate the two parameters b and c at the actual time of observation? Would that be a good plan?

T.J. Kukkamäki: We can do it, but we cannot get any dependable result from that. We cannot get c from those two differences, as it is changing so much that there is no sense to compute it. Already in the beginning of our work we saw this, and therefore we used recordings of Best and we determined the exponent c statistically. Later we saw that even that was not necessary, but we can use an average value for c.

B. Garfinkel: What do you do about the parameter b? Is that also a fixed number?

T.J. Kukkamäki: The parameter b must be determined along the measurements. After some minutes we will hear the presentation of Holdahl about a method to compute b for old measurements, where no gradient thermometers are used, and only information about cloudiness and direction of the slope is available. When you are making new levellings you have the possibility to make these direct measurements, please do that. It is easy, simple and not at all expensive.

E. Tengström: Professor Kukkamäki knows that there are plans going on to make a connection between the precise levelling of Finland and the corresponding of Sweden. And, of course, we have to cross the Gulf of Bothnia. You have, as I understand, done levelling through islands up to the lighthouse Märket. And we have carried out our levelling on to the lighthouse Understen. When I talked to Kakkuri about your levelling across the Finnish islands, he told me that this type of topography, that is precise levelling over water surfaces, is extremely nice to make, with very small errors in summertime. And he observed, that during this time it was good to be as near the surface of the water as possible, in order to have the turbulence effects not so big, because of the constancy of the temperature of the water and air. Now, you know we are talking about the two-colour method. The rest, that is the distance Understen on to Märket, cannot be hydrostatically levelled because of the ships.

T.J. Kukkamäki: And because of the lack of money!

E. Tengström: OK, can be. So we thought that the first thing we should do, after these experiments along our testbase in Uppsala, was to try trigonometrical levelling and correction for refraction over the remaining distance mentioned. So if we have our instruments rather near the surface of the water, which means much better conditions than we have here at the testbase topography, what accuracy shall we aim at in order to make a positive contribution to the connection between our two levelling nets? If the conditions are better than here, perhaps we may not be pessimistic in hoping to get an accuracy of $0".5$.

T.J. Kukkamäki: I can tell, that when we measured from the Finnish mainland to lighthouse Märket, we made about 100 watercrossings, and the accuracy was ±25 mm. So to make a significant contribution to the connection Märket - Understen, the accuracy must be something like ±25 mm, not essentially worse. The accuracy of a levelling around the Gulf of Bothnia is of that magnitude, also.

E. Tengström: The distance between the islands of Understen and Märket is 12 km. So I believe then it will be possible.

L. Hradilek: We are performing a spirit levelling in the valley of the High Tatras, with the elevation difference 700 m over 15 km distance. What accuracy is attainable when doing our best and applying all recent improvements?

T.J. Kukkamäki: You are making very careful work in Czechoslovakia, and your random precision must be high. But if the roads are gently sloping, and you are able to use 50 m sightlengths, the systematic levelling refraction can be rather dangerous. In our country, where slopes are very gentle, the systematic error caused by refraction is 0.06 mm per 1 m elevation difference on 1000 m high mountains a 60 mm levelling refraction is possible. So you can imagine what is the magnitude of that correction. We may expect a real accuracy of some centimeters.

THE DIRECT EXPERIMENTAL DETECTION OF THE SYSTEMATIC ERROR OF PRECISE LEVELLING

O. Remmer
Geodetic Institute,
Charlottenlund, Denmark

ABSTRACT

A method is described for determining one and only one parameter governing the refraction effect of levelling. The central idea is that this parameter is determined <u>not</u> by (temperature) measurements during the observation but by adding one more element in the Least Squares Adjustment of the levelling network.

Furthermore is described an experiment where it was found that this parameter differed significantly from zero and a tentative first value is suggested.

REFERENCES

Remmer, O.: 1977, "The Direct Experimental Detection of the Systematic Refraction Error of Precise Levelling", Geodaetisk Instituts Meddelelse no 52.

DISCUSSION

T.J. Kukkamäki: Dr Remmer mentioned that is is not possible to determine the difference δt. We know that the air temperature is vibrating, but it is possible to determine an average value of δt. You must not use too sensitive thermometers, but thermometers which are integrating for some seconds period. In the morning the observed temperature difference from 0.5 m to 2.5 m height is zero. During daytime it is something like $0°.5$ or $1°$, and then it is going down toward the evening. We can compute a correction. This correction is not exact for one instrument station, but for 1 km is more representative and for 1000 km still more real. In that way we get average correction, which improve our results.

O. Remmer: I don't think we disagree that much, professor Kukkamäki.

Of course, you should have integrations of temperature differences. It is a matter of experience of what you get when you measure things, what are the results when applying these corrections. What we have found is that there is indeed some kind of systematic error which may be attributed to refraction, but maybe there are also other sources. Because if you apply the method I advocate you get hold of all kinds of systematic errors which follow this law, giving proportionality to the square of the distance and direct proportionality to the height difference. Then, of course, you will get every other effect which have this kind of law.

T.J. Kukkamäki: Yes, I agree there are many other sources of errors which hide the refraction effect in regular levelling procedure. But when we exaggerate the refraction effect by using 100 m sightlength, as I described in my paper, we found an effect of 10 mm. The directly observed differences agree with the corrections computed from the temperature observations very well.

O. Remmer: I am relying on older Danish experiences. There has also been some small difference of opinion between the Danish Geodetic Institute and the Finnish on this point. Dr Simonsen was of the opinion that you shouldn't try to find the refraction correction. You should only measure at a time when it was as small as possible. Because he made some experiments where he couldn't find a number, which he could use for these corrections. So the work I have done is more or less a repetition of those. We do not believe so much in applying directly the corrections. We try to estimate them afterwards in some way.

B. Garfinkel: I have only a minor question here. At the distance of 100 m the refraction will contain much noise. And you suggested a distance of 1 km, and there you say that systematic errors are about twice as large as the noise. Now this kind of ratio is still not quite convincing. I suppose if you make it more than 1 km you will get better results.

O. Remmer: Well, this is not a sidelength of 1 km. It is a line composed by ten individual set ups of 100 m distance. It is just a slow accumulation of the systematic error. I mean, the systematic error accumulates by adding linearly, while the random errors accumulate according to the square root of the distance. I am not looking through 1 km of air. I am all the time looking through 50 m of air, and I am doing it ten times. It is only the order of magnitude I have reported.

P.V. Angus-Leppan: I agree with the author that individual measurements of temperatures and temperature gradients don't mean very much. One has to allow them to accumulate before they have any significance. It appears that you are assuming all the systematic errors are due to refraction.

O. Remmer: Yes!

E.G. Anderson: Variations in the temperature with time, particularly during short time intervals, have been mentioned, for instance by professor Kukkamäki. Surely there are also considerable variations in the temperature gradients horizontally along each line of sight. It seems that individual temperature measurements will not be capable of modelling these horizontal variations and we can only approach that problem with some kind of overall, averaged effect. This surely would support Dr Remmer's statistical approach. There seems to have been rather little attention paid to these horizontal variations. Professor Angus-Leppan has shown that this is also a considerable problem, mainly because of the variation of surface material along the line of sight.

O. Remmer: Well, I should like to comment that I have had some experience trying to do as the Finns. And as professor Kukkamäki said, we are more pessimistic in Denmark. We have not been able to get numbers which could be used. I have always had more intuitively than scientifically this feeling. I was measuring the wrong things. I was measuring temperatures which I really could not use.

T.J. Kukkamäki: When we are determining the bending of the light beam, we should measure the gradient on every mm along the line. That would be very good, but in practice not possible. So we measure the gradient only close to the instrument. That is not representative for the whole sightlength of this instrument station. But when we measure a 1000 km long line, we make 10.000 gradient measurements. I think all these measurements will give a rather good average value.

J. Saastamoinen: Maybe the Danish pessimism somehow depends upon the topography of Denmark. What is the height of the highest mountain in Denmark?

O. Remmer: I am not going to speak about high mountains in flat Denmark. Our experiences are based on 20 m height differences.

J. Saastamoinen: The point I wanted to make is the following. Maybe the effect of refraction is quite small even for 20 m height differences. So you are likely not to put too much weight on your results. Finland is also a flat country, but not that flat, and the corrections are much greater.

O. Remmer: I think neither Finland nor Denmark have been noticed for being very hilly. What we have been investigating is a special case. Of course, in a mountainous country - I am sure professor Kukkamäki and I agree - you get a very large error on top of the mountains. That was what professor Kukkamäki said to professor Hradilek. You get the maximum errors at the top of the mountains, of course.

T.J. Kukkamäki: I was just going to say, that after some minutes we will hear something about experiences in the U.S.A., where the gradient is much larger than in these Nordic countries.

REMOVAL OF REFRACTION ERRORS IN GEODETIC LEVELING

Sandford R. Holdahl
National Geodetic Survey
National Ocean Survey/NOAA
Rockville, Maryland 20782

INTRODUCTION

Refraction error is often regarded as the most serious problem with geodetic leveling. Its accumulation depends on the slope of the terrain being leveled, length of sight, and the vertical temperature gradient. Refraction error can be minimized by limiting and balancing sight lengths, and by not reading the portion of the level rod which is within 0.5 meter of the ground where air density changes most rapidly. The remaining refraction error should be removed by application of a correction to leveling data, otherwise heights and crustal motion may be determined so weakly that meaningful conclusions or interpretations cannot be inferred from the data by geophysicists.

A refraction correction for leveling was first developed by T. J. Kukkamaki (Kukkamaki 1937). Ironically, Finland is not one of the countries that most needs a remedy for refraction error. The vertical temperature gradients in Finland are not as large as those found to the south, terrain relief is not severe, and sight lengths used in Finland are limited to a distance, which is shorter than the limit used by most countries. However, Finland is noted for the high quality of its leveling. The leveling data are used to resolve geophysical problems and to estimate post-glacial uplift, and, therefore, must be very accurate. Finnish geodesists developed instrumentation to measure vertical temperature differences while leveling is underway, and gained confidence in the temperature function which is the basis for the correction.

Most countries have been reluctant to utilize the refraction correction. This is probably due to several factors: (1) extra instrumentation and computational effort are required, (2) disbelief in the idea that temperature in the lowest 3 meters can be represented by a single temperature function, and (3) belief that refraction error is small. The full effect of leveling refraction is normally not seen in misclosures between forward and backward levelings because it is usually common to both in approximately the same amount. Similarly, circuit misclosures do not reveal the total refraction error, consequently it is sometimes underestimated.

It has also been thought that only mountainous countries suffer from significant accumulations of refraction error; and then, only the heights of the mountain peaks would be in serious error.

A recent experiment was conducted in California by the author to determine whether refraction error in locations with low latitudes and arid climates might be considerably greater than in Finland or England. The experiment was performed in December when temperature gradients should be smallest. Observed vertical temperature differences averaged 4-10 times as large as those predicted by a table developed by A. C. Best (Best 1937). The Best table gives hourly values of Δt, for each month, between the heights of 30 cm and 120 cm, and is based on means of 2 years of continuous observation in England. The high Δt values observed in California are alarming because they indicate, provided the existing refraction correction is valid, that refraction error in California and probably the remainder of the United States is much greater than was assumed previously.

A further objective of the experiment was to see whether vertical temperature gradients were larger on south slopes (i.e. slopes facing down to the south) than on north slopes (i.e. slopes facing down to the north). Simultaneous observations on north and south slopes revealed that south slopes gave vertical temperature differences, between heights of 50 cm and 250 cm, that averaged .89° C higher than north slopes. This result is disconcerting because it means that refraction error accumulated while leveling up the south slope of a topographical feature will generally not be canceled adequately by refraction error of the opposite sign accumulated while leveling down the north side. In mid-latitudes, refraction error will systematically accumulate in the north-south direction as each topographical feature is traversed by leveling.

The vertical temperature difference, Δt, is an essential parameter in the refraction correction. Because present methods for estimating Δt do not consider the angle at which the sun's rays strike the ground, or consider regional and seasonal variation in cloud cover or turbidity of the atmosphere, the author suggests the use of solar radiation measurements as a basis for estimating Δt. The formulas in the latter section of this paper are used to compute the orientation and slope of the terrain surface relative to the sun's rays.

THE REFRACTION CORRECTION

The refraction correction, R, in mm, for a single setup of the instrument is given by:

$$R = -10^{-5} \cdot A \cdot \left(\frac{L}{50}\right)^2 \cdot \Delta t \cdot \Delta h \qquad (1)$$

A = a function tabulated by Kukkamki for each month, each hour and the respective geographic latitude, depending on the exponent c of the used temeprature function $t = a + bz^c$, the altitude z of the temperature measuring places, and the height of the instrument above the ground. Mean values for c were derived from temperature measurements by Best. A is frequently assumed constant, and equal to 70;

L = sighting distance at station in meters;

Δt = temperature difference $(t_2 - t_1)$ between the heights of the temperature measuring stations z_1 = 50 cm and z_2 = 250 cm in degrees celsius;

Δh = measured height difference in 0.5 cm units.

When the temperature difference, Δt, is observed, the correction is determined at each instrument station and accumulated over the whole section. This is the most rigorous application of the refraction correction.

When used to refine historical leveling data, the correction relies on estimation of Δt because air temperature measurements were traditionally not made at more than one height. The Δt is assumed to be constant for each setup of the instrument in the leveled section. Sight lengths and measured height differences are also assumed to be the same at each setup. Average values of L and Δh are calculated by an algorithm. The correction for the average setup is then multiplied by n, the number of setups, to obtain the correction for the whole section.

$$R = -10^{-5} \cdot A \cdot \left(\frac{\bar{L}}{50}\right)^2 \cdot \bar{\Delta t} \cdot \bar{\Delta h} \cdot n \qquad (2)$$

where $\bar{L}$, $\bar{\Delta t}$, and $\bar{\Delta h}$ are the average setup values for sight length, vertical temperature difference, and observed height difference. The number of setups, n, is available from the leveling records.

It can be seen from equation (1) that refraction error would be negligible on flat terrain or when cloud cover minimizes Δt. Conversely, on clear days, large refraction errors will accumulate when leveling on sloping terrain. It is also important to note that the refraction correction is proportional to the square of the sighting distance. This implies that the leveling specification, which defines the maximum sight length will have considerable influence on the amount of refraction error that may accumulate in leveling. The poorest conditions for leveling, from the viewpoint of refraction, will generally exist on clear days, at noon, when leveling with long sight lengths on a gentle slope.

In the temperature function used by Kukkamaki, given below,

$$t = a + bz^c \tag{3}$$

where, a and b are constants, and z is height above the ground, the coefficient c has been determined by harmonic analysis using two years of data obtained in England (latitude 51. 2) by A. C. Best.

$$c = f + g (T - 12^h), \tag{4}$$

where

$$f = -0.14 + 0.10 \sin\left(\frac{360°}{365} D - 61°\right), \tag{5}$$

and

$$g = +0.016 - 0.015 \sin\left(\frac{360°}{365} D + 95°\right) \tag{6}$$

D is the number of days elapsed since the beginning of the year, and T is the hour of observation.

Equation (4) is used, starting at 2.5 hours after sunrise and ending at 0.3 hour before sunset, to calculate the value A in equation (1):

$$A = \frac{5.95}{z_2^c - z_1^c} \left[\frac{1}{c+1} \left(z_1^{c+1} - z_2^{c+1} \right) + z_0^c \left(z_2 - z_1 \right) \right] \tag{7}$$

z_0, z_1, z_2 are heights of the instrument, the low temperature probe, and the high temperature probe, respectively.

EXPERIMENTATION

An experiment in California was motivated by concern that Δt values interpolated from the English data would be too small. Further, it was hypothesized that Δt values on slopes facing down to the south should be larger than those occurring simultaneously on nearby slopes which face down to the north. The results of the experiment showed that both of these concerns were justified.

The California experiment was conducted at several locations for two weeks in December 1977. Temperatures were measured at several heights (30, 50, 120, and 250 cm) above ground. During most of the experiments measurements were made only at heights of 50 cm and 250 cm which correspond to z_1 and z_2 of equation (7). In precise leveling, the sighting path will normally intercept the leveling rod between these two vertical limits.

The magnitude of the observed vertical temperature difference frequently exceeded 2° C on south slopes, and reached a maximum of 3° C on a clear day with light breeze. The maximum value that can be obtained for December from interpolation using the Best Table is 0.3° C. Throughout most of the experimentation observed values exceeded the corresponding tabular values by a factor ranging between 5 and 10.

On seven days of the experiment, two observers worked simultaneously on opposite sides of a mountain or valley. Comparison of Δt values, observed at identical times, showed that values from south slopes averaged 0.89° C higher. The significance of this result is that refraction can induce a systematic accumulation of error in the north-south direction. High refraction error accumulated while going up the south face of a mountain would be insufficiently canceled by lesser refraction error of opposite sign accumulated while leveling down the north face. This accumulation would be almost as great with rolling terrain as it would be for mountains. The sign of the accumulated error is such that it would cause north-directed leveling to yield heights which became increasingly too low as the survey progressed over several hills or mountains.

Figures 1 and 2 are computer plots that depict observed values of Δt, incident light (scale on right vertical axis), and recommended tabular values of Δt. The values plotted in figure 1 were observed on a slope which faced down to the south; the values in figure 2 were observed simultaneously on the opposite side of a small valley which sloped down to the north. The sun's rays struck the down-to-south slope much more directly and, therefore, produced steeper temperature gradients. A similar result was observed at several other locations as well. This implies that the method of estimating vertical Δt needs to be modified to reflect the incidence angle of the sun's rays.

Figures 1 and 2 also indicate that the tabular values (shown in figure 3 and developed in England by A. C. Best) are not representative for California, and are probably not representative for most of the United States. The depicted tabular values have been used in Finland with some success, but they are much too small for latitudes below 45° where cloud cover is not prevalent.

The development of an improved table of Δt values is a very important undertaking. Vertical crustal movements determined by analysis of old levelings will be influenced by its quality.

SOLAR RADIATION

The amount or quantity of the sun's radiation reaching a horizontal unit of the earth's surface depends upon a number of factors. These include the intensity of radiation emitted by the sun, astronomical considerations determining the position of the sun, and the transparency of the atmosphere.

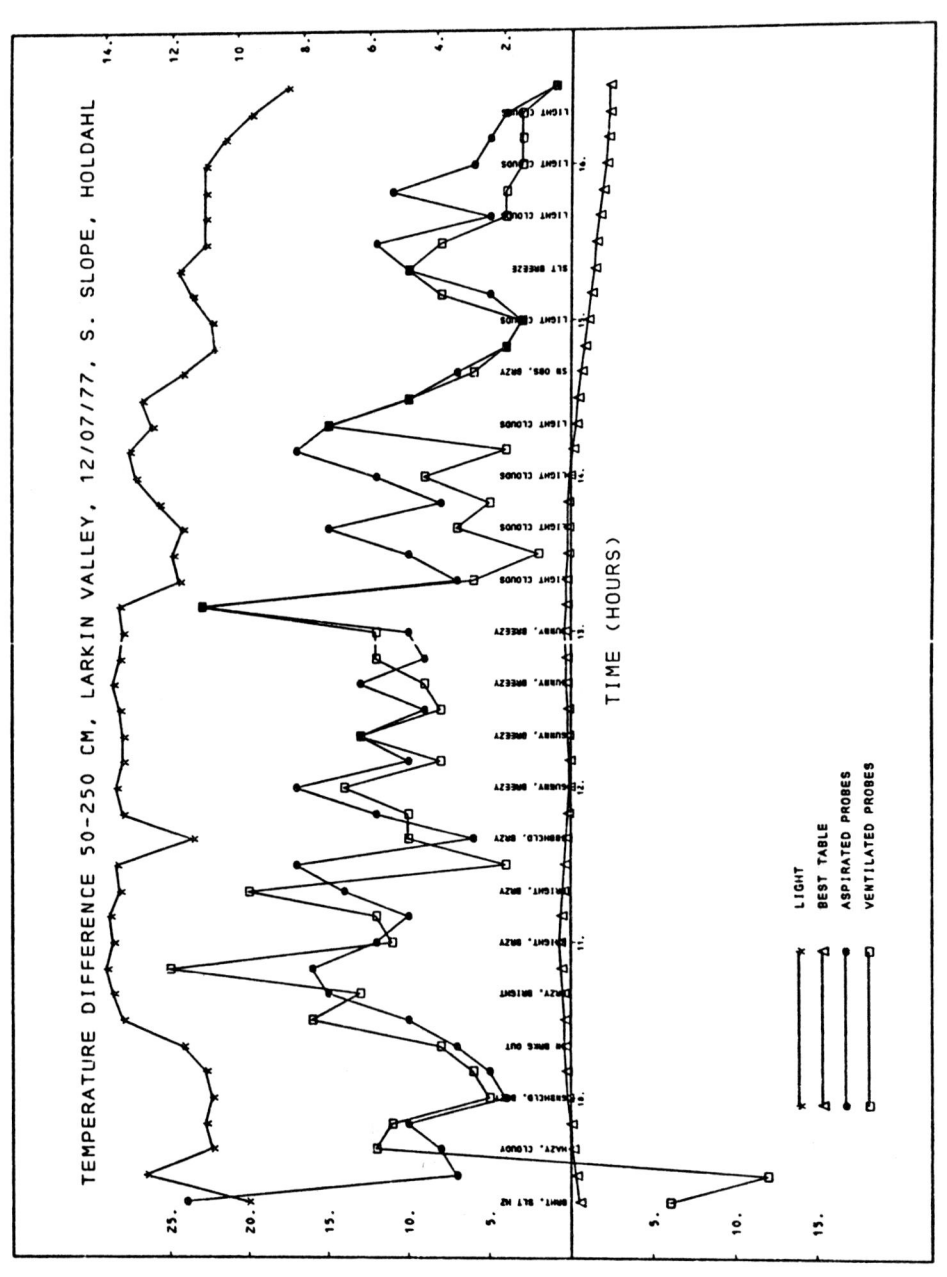

Figure 1

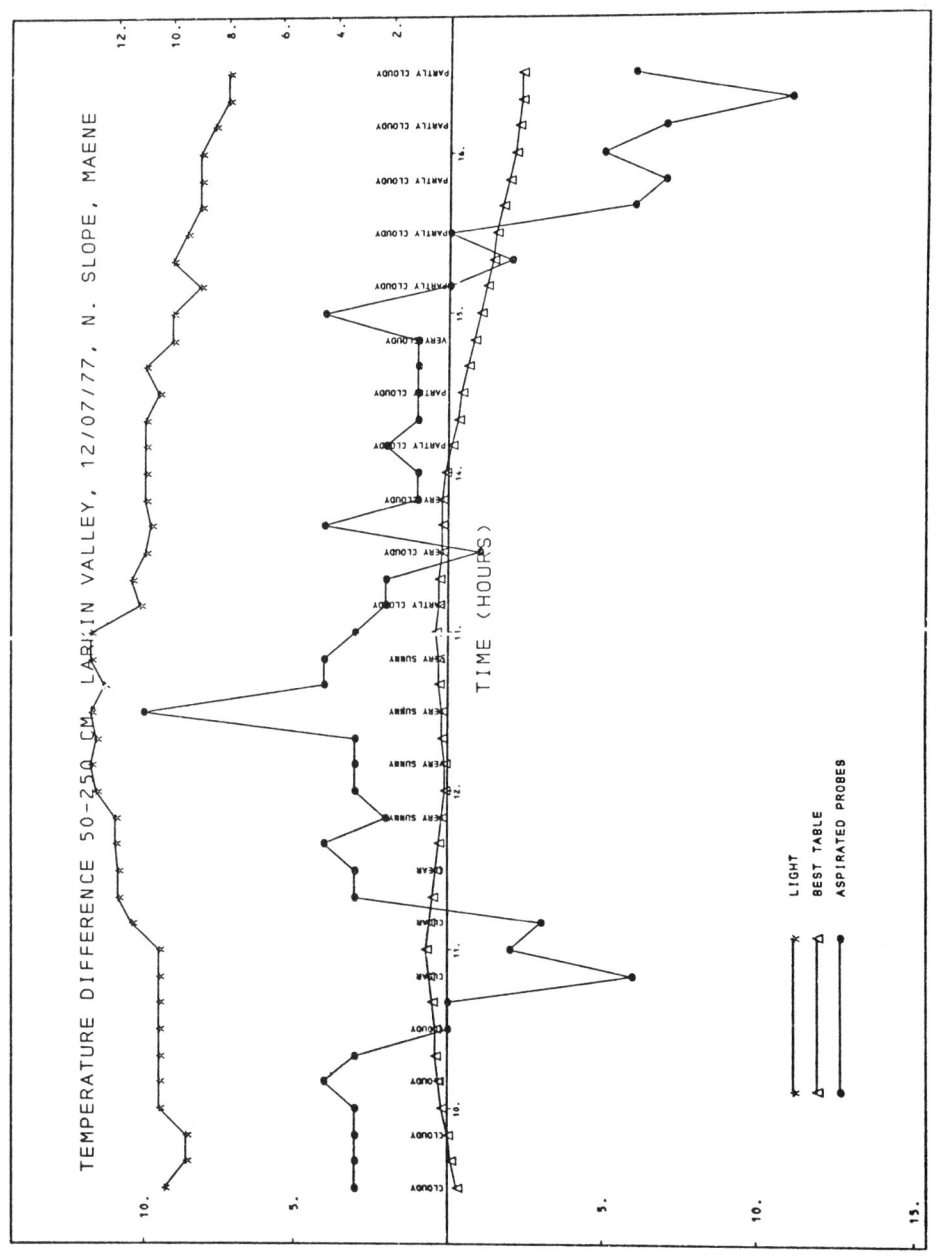

Figure 2

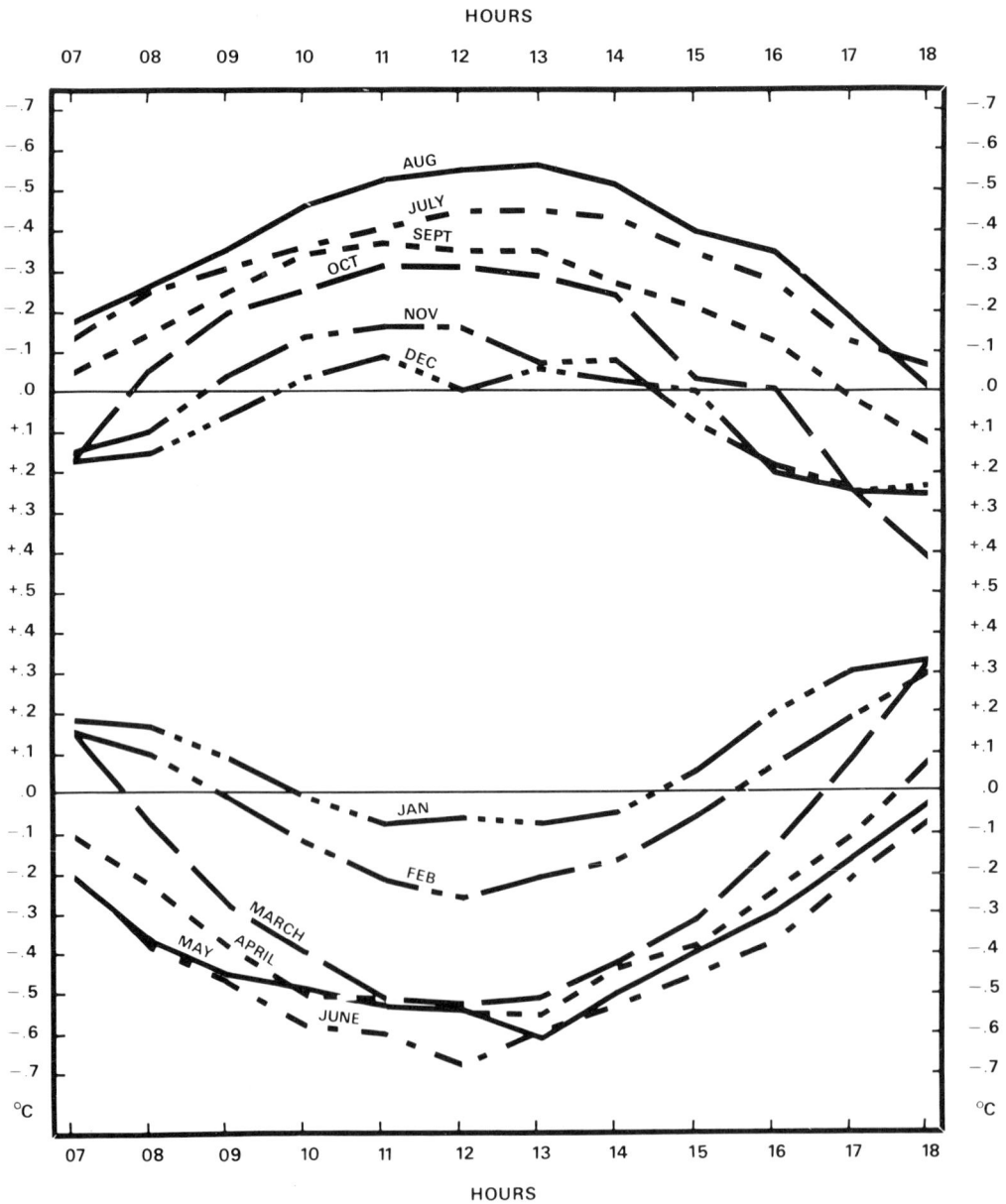

Figure 3

Vertical temperature gradients are influenced by the amount of solar radiation received by the ground. Vertical temperature gradients, therefore, are governed to a large degree by Lambert's Cosine Law which states "the intensity (of the sun's radiation beam) on the ground varies as the cosine of the angle between the normal to the terrain surface and the beam. This means that a solar beam spreads out on a slanted surface and, therefore, the energy received on a unit area of that surface must decrease as the incidence angle decreases from 90°.

The sun's azimuth and zenith distance also determine how much atmosphere a solar beam must pass through before reaching the ground. The atmospheric gases surrounding the earth absorb considerable portions of the direct solar beam.

For the above two reasons, the solar radiation received simultaneously at two different latitudes, or at two locations with the same latitude but different ground slopes, would not be the same. Consequently, any method of estimating values of the vertical Δt must consider latitude and ground slope. Latitude, longitude, height difference, distance leveled, and time have traditionally been recorded; and can be used to calculate incidence angle of the sun's rays, zenith distance or declination of the sun, and the azimuths of the sun and level line.

An attractive model for interpolating Δt, which considers latitude, clouds, and turbidity of the atmosphere can be developed from averages of actual measurements of solar radiation received at the earth's surface. The climatic records of the United States give mean daily totals of solar radiation for each month, determined at 192 stations distributed throughout the United States. Measurements have been made for as long as 46 years at one of the locations, and the average measurement history extends more than 20 years.

A mathematical fitting procedure can be used to find an expression for seasaonal and regional variations in solar radiation. Measurements of solar radiation from Canada may also be used to obtain a result which would be applicable throughout most of North America.

The fitting procedure can proceed by first expressing solar radiation as follows:

$$S(x,y,D) = P(x,y) + Q(x,y) \sin(2\pi D/365 - \pi/2) + R(x,y) \cos(2\pi D/365 - \pi/2) \qquad (8)$$

where D is the number of days since December 21. $P(x,y)$, $Q(x,y)$, and $R(x,y)$ are polynomials whose coefficients can be determined by a least squares fit to monthly means of observed daily totals of solar radiation.

It will be helpful to adopt the following notation:

S = mean daily total of solar radiation, measured between sunrise and sunset, which is incident on a horizontal surface;

S' = instantaneous solar radiation on a horizontal surface;

S'' = instantaneous solar radiation on an inclined surface;

S_{max} = maximum value of solar radiation on a given day

The mean daily total solar radiation, measured on a level surface, must be converted to instantaneous solar radiation on a surface that is usually inclined.

The plotted diurnal variation of solar radiation, between sunrise and sunset has the shape of an inverted parabola, peaking at noon, and zero at approximately one-half hour after sunrise and before sunset (see figure 4 below).

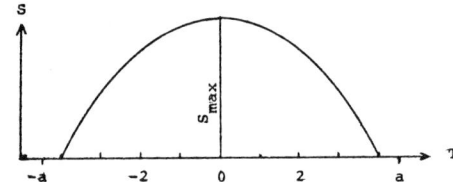

$$S' = S_{max} - \frac{T^2 S_{max}}{(a - .5)^2} \qquad (9)$$

T is the hour angle of the sun; and, a is half the number of hours between sunrise and sunset.

The value, a, can be calculated according to

$$a = \frac{1}{15} \cos^{-1}(-\tan \delta \tan \phi) \qquad (10)$$

where δ is the declination of the sun, and ϕ is the latitude of the location where releveling is being done.

We can integrate (9) to obtain the daily total of energy received, and get

$$S = \int_{-a+.5}^{a-.5} \left(S_{max} - \frac{T^2 S_{max}}{(a-.5)^2} \right) dt = \frac{4(a-.5)S_{max}}{3} \qquad (11)$$

Rearranging,

$$S_{max} = \frac{3S}{4(a - .5)} \qquad (12)$$

Substituting (12) into (9) allows us to calculate S', the instantaneous solar radiation on a level surface:

$$S' = \frac{3S}{4(a - .5)} \left[1 - \left(\frac{T}{a - .5}\right)^2 \right] \qquad (13)$$

The following equation uses a variation of Lambert's Cosine Law to convert the instantaneous solar radiation, S', on a level surface to solar radiation, S", received on an inclined surface:

$$S'' = \frac{S' \sin B_1}{\sin B_0} \qquad (14)$$

B_0 is the incidence angle between the sun's rays and a level surface, and B_1 is the incidence angle between the sun's rays and the ground surface.

$$B_0 = 90° - \gamma \qquad (15)$$

where γ is the zenith distance of the sun, and

$$\sin B_1 = \cos \gamma \cos |\alpha| + \sin \gamma \sin |\alpha| \cos (DAZ)$$
$$\text{if } \alpha \leq 0, \, DAZ = A^* - A'$$
$$\text{if } \alpha > 0, \, DAZ = A^* - A' - \pi \qquad (16)$$

where α is the slope of the terrain surface, and is given by

$$\alpha = \tan^{-1} \frac{(\Delta h)}{2L} \qquad (17)$$

Δh is the observed height different, and L is the sight length. A^* and A' are the azimuths of the sun and level line, respectively.

THE VERTICAL TEMPERATURE PROFILE

A vertical temperature profile can be constructed from a knowledge of solar radiation by first converting solar radiation to net radiation, S_n, using the following expression (Rosenberg 1974):

$$S_n = 0.85 \, S'' - 0.14 \qquad (18)$$

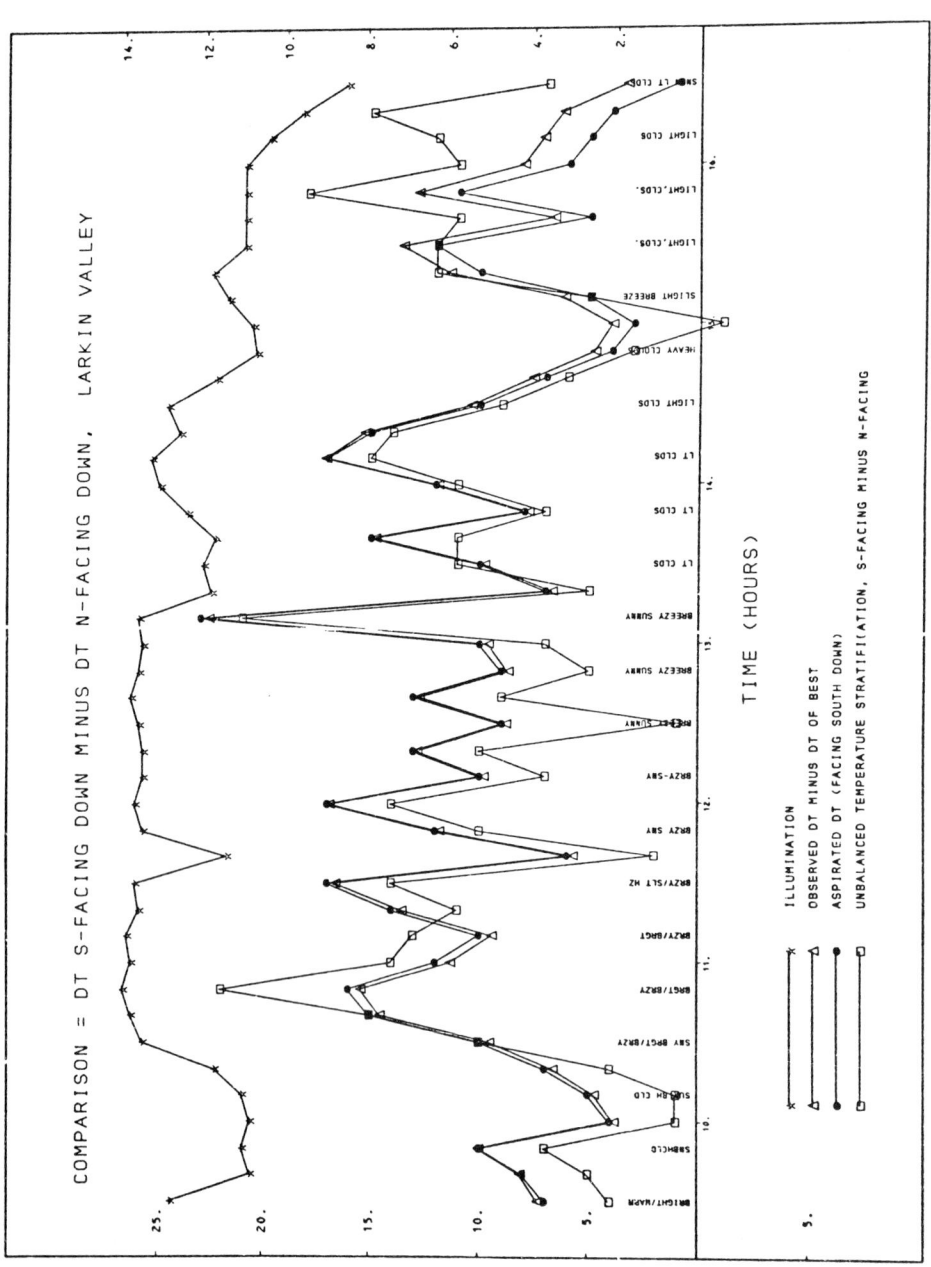

Figure 5

Net radiation combines with heat flux into the ground, G, to yield upward sensible heat flux, H:

$$H = (S_n - G) - \lambda E \qquad (19)$$

where λE is the evaporation flux with E the evaporation rate and λ the latent heat of vaporization of water.

Heat flux into the ground is estimated using the following equation:

$$G = A_0 K_0 \sin(wt + \pi/4) \qquad (20)$$

A_0 is the amplitude of the daily surface temperature, K_0 is the thermal conductivity of the soil, and t is the time in hours from the time of maximum temperature, and $w = 2\pi/24$, the period of the cycle being 24 hours.

The temperature T_{h_1} at height h_1 can be calculated with the following equation:

$$T_{h_1} = T_h + 3\left[\frac{H^2 T_h}{(C_p \rho)^2 g}\right]^{1/3} (h_1^{-1/3} - h^{-1/3}) - .0098(h_1 - h) \qquad (21)$$

where C_p is the specific heat of air at constant pressure
  ρ is the density of air ($C_p \rho = 1200$)
  h is the height at which T_h is measured
  g is the acceleration due to gravity, m/sec^2

To obtain the temperature difference between two heights equation (21) is applied twice to obtain:

$$\Delta t = T_{h_2} - T_{h_1} = 3\left[\frac{H^2 T_h}{(C_p \rho)^2 g}\right]^{1/3} (h_2^{-1/3} - h_1^{-1/3}) - .0098(h_2 - h_1) \qquad (22)$$

In the above two equations T_h is the air temperature in °K and can be obtained from the old leveling records where it was needed to correct for expansion or contraction to the graduation invar strips of the level rods. The height, h, would ordinarily be 1.5 meters.

Equations (19), (20), and (21) were presented with more detailed discussion by Webb (1969), and later by Angus-Leppan (1970; 1971) and Angus-Leppan and Webb (1971). The equations were suggested for reducing EDM measurements, being applicable for heights from less than a meter above the ground, up to tens of meters. Consequently, they should also be suitable for estimating Δt for input to the refraction correction for leveling.

If we ignore for the moment the need to know the evaporation flux, λE, in equation (19), we can say that a method for estimating the vertical temperature difference, Δt, is possible even when only one air temperature, T_h, has been measured. However, λE can be very important in lower latitudes where there is normal or above average precipiation. It is less important at higher latitudes or in arid climates. In moist areas, λE can range between 25-50% of s_n, and range between 5-25% of s_n in arid regions. It may be possible to model λE using weather data obtained over a period of years, as has been done with solar radiation data. The author has not yet explored this and other possibilities.

As of this writing, the above method of estimating Δt has not been tested. The described method is regarded as tentative and is being presented for discussion. Testing will begin in the coming months.

REFERENCES

Angus-Leppan, P.V.: 1970, *Proc. Conf. on Densification of Geodetic Networks*, Budapest.
Angus-Leppan, P.V.: 1971, *Commonwealth Survey Officers Conference*, Paper No. 85, 9.
Angus-Leppan, P.V. and Webb, E.K.: 1971, *General Assembly IUGG*, Moscow, Section 1, 15.
Best, A.C.: 1935, *Geophysical Memoirs No. 65*, London.
Fraser, C.S.: 1977, *Uniserv G 27* (1977), p. 42-51, Univ. NSW, Sydney.
Geiger, R.: 1975, *Harvard University Press*.
Hytonen, E.: 1967, *Pub. of the Finnish Geodetic Inst.*, No. 63, Helsinki.
Kukkamaki, T.J.: 1936, *Verh. d. 9. Tagung d. Balt. Geod. Komm.*, Helsinki and Helsinki 1937.
Rosenberg, N.J.: 1974, *John Wiley & Sons*, New York.
Webb, E.K.: 1969, *Proc. of REF-EDM Conf.*, Univ. NSW, Sydney (1968), 1-9.

DISCUSSION

D.G. Currie: Have you compared this effect of actual lumination on the data? It would appear, that your 0.89 degree difference may have been explained by the fact that you had a different lumination on your north and south slope, and therefore have become a direct confirmation of the other things you were talking about.

S.R. Holdahl: The 0.89 degree difference is well explained by Lambert's Cosine Law. The radiation on south slopes is more intense because the ground is more normal to the sun's rays at latitudes higher than $23°$.

O. Remmer: I only have a simple question. I am planning to make test measurements in small networks, where I can use your formula for significantly reducing the mean errors. Do you think it is worth while to do that?

S.R. Holdahl: No. I suggest that you measure Δt and plug it into Kukkamäki's formula for leveling refraction. My method of predicting Δt should be used for removing the influence of refraction from old leveling measurements obtained without measuring Δt.

J.C. de Munck: In the United States there is a systematic difference between the levellings near sea-level and those on land. Can the mentioned effect be an explanation for that?

S.R. Holdahl: Unfortunately, correcting for differing amounts of radiation on north and south slopes will make a larger disagreement between the older levelings and the oceanographically determined sea slopes. The most recent coastal leveling in California would be improved by it.

P.V. Angus-Leppan: This is an interesting approach, which could be used for future levelling as well. Do you think, that you should measure the actual temperature differences? Would it not be better to use the heat balance approach, in which case the measurements would be qualitative ones, e.g. cloudiness type of surface, moisture, and perhaps wind?

S.R. Holdahl: I don't think that modeling of Δt could be as reliable as measuring it, but it would be more convenient. It would be necessary to record all the parameters you just mentioned before the modeled Δt is comparable, and it is probably easier to just measure Δt.

T.J. Kukkamäki: Remmer asked, whether you are making some test measurements. I understand that you can test your method easily. You have the direct measurements of temperature differences, and you have recordings of cloudiness, and you can compare your theoretical values with the directly observed values. If you see - and that's an answer to Angus-Leppan - that the theoretical computation gives good results enough, then you stop your observations as unnecessary ones.

S.R. Holdahl: Exactly!

REFRACTIONAL EFFECTS IN PHOTOGRAMMETRY

J Larsson
Department of Photogrammetry
Royal Institute of Technology, Stockholm

ABSTRACT

A brief description of photogrammetric procedures are made to show how it has been and today is possible to make correction to photogrammetric measurements for the effect of refraction.

Then a review of how refractional effects have been calculated is made.

Finally some suggestions are made of how to meet the demands for improved accuracy in photogrammetric measurements.

WHAT IS PHOTOGRAMMETRY ?

Photogrammetry is a tool for determining 3-dimensional, object-coordinates from two 2-dimensional photographs, preferably taken with a metric camera.
A metric camera is a camera that has a fixed and known inner orientation which means that the principal point and principal distance must be determined. This is done in a calibration. The basic unit for measurement is the photograph coordinates x' and y'.
From the photograph coordinates and the inner orientation we then try to reconstruct the rays between the projection centre and the objects at the moment the photograph was taken. These angles or rays are used to make a resection of the projection centre in space in the first place. This is a standard method for two overlapping photographs which is performed with a relative orientation and an absolute orientation. When we know the exact situation for the projection centre we can continue to compute coordinates for all other points using intersection. One big difference between photogrammetric and geodetic measurements, however, is that we only get one observation for our computations. This means that we must have a very good control over all possible errors that affect our observations if we want to achieve a good accuracy in our final results. Calibration of photogrammetric instruments and procedures leading to accurate adjustments and good correction-functions is therefore of great importance.

Such error-functions are for example
- radial and tangential lens distortion
- film shrinkage
- refraction and earth curvature.

HOW IS PHOTOGRAMMETRY USED?

Photogrammetry is used for many different purposes. The classical area is of course mapping from aerial photographs, from the beginning only for small scales 1:50 000 - 1:20 000 but today also for large scale mapping up to 1:400.

In ordinary civil engineering photogrammetric methods also find a place for deformation measurements and volume computations etc. Here terrestrial cameras are used and the distance between camera and object is much smaller than in aerial photography.

Another increasing area is the densification of the net of known ground points used for absolute orientation of stereomodels. It is in this field of photogrammetry most efforts are spent to get the highest possible accuracy today.

WHEN IS CORRECTION MADE FOR REFRACTION?

Photogrammetric computations can be made with analogue or analytical methods.
In the first case a stereoplotting instrument is used and we get a drawn map or 3-dimensional coordinates as primary products. Because we do not measure the coordinates of the picture directly we have very limited means of introducing corrections.
One way is to include corrections plates in the copying process from negative to diapositive or put correction plates in the image-holder in the stereoplotting instrument. These glass-plates can only be made for standard heights and standard atmospheres.
Antoher way is to compute the effects on the model-coordinates and make corrections to these. (Tham)
But with analytical methods we have other possibilities. Here we use comparators to directly measure the photograph-coordinates and numerical methods for reconstruction of the central projection and both relative and absolute orientation.
The introduction of any type of correction to photograph coordinates can of course be done very easily in the computations.

HOW HAS REFRACTION BEEN TREATED?

The influence of refraction on photogrammetric measurements was outlined in a basic work by Leyonhufvud in 1950.
In this work the following assumptions were made
- only rays that have nadir-distances between 0 and 60 degrees are treated
- atmosphere consists of concentric shells

- flying heights lie between 0 and 10 km
- ICAN standard atmosphere
- linear temperature gradient.

The following conclusions are drawn concerning the use of a standard atmosphere:
- variations in absolute temperature, humidity and pressure is of minor importance
- only variations in temperature gradient can be of importance
- linear temperature gradient can be assumed since photogrammetric flights always take place on a clear day below a thin cloud-cover.

Errors caused by using the observed angle as true and by assuming a flat earth is neglected.
The astronomic refraction is computed for different nadir-distances and from that the photogrammetric refraction is computed for different flying heights and terrain heights.

For near vertical photographs the influence on coordinates can be assumed as radialsymmetric around the principal axis.

Later works in this field by for example Bertram and Schut do not add much new to solve the problem. They make the same assumptions of radialsymmetry and only tabulated densities for other standard atmospheres are used that extends up to 80 km.
Bertram states that refraction must be corrected for, an example is given where it causes a distortion of 10 microns which is very much compared to the measuring accuracy of 1 micron for a comparator.

PHOTOGRAMMETRIC HANDBOOKS

If we look into different handbooks for photogrammetry there is not much more to be found.
In Handbuch der Vermessungskunde methods are given for computation of the refraction angle using numeric integration or a simplified summation.
The values of n as a function of height H must be given and an example is shown where $n(H) = 1 + 0.000226\sigma(H)$, where the σ-values are taken from the ARDC-model-atmosphere.

The Photogrammetric Guide by Albertz/Kreiling gives a method for computation of refraction using the ICAO-standard atmosphere with temperature-gradient of 0.65 K/100 m up to 11 km and then constant temperature up to 32 km and pressure $p = 1013.25(1 - 0.022557696 H)^{-5.25588}$ mb between 0 and 11 km and $p = 226.3204\ e^{0.15768852(H - 11)}$ mb up to 32 km.
The radial distortion due to refraction is computed with observed nadir-distance and astronomic refraction angle
$\Delta \tau_O = T - 6366200(Q_O - Q_p)/(H_O - H_p)$.
T is derived from a 4-degree polynomial in H_O and Q_O, Q_p with a 5-degree polynomial in H_O, H_p respectively. This is perhaps the most commonly

used formulae.

In Ackermann/Schwidefsky are references made to Leyonhufvud and Schut for the computation of the refraction angle.

HOW IS ACCURACY BEING IMPROVED TODAY?

The results achieved with numerical photogrammetry today are very excellent, however, methods are tried to further improve accuracy.
In block-adjustment, for example, parameters for the inner orientation are included as unknowns in the computations, this is called self-calibration.
If parameters for radial lens-distortion are included the effects of refraction must to some degree be included in these.
This could be one way to compensate for the actual atmosphere.

Another method to improve accuracy is to use the discrepancies in known points after block-adjustment to compute parameters in a correction polynom separate for x- and y-coordinates. These polynoms are then used to compute corrections for all new coordinates.

CONCLUSIONS

- Today only model-atmospheres are used to compute the refraction for aerial photogrammetry.
- Traditional methods for photogrammetric measurements offer limited possibilities for corrections of photograph coordinates.
- There is a demand for more accurate photogrammetric measurements for point densification.
- With modern computing methods where comparator measurements are used it is easy to introduce corrections.
- Refraction, among other errors, must be fully controlled using inflight registrations of actual conditions if accuracy is to be improved.

SELECTED REFERENCES

Ackermann/Schwidefsky. Photogrammetrie. Grundlagen, Verfahren, Anwendung. Stuttgart 76

Albertz/Kreiling. Photogrammetrisches Taschenbuch 1975

Bertram, S. Atmospheric Refraction. Photogrammetric Engineering no 1 1966

Jordan/Eggert/Kneissl. Handbuch der Vermessungskunde

Leyonhufvud, A. On Photogrammetric Refraction. Photogrammetria 1952-53

Schut, G H. Photogrammetric Refraction. Photogrammetric Engineering no 1 1969

Tham, P H. Addition to Leyonhufvud's Article: On Photogrammetric Refraction. Photogrammetria 1952-53.

DISCUSSION

J. Saastamoinen: During this elapsed 19 years period, there have been two shocking things for photogrammetrists. One was the earth's curvature, and the other had to do with refraction. I don't remember exactly - was it 15 years ago or so - there was an international test of photogrammetry, and our office was trusted the treatment of the results. Everybody was extremely surprised when at the center of the picture the measured elevations were very good, but going off the center, a few kilometers, the elevations were very bad. And someone said that maybe the earth is curved. Now, the second shock came in analytical photogrammetry with these refraction corrections. There was a Russian paper, written by - I think - Dr Kushtin. He was the first to point out that in modern airplanes, when the camera is mounted behind a port-glass in a heated and pressurized cabin, the optical system is no longer an ordinary lens but there will also be refraction between layers of different densities separated by the port-glass. And these effects happen to have the opposite algebraic sign to the ordinary refraction correction. At a certain flying height - about at 6000 meters - the effects cancel out each other. So, in analytical photogrammetry, it might be better not to apply any refraction corrections at all. A third remark is on the papers by Bertrand and Schut in "Photogrammetric Engineering" - if you look a couple of years later, you will find that the solution for the actual atmosphere is very simple, as simple as the refraction correction computed from standard atmosphere.

J. Larsson: I would say, that in standard procedures of photogrammetry of today we don't try to measure the actual atmosphere. We don't compensate for it. But we should correct for the actual atmosphere to further increase the accuracy of the analytical photogrammetric procedures.

J. Saastamoinen: You need only the pressure at the ground and the pressure at the camera.

J. Larsson: These measurements should be made at all flight missions when analytic procedures such as a block-triangulation shall be used.

J.C. de Munck: In photogrammetry the scale is normally kept free. So the greater part of the refraction is not important. Only second and higher order terms are of interest for the photogrammetrist.

J. Larsson: That is certainly true. The greater part of the refraction may be thought of as a scale difference, which is eliminated in the absolute orientation.

THE USE OF MULTI- OR SINGLE WAVE METHODS TO ELIMINATE TERRESTRIAL
REFRACTION FROM GEODETIC MEASUREMENTS

J. C. de Munck
Geodetic Department of the Delft University of Technology
Delft, The Netherlands

ABSTRACT

Refraction is a very important aspect of geodesy and astronomy: it is
one of the main sources for troubles, papers and dissertations, at
least in geodesy. Particular important is the vertical refraction because of the up- and down-asymmetry in the atmosphere. However, we
must not forget that gravity does not only cause the asymmetry of the
atmosphere, also the asymmetry of the mechanical parts of the instrument (bending of the tube, deformation and excentricity of the axis,
etc.) and even an asymmetry of the mind of the observer. See figure 1.
If it becomes possible to measure the refraction these effects will
often become limiting factors for the accuracy.

However, at the moment the refraction is still (?) a problem and
there are a number of ways to attack the difficulties. Four years ago
Prilipin gave a survey of the different methods. He distinguished
direct methods and indirect methods.

The indirect methods consist of intelligent averaging over different
weather conditions and avoiding bad conditions. These are the methods
currently in use; they might be improved by studying direct methods.

The direct methods, where measurements are used to determine the refraction, may be divided into four types (see figure 2):
A. Simultaneous meteorological measurements
B. Symmetrical measurements
C. Other single wavelength methods
D. Multiwavelength methods

A. Simultaneous meteorological measurements

The refraction may be calculated from measurements of temperature,
temperature difference, humidity, humidity difference, radiation, wind
observations, etc. For practical reasons such measurements can only be
performed on a very small number of sites. Because interpolation and

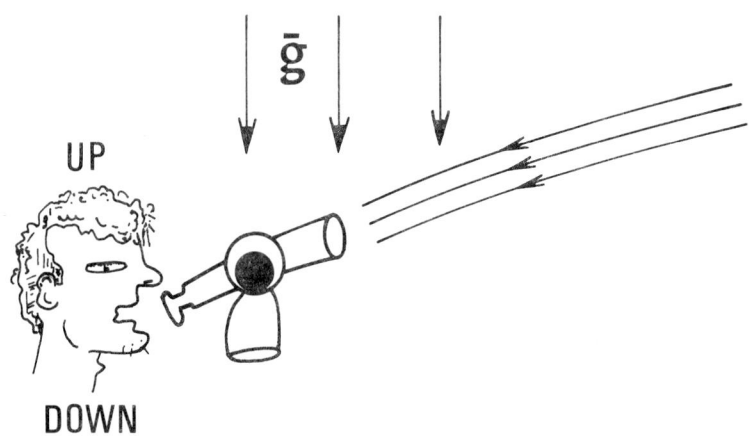

Figure 1. The influence of the gravity on the atmosphere, on the instrument and on the observer.

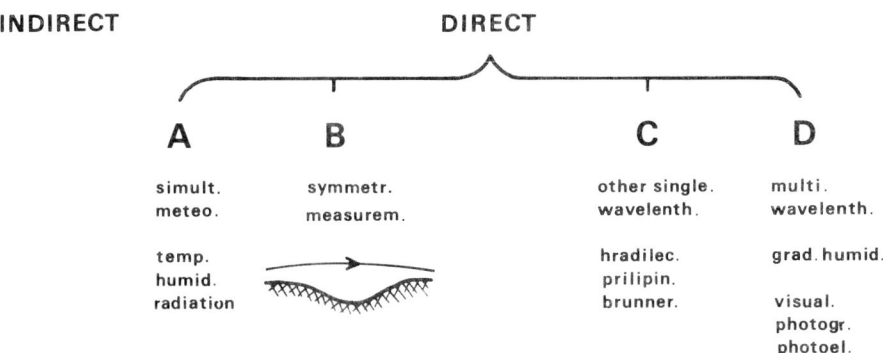

Figure 2. Methods to eliminate the terrestrial refraction.

extrapolation are necessary, studies of atmospheric models are urgently needed. Also the type of the terrain should be considered.

B. Symmetrical measurements

Angle measurements from both endpoints, measurements of angles from the midpoint towards both endpoints and other symmetrical measurements belong to this type that eliminates the symmetrical part of the refraction, i.e. the refraction of the symmetrical component of the curvature profile or of the "gradient n" profile.

Such symmetrical measurements will give good results for horizontal paths over a horizontal homogeneous terrain or for horizontal paths over a symmetrical valley. For sloping times and for asymmetrical profiles, however, a systematic effect of refraction will not be eliminated.

C. Other single-wavelength methods

Besides the above mentioned methods and the use of dispersion a number of procedures have been proposed or are in use:
1. The adjustment of the refraction coefficient for each site, including distance measurements over steep slopes. This method, used by Hradilec, will only work for great height differences.
2. Another method was proposed by Prilipin in Bulletin Géodésique. A special interferometer is used with two mirrors being moved along two bars making different angles with the wanted direction. If no feed-back is used the distance is limited to less than one kilometer by consequence of air turbulence.
3. This week the ideas of Brunner have been announced. He tries to find the refraction angle from the variations in the refraction. A thorough study of the instabilities of the atmosphere and a lot of experiments are still necessary to find the circumstances under which this method will work.

D. Multi-wavelength methods

When two wavelengths are used it is possible to eliminate the refraction angle more or less for an atmosphere of constant composition. The only relevant variation in the composition of the air is the variation of the humidity. It may be expected that the humidity gradient limits the accuracy of a vertical angle generally 0.02 seconds of arc per km.

It is not possible to eliminate the humidity effect in the refraction angle by using three wavelenghts because the required accuracy for the measurements and also for the dispersion functions for dry air and for water vapour is much too high if optical- and near optical wavelengths are used and with radio-waves precise angle measurements are impossible for terrestrial use.

In order to reach higher accuracies with multi-wavelength methods or to avoid risks of greater errors in unfavourable circumstances, better atmospheric models should be found: in particular relations between humidity variations and other atmospheric parameters like variations of temperature. An example of such a relation is given by the invariability of the relative humidity under certain conditions.

(During the symposium I learned that other important items to be studied are the fluctuations in the air and the distance between the paths for the two wavelengths.)

As to the instrumental realization three methods are used: visual, photographic and photoelectric. The visual methods are very restricted in the separation of both wavelengths.
The photographic method has the advantage that the averaging of the signals over longtime intervals is very easy.
The photoelectric method is maybe the most complicated one, but the signal processing is possible in an almost optimum way.

DISCUSSION

E. Tengström: Thank you, Dr de Munck. This review was a very interesting complement to that one given by Prilepin in Stockholm 1974. It contained also new details and fresh points of view concerning these important problems.

ROUND TABLE DISCUSSION ABOUT THE USE OF DISPERSION METHODS FOR DETERMINING REFRACTIONAL EFFECTS IN ASTRONOMY AND GEODESY

Chairman: E. Tengström

E. Tengström: During the discussion today I like to hear the opinions not only from people dealing with observations of geodetic nature, but also from people who are astronomers or meteorologists, concerning their ideas of eventual applications of the multi-wave method in their investigations. We may start with meteorology. We have unfortunately no pure meteorologists here, but some of you are experienced in what we may call geodetic meteorology, a name for an area, which was introduced already by Brocks, and dealing with wave propagation, turbulence and refraction. The meteorologists have used sound waves to study various problems. Perhaps there may be someone who knows what has been done to study the propagation of sound through the atmosphere, in order to correlate the refraction of these waves with various meteorological models and conditions. As to my knowledge, no attempt has been made to use dispersion for sound waves of various frequencies, up til now. I like to ask some representative here, if they have any experience of success in studying these problems with sound waves. Is there anyone familiar with this area of research? Professor Liljequist and his assistant Dr Israelsson will join us during the next session, when we are going to discuss the cooperation between astronomy, meteorology and geodesy. And they will probably be able to inform you further about any existing contributions from the side of meteorology concerning refractional problems.
So, having no meteorological contribution from the floor at present, I like to direct myself toward the astronomers. At first I would like to address myself especially to Dr Teleki, Dr Hughes and Dr Sugawa, asking them if anything has been done to these problems of eliminating refractional effects, before Dr Currie started his important investigations, which he told us about the other day. Or is this the first time the two-colour principle has been applied in astrometry?

G. Teleki: I have never been informed about any investigations of such kind before. I have only heard about your suggestions to investigate astronomical refraction with multi-wave methods. Yesterday we listened at Dr Currie's interesting paper about realistic research in this field. Before that, no other investigations in astrometry.

E. Tengström: As I know from Dr Brein, Hertzsprung started to measure directly the atmospheric dispersion at stars of different elevations, by applying an interferometric method, using a diffraction grating in front of the telescope objective. Method and results were published in Astronomische Nachrichten, Vol. 192, 1912, under the title "Photographische Messung der atmphärischen Dispersion".

G. Teleki: But he did not use the results for astrometry.

E. Tengström: No, but it was the first time an astronomer tried to determine the actual atmospheric dispersion, which could have been used for deriving astronomic refraction. His work might be regarded as the origin of a research into a direct determination of astronomic refraction. The origin of research into geodetic refraction was the work by Näbauer. He was the first one to derive a formula for the relation between refraction and dispersion. His technique, using pointing with alternating read and blue filters, was however too inaccurate.

D.G. Currie: I think there was an intention to use this method at the automatic transit circle of the U S Naval Observatory, for recording in two colours with alternating filters in red and blue.

J.A. Hughes: It has been a practice to use yellow and blue plates in photographic astrometry for purposes other than refraction. I should mention that the automatic transit circle is now back in the factory being reworked because it did not perform satisfactorily. The filters which were, at least provisionally, included in the design were not directed primarily towards refraction. I agree with what you said about Hertzsprung.

E. Tengström: I believe, it is not so important just to give a review of ideas we should already know of, and which are rather new as to their practical applications, but such a review may form a good background for starting to think of some proposals for the future use of this method in astronomy and astrometry. The refraction problem is, of course, very important for our star catalogue work, which we have touched, but also for geodetic astronomy, which may be regarded as a part of astronomy, as it always was. In the case of the two-colour approach, I think, wouldn't it be possible to construct such an instrument, with which we can measure the absolute angular distance between two lines in a star spectrum and compare this measured angle with the angle calculated from the exact properties of the reflecting optics? E.g. having a Rowland grating system, which gives you two indications, one for the atmospherically transferred spectroscopic beam from the star, and the other from the corresponding spectral lines observed by the same instrument, in the laboratory, using the exactly known mathematical properties of the receiving system and earth bound sources. I don't talk about the technical possibilities, but only about principle of application. As we have a round table discussion about the two-colour method, I thought I should also take this idea up. The difficulty, from the theoretical point of view, would be that Edlén's formula, which seems to be slightly better than Barrel and Sears' formula, both used in geodetic work, will not necessarily be applicable to the problems of astronomy because the composition of the atmosphere ought not to be the same as in the laboratory, where the formulas have been derived. I have talked to Dr Bender about this problem, and he was pretty sure that the percent of different gases, even CO_2, was rather constant up to at least 500-600 km. So that the contribution to the total astronomical dispersion coming from this part of the atmosphere should be a very essential part. Is there any objection against my trying to investigate such a method of deriving astronomic refraction? If you are not totally negative, I shall start a cooperation with the people at the Schmidt observatory here to see if we can do something together. Of course, the humidity is again a problem, but the content of water vapor goes rather quickly to zero and its gradient profile would be possible to obtain by meteorological soundings. Unfortunately we can't do as with the distance measurements, eliminating the humidity term by using an additional frequency in the microwave region. The humidity gradient integral cannot be eliminated by using a microwave frequency, because the corresponding dispersive effect for the optical waves and the microwave is too small. In astrometry, by the way, we have probably to stick to optical frequencies only. Now I ask the astronomers if they like us to help them to investigate the possibilities to apply the two-colour method for the determination of the astronomical refraction, or if they think they do not need such a help.

J.A. Hughes: Honestly I must say that astronomers appreciate all the help they can get.

E. Tengström: Also in this particular problem?

J.A. Hughes: Yes. Regarding the humidity; I have taken part in LIDAR measurements which work fine for humidity profiling, so I don't consider that a problem. Well, it might be a financial problem. In any event, since we only have to probe 2 or 3 KM for humidity, a modest laser, say 1 millijoule or so, does it easily. There are radiometric methods as well, so let's assume, for discussion purposes at least, that the humidity is known. I do have a question though. Earlier, Dr Teleki read a Soviet paper having to do with the chromatic effects upon refraction. In the past the refraction formulas, or tables based upon them, have been based upon some particular wavelength. Departures from this wavelength are then tabulated in some way, for instance as in Pulkovo III. It is not clear to me in the case of the multi-wave approach exactly what wavelength one is effectively using. Is it some integrated visual wavelength, or some weighted results which depends upon the details of the dispersion curve? What precisely is it? It could be an important point, or it may not be a problem.

E. Tengström: You know you have one equation for the deviation of one line, another equation for the deviation of the other. In the first you multiply the unknown atmospherical integral with the calculable n-function. If you forget about the humidity, the measured dispersion between the lines is equal to the atmospherical integral times the difference between the n-function values. From this dispersion equation you can calculate this unknown atmospherical integral. With this you can then derive the refraction for any desired wavelength, having its appropriate and calculable n-function. In the case of a star I think the desired one has to be the effective wavelength of the incoming spectrum. The assumption I have made, namely that the atmospherical integral is the same for every considered wavelength, is of course a weak point, but from measurements along our 20 km test line we know that existing refractions of 3' or more in this complicated atmospherical region gives a maximal distance between the UV beam and the red beam of only 0.5 dm. So we are convinced, that in dry air the two atmospherical integrals are identical, especially as the effective total spacing must be much smaller. Perhaps the assumption is more dangerous in astronomical refraction?

J.A. Hughes: Yes, I think it's alright. I just want to make sure that we are coming up with results which are well defined in the sense that they represent a standard. For example, a situation where model number one of a multi-colour device using some particular wavelengths gave you something referred to one thing, and model number two gave something else, would be bad. Apparently that's not true, and I am happy to hear that.

E. Tengström: And you may imagine, that we liked to check the consistency, using various wavelengths. Of course, two colours could be replaced by three or four, and we have done that for three. Mårtensson has proved that even if we have He-Ne red, Argon blue, and He-Cd UV, and we use blue - UV to compute the refraction, or we use the red - UV, we get the same result within the accuracy we have predicted.

D.G. Currie: With regard to Dr Hughes' question, as far as it was presented here for the two-colour refractometer, the wavelengths mentioned are 3400 Å and 6000 Å and considered as the effective wavelength for the two relatively wide band filters. And that is what I have mentioned earlier about the effective wavelength, perhaps changing as you absorbe on the blue edge due to increased zenith distance. The result, which was the number which converted the apparent separation of those two effective wavelengths into what I call the refraction. On the listed equations I had, was the conversion 30.55, which was the conversion to a visual D line, mentioned by professor Tengström, that is to 5600 Å. So the conversion from the effective wavelength, which are instrumental statements in my case because I am not using lasers, I presume to be a reasonable standard, but which perhaps might be discussed of as to the standard used.

E. Tengström: There is another thing also. Even if you don't say that you define the refraction for the D line, that is if you use something in the neighbourhood of that which has to do with our visual of photo visual experience, it does not matter, because the dispersion change is extremely small for small changes of the wavelength.

J.A. Hughes: Yes, that's true.

G. Teleki: Can I continue this discussion about the red and the blue problem? I asked yesterday Dr Currie how to select the stars for your investigations. Because, for one star you have a very intensive red line and a very weak blue line. Therefore I would like to ask you how this difference between the intensities in the lines influences the refractional determinations?

D.G. Currie: I believe, and this has to be a statement of prediction not measurement, that this will not resolve in a systematic error. It will resolve in a lower signal. Therefore we require a longer integration to a given accuracy. That means, if you look at a M star, you must either look at longer, or except in the standardized programmes, larger error barriers. But in the analysis, nothing has arisen which indicates that there would be a systematic effect, apart from what I mentioned about having to calibrate where the effective center is. For a M star it would not be 6000 Å, it would be 6020 Å. That change you would have to calibrate. And I believe, at that point we will not have a systematic error, or it might be ignored.

J.A. Hughes: I think that out of this discussion I have gotten enough information so that I can phrase my question better. The thing is, I am concerned about whether or not atmospheric attenuation on the one hand, coupled with the spectral type of a star on the other, could possibly give us a zenith distance dependence of the refraction measurement. This is exactly the kind of thing that plagues us now. I am willing to admit it may not be a problem at all, but that is my question, and I think it is being answered.

G. Teleki: It would be very good to have the refractional correction for the moment of observation of the star. Connected with this, I ask for some information about the difference between Tengström's proposed method and Currie's method for the determination of this effect.

E. Tengström: I think, in principle the methods mean the same thing, but with the big resolution we can get from the spectra, and knowing the real wavelengths we can perhaps achieve more. The effective wavelength which is responsible for the position of the exposed image, might easily be determined through photometry of the incoming spectra.

G. Teleki: So there is basically no difference between your idea and Dr Currie's?

E. Tengström: An attempt to increase the accuracy of dispersion by having a very big resolution from the reflecting optics. Then it is not necessary to use filtering separation. The separation is made in the spectrograph, and you measure with the same spectrograph, in our case the Rowland grating, the distance between well defined spectral lines. It is not any interferometer or any special instrument of other type necessary. I think there exist for solar investigations already

Rowland gratings which can be used for our purpose.

B. Garfinkel: I learnt about this two-colour method just a few days ago, and I have only a vague recollection of the principle involved here. Was it not assumed that the refraction in a given wavelength is proportional to the refractive index in that wavelength minus one or something like that? That is certainly true for the first order in the refraction. But there are also higher order terms.

D.G. Currie: That is correct, but the accuracy increase by including the higher order terms is well beyond our needs in the optical region.

E. Tengström: I agree with that. I also like Dr Milewski to say something about the behaviour of the refractive index in different parts of the spectrum, related to the atomic structure of the atmosphere.

J. Milewski: I think that from the physical point of view we are in a favourable position. The region 2000 Å to 7000 Å is without great resonance influences with the atomic and molecular structure of the atmosphere. This is important, because for frequencies very near to the resonance frequencies of atoms and molecules, we have extremely great difference in refractive index. Great molecules have some resonance effect beyond infrared, but also here it is not dangerous because the content of great molecules in the atmosphere is very small, in any case in the used spectral region of the multiwave method. This method is very accurate everywhere. It is a direct method and a very objective one. It is only the question of how good the Edlén's formula is for our purpose, being derived by physicists in the laboratory. I believe that we now know that the formula works with an accuracy of, say 4 parts in 100 millions, and this is in astronomical and geodetic practise a very good accuracy. The multiwave method ought to be a fine and objective method, at least in the optical region. There are however some technical problems and difficulties, but this is another question.

E. Tengström: Thank you Dr Milewski. In the Association of Geodesy we have since long time agreed upon that we need a certain formula for cm-waves and one for the optical region. Your information has now told us where we can be safe. And I believe that not only the geodesists can use Edlén's formula, or Barell and Sears', but also the astronomers when studying refraction by means of the multiwave method in the whole optical region including UV.

J. Milewski: Yes, down to about 2000 Å which is a very deep UV. Dr Currie, do you agree with me?

D.G. Currie: Yes!

G. Teleki: I don't think we can make some definitive conclusions out of this discussion, which is mainly an exchange of informations. I can say, that we from astrometry support this kind of investigation without

ROUND TABLE DISCUSSION

any further discussion. We expect with great interest the results of these experiments. Therefore we propose this kind of investigations to be put in the resolutions.

E. Tengström: Thank you Dr Teleki. We have previously been talking very much about tilt of the isopycnic layers and its effect on astrometric work, catalogue work and geodetic astronomy. This problem contains the zenith refraction question. I think we can master this question by using two-colour devices as Dr Currie has proposed, or looking at star spectra with high resolution near the zenith. The deviation from the theoretical shape of the layers in the models we use and the real shape would perhaps be possible to study also using greater zenith distances, though that is not yet proved. On the other hand the refraction at any elevation and the tilt near the zenith might be studied also by simultaneous, or almost simultaneous, latitude or longitude observations, carried out separately with transit observations and elevation observations, e.g. in the case of zenith refraction with latitude out of Struve observations and Horrebow-Talcott observations. Dr Teleki, I wrote to you about such attempts we plan to make, and I know that you considered them realistic long time ago. What is your attitude today? Have you done any observations using different methods of latitude and longitude determination to study these problems?

G. Teleki: I do not understand the question.

E. Tengström: I can take an example. You make a latitude observation by means of Struve's method, and you make a latitude observation by means of Horrebow-Talcott's method. In the last case you have a zenith refraction. That is you will have a shifted zenith, and you will have half of the zenith refraction which appears as a term in the difference between the result of Struve and the result of Horrebow-Talcott. There, I mean, we have a possibility to study the zenith refraction. About the attainable accuracy I cannot tell you, but I have a feeling that the internal accuracies of the two methods permit us to solve for an eventual zenith refraction with reasonably small error.

G. Teleki: Perhaps you are thinking of my own and professor Shevarlich's proposition of determining the anomalous refraction in the zenith zone (Publ. Dept. Astron. Belgrade, 3, 1971, pp. 5-16), maybe? Because it is a comparison between Horrebow-Talcott and Struve methods.

E. Tengström: I refer to a letter from me to you, and an answer from you to me. I do not by references know about your investigation. We have been thinking of this ourselves for many years, but we have not yet been able to start any observations along the line. The investigations could be done with longitude observations also.

G. Teleki: You know, we proposed measurements in the zenith zone in the meridian and in the prime vertical for the same star, using Horrebow-Talcott and Struve, respectively. However, with this method we had special problems, especially instrumental problems. You need very good

instruments for these observations. It is a basic problem. The second
problem is the symmetry. And you have no observation at the same time.
You need, but I don't exactly know, something like half an hour for the
observation of a complete set, that is from one point to the other.
Therefore, separating out from these observations the anomalous refraction is a very problematic question. For this reason I expect a new
type of investigation, yours or Currie's investigations, using the
same moment of observation. It is very important for us. Because we
need a correction just for the moment of observation. It would be very
useful - this might be a strange idea - to use the same star for the
determination of the refraction as you use to determine the declination.

E. Tengström: Thank you very much for this information about your important work already done to try to solve this problem.

K. Ramsayer: We have developed some devices for automatic star tracking. With them you can measure simultaneously the vertical and horizontal direction to a star. If we observe a set of stars we can compute
latitude and longitude from the observed vertical angles alone or from
the horizontal angles alone. The results will be different because of
the errors of the measurements and the different influence of the vertical and lateral refraction. If there is a difference larger than tole-
rated we can say we have a refraction anomaly, and we have an indication that the measurements are to be repeated. Some preliminary theoretical investigations of the influence of the tilt of optical layers
to the vertical and lateral refraction have shown, that in general both
components of refraction have different influences on the simultaneous
determination of latitude and longitude, that this influences are proportional to the square of the secant of the zenith distance and that
they get smaller the more you come to the zenith. The investigations
will be continued.

E. Tengström: Your important investigations of anomalous refraction in
the zenith zone have been highly appreciated by us geodesists during
many years. You are the first who could talk realistically about amounts
of zenith refraction to be expected at sites like yours. You have been
optimistic in this respect til now, relying upon your own experiences,
but I think that every observer has to make observations by himself to
get results which could be applied to that site where he is working,
and the time of his observations at that site. Your method, I think,
however, must be very valuable to use for all observers concerned,
especially because it seems to have solved the simultaneity problem.

J. Dommanget: I think that the position for the astrometrists in general
is somewhat different and is fundamentally connected with the field in
which they are working and the equipment they have at their disposal.
Astronomers are not really interested in the atmospheric refraction itself but only in the way of how to get rid of it. So practically speaking, when the astrometrist is in his observing room, the point for him
is to have at his disposal a method of observation (or of reduction)
such as to free the observation from the refraction effect. The best

way to do this should be of course to use a technique giving, in the case of photographic astrometry for instance, the importance of the refraction all over the plate's field. The measurement of the refraction effect should be made and the photographic plate should be taken simultaneously. This is unfortunately not easy to do at least with the existing equipment: one cannot pursue two programmes simultaneously at the same instrument. It may be different for Dr Teleki when he is observing at the meridian: he may perhaps determine by the same programme of observation, the latitude, the declination of the star and some information on the refraction. So, concerning photographic astrometry, my point of view is that two kinds of observers should be considered. On one side, a very small group which is interested in the study of the refraction phenomena and who should try to fine correlations between refraction effects and atmospheric characteristics for improving our refraction model and our refraction tables. In that respect, the formulation I have proposed for the refracted X and Y coordinates (in my paper presented earlier) may be of some help. The other astrometrist who have to use the improved tables on the basis of any information concerning unfortunately ground level meteorological observations only.

E. Tengström: What you are talking about, as I understand, is the difficulty of having simultaneous information. This is one thing. But when Teleki started to try to improve the refraction tables he was probably not aware of the possibility to determine simultaneous effects. So there is just a way of improving the whole thing. That is, with refraction tables, dynamical atmospheres included, we can nowhere get the correct refraction values. I think the astronomers should be grateful if any instrument could be constructed which eliminates the refraction from the data observed or gives a simultaneous information about it, which can be used for correcting the observed astrometric results. I think we should work on this technical problem. As Hugget has been able to construct a distance meter for refraction, it would perhaps be possible to do something in the astrometric work, photographing the star spectrum for instance at or almost at the same time as we photograph the star. I believe this would not be impossible.

D.G. Currie: For certain areas in astrometry, that is PZT, transit circle and astro geodesy where one observes the star, and may observe it photoelectrically, the ability to do it simultaneously is possible. The brightness of FK 4 permits you to make these measurements. In the case of the PZT, when one extends the observations beyond the FK 4 one has to work with a larger aperture. In order to make an instrument like the two-colour geodimeter, which does both astrometric work and refraction, will be a new astrometric instrument. New astrometric instruments are very expensive. So what we are doing first is to make an observing programme to determine what the time spectrum and the magnitude of the fluctuations are, and how they change across the sky. If we see that the refraction remains constant to a 20th of an arc second - I am talking in extremes now - one would not need to make it at the same time. If it changes rapidly, in a spatial sense, then obviously we have to progress to a joint instrument. This would probably from some initial

studies be more similar to an astrolabe than a transit circle, because
of questions of support of equipment and calibration. You do have enough
brightness to look at the refraction, that is the dispersion, at the
same time as you look at the positions. The initial programmes with the
PZT that we would do will be two separate telescopes, and we would have
looked at whether room refraction is bad or not. Another thing is, you
have a joint instrument, you don't care about the room refraction, because everything comes through the same telescope. So, yes it is possible, but I hope we will not go into the expense of doing that.

J.A. Hughes: There is one great difference between fundamental astrometric observations and astronomical geodetic observations, and that is:
one sets up a transit circle for example, and it stays put, observing on
a single program night after night for five or ten years at that spot.
So one is immediately concerned with how the refraction is systematically changing. But I think that this precise simultaneous determination
is much less necessary than you might think. At least in the transit
circle case. What I mean gentlemen, is that when a geodesist is out on
a Laplace station he doesn't stay there for five years, I am sure you
are much more efficient than that, but the astronomer must stay put if
there is ever to be an FK 4 of FK 5. That's the only way it can be done.
The point is that the astronomer does integrate and sample over a great
many different atmospheric conditions. So the short range of variability
does not worry me nearly as much as the isopycnic tilt that sits over
your site for five years, at least statistically, and every time you
observe it's there, or perhaps it changes annually. That's the kind of
thing we are concerned about. The shorter term things, with which the
geodesist is very definitely concerned, is not quite the problem in the
astrometric case. This is perhaps one small advantage we do have.

E. Tengström: Thank you very much. That was a clear statement of the
difference between the geodetic astronomy and astrometric work.

C. Sugawa: At Mizusawa, the international latitude observatory, we are
now carrying on simultaneous observations with a visual zenith telescope,
a photographic floating zenith telescope, a PZT and a Danjon astrolabe.
At almost the same parallel of latitude we are simultaneously observing
time and latitude every night. The refractional influences in the free
atmosphere should be the same, but the mutual comparison indicates fairly
peculiar differences. Dr Currie says that these might be due to different
room refraction for each instrument. But for explaining these complicated
differences, I think that special experiments must be carried out using
various wavelengths.

E. Tengström: Have you not detected any seasonal variation in the zenith
tilts?

C. Sugawa: No seasonal variation in the zenith tilts has been found from
our observations.

G. Teleki: We permanently discuss about the multiwave observations, but you remember that we have many problems with chromatic refraction. Connected with this it is possible to propose (A. Kubichela suggested this idea) to observe only in one colour, for instance in red, all stars in red colour, or in blue. What will happen in this case with refraction? We can eliminate chromatic refraction, maybe. Maybe because the intensity of the line can influence our observations. Suppose there are no influences of this kind, in this case no chromatic influences. But the normal refraction exists. Anyway, this idea of an one-colour instrument is interesting, and must be investigated.

E. Tengström: I think we have to close this session now. But if there are additional questions about this we can continue during the next session.

NOTES ON SOME ABNORMAL REFRACTION PHENOMENA IN THE ATMOSPHERE

G.H. Liljequist
Meteorological Institute
Uppsala University
Sweden

To geodesists and astronomers atmospheric refraction is a phenomenon, which makes it necessary to introduce corrections to the observations of carefully performed measuring series. Such corrections are generally obtained from tables giving values, which are applicable under "normal" or standard conditions. These conditions may be taken as more or less obtaining with a mean variation of temperature with height in the atmosphere - in the troposphere mainly a decrease of 0,6 - 0,7°C per 100 m.

However, in the low layer of the atmosphere there often occur situations, which deviate markedly from such standard conditions. In some specific weather situations and in some geographical regions the deviations may become especially marked. The influence of atmospheric refraction upon astronomic and geodetic measurements may then become exceptional and, besides, almost attain a random character.

In the winter, with strong cooling of the earth's surface, so-called surface inversions are common. Temperature then increases with height in the low layer, from the surface and some distance upwards. Such situations are common in fine-weather situations, especially when the ground is snow-covered. They are thus common at high latitudes in the winter. They are also found in late spring and early summer over the sea, i.e. at a time when the air is heated over the land, but the sea is still cold.

In summer - or in the warm zone - the heating of the ground in daytime gives rise to an abnormally strong temperature decrease with height in the layers close to the surface. Such conditions will not be considered here.

The meteorologist is mainly concerned with the atmospheric refraction phenomena as such and their varying appearance under different meteorological conditions, especially in extreme situations. To geodesists and astronomers the abnormal refraction phenomena are, however, a mere nuisance. Due to the stochastic character of the ensuing "errors" these cannot be eliminated properly by introducing corrections.

Let us have a closer look into the matter.

The refraction index n of air is a function of air density ρ and can be written

$$n = 1 + k \cdot \frac{\rho}{\rho_o} \qquad (1)$$

where $k = 2{,}93 \cdot 10^{-4}$ for "white" light,

ρ_o = the density of air at NTP (= 1013,2 mbar and $0°C$).

With a temperature inversion - i.e. with a temperature increase with height - density ρ thus decreases more rapidly with height than in the normal case.

With refraction phenomena located to the surface layers, e.g. in connection with a marked surface inversion some 5-10 m thick, pressure p may be considered as constant within the layer. Density ρ and refraction index n are then for all practical purposes determined by temperature alone. The surfaces of equal refractive index n therefore run parallell with the isothermal surfaces.

Analogous conditions are found within inversions situated at higher levels in the atmosphere. In some cases, however, the inversions are so thick, especially in the polar regions, that the vertical variation of pressure p must also be considered.

The paths - or rays - of light can be derived from Snell's equation, viz.

$$n \sin \alpha = \text{const} \ (= n_o \sin \alpha_o) \qquad (2)$$

or in the modified form valied for spherical equidensity surfaces

$$rn \sin \alpha = \text{const} \ (= r_o n_o \sin \alpha_o) \qquad (3)$$

The index "0" marks values at a certain point of reference along the ray, often at the surface. The denotations are given in Fig. 1. Generally the surfaces of equal density are supposed to be strictly horizontal.

By using the first equation the equidensity surfaces are thus supposed to be horizontal planes; by using the second equation they are supposed to be spherical, r being the distance from the considered point of the ray to the earth's centre.

Knowing the vertical variation of density it is thus possible to compute in detail the rays of light, which originate from a certain object, or those rays which reach our eye. We have here only to measure the vertical variation of the temperature with height and if necessary, besides, pay attention to the vertical decrease of pressure, compare above.

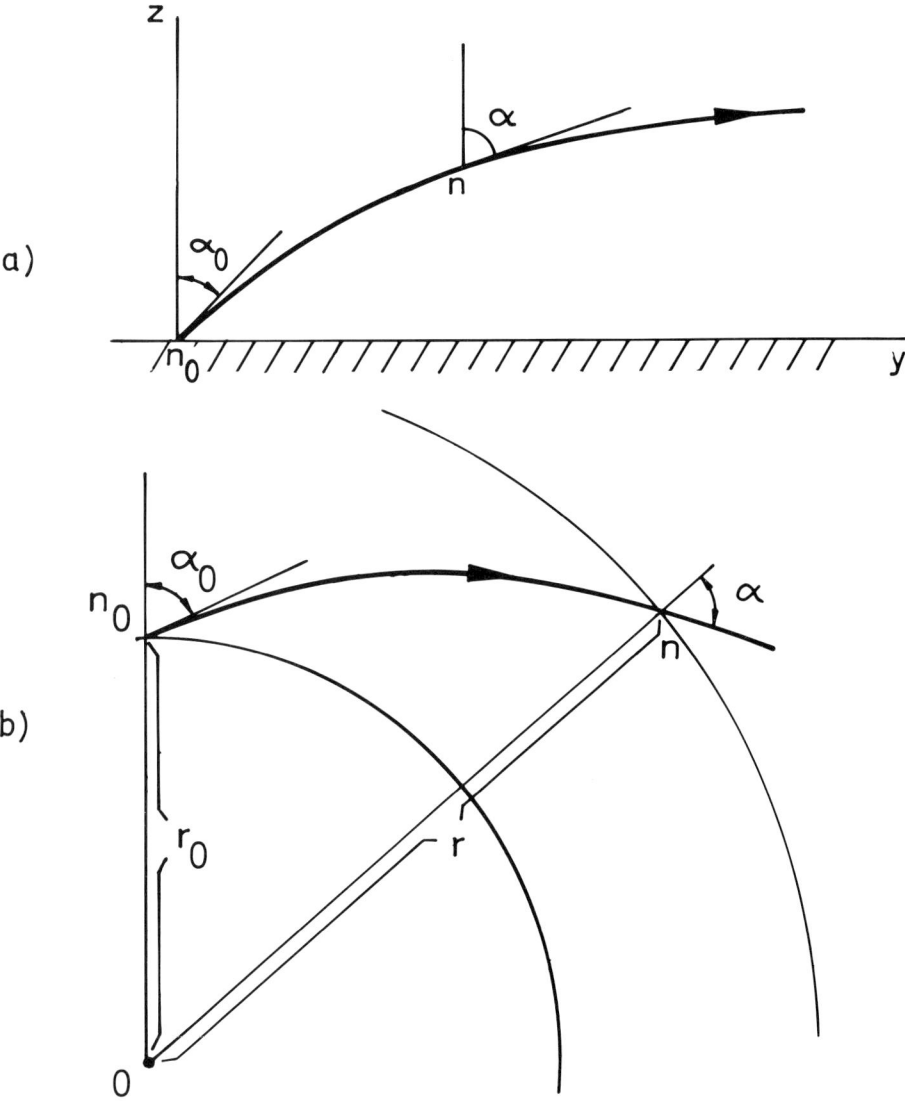

Fig. 1 Atmospheric refraction:
a) the surfaces of equal refractive index n are supposed to be horizontal planes;
b) the same surfaces are supposed to be spherical.

With marked refraction phenomena in the lowest layer - say up to 10 m - we can generally use the first equation; with refraction phenomena within thicker layers we must use the second equation, i.e. pay attention to the curvature of the equidensity layers.

The vertical variation of n is obtained from measurements of the temperature at different heights. In the lowest layer temperature is obtained from thermometers (preferably electric thermometers) mounted at a number of heights along an observation mast. If the temperature variation with height is wanted in more detail we can besides use a ventilated thermometer, which can be moved upwards and downwards along the mast. For thicker layers - say up to 500 m - it is possible to obtain detailed information from a tethered balloon equipped with suitable instruments. Such a balloon can analogously be moved upwards and downwards. If still thicker layers must be considered, data can be obtained from radiosonde ascents.

It is thus possible to obtain the refractive index and its vertical variation quite easily at the place of observation and at the time of recording.

A temperature inversion is, however, not a phenomena which remains more or less unchanged with time and in the horizontal. Internal waves are generated in the highly stratified air. Also more sudden and unexpected changes do occur. An inversion is also affected by the topography of the underlying ground. This means that the equidensity surfaces may not be horizontal, nor may they remain unchanged with time or in the horizontal.

This means that the atmospheric refraction and the optical phenomena coupled to it often change with time in a highly unexpected manner. Refraction and optical phenomena may also be different in different directions. In such cases atmospheric refraction corrections cannot be obtained from mere tables.

In some situations the atmosphere offers more than one path to light to travel from object to eye - each one giving rise to a separate image of the object. The images are then observed in the vertical, one above the other. In some cases all rays from the object, within a certain angle, can pass into the eye of an observer. The atmosphere then behaves in principle like a cylindrical lens. The image of a dark object situated at a critical distance is seen like a vertical dark pillar. The "points" constituting the ground - or snow surface - are then drawn out vertically to present an image of the surface similar to a gigantic wall surrounding the observer at a distance of 1/2-5 km or more depending upon the value of the vertical gradient of the temperature, or rather, of the refractive index. Objects at greater distance may disappear from view. Upper inversions may in analogous situations make distant mountains appear like gigantic pillars.

A person walking away from the observer may in such situations suddenly appear double and – as suddenly – transform into a dark pillar and then disappear from view. The rays of light, which originate from him, cannot then reach the observer as they are bent downwards towards the surface before then.

One way to study the special type of mirage, which is characterized by total reflection is to derive the so-called vertex curve, i.e. the locus of points situated on the top (vertex) of the different rays, which "leave" the object and then become total-reflected on the way, Fig. 2. Knowing the refractive index and its vertical variation, the vertex curve is easily obtained.

With object and eye situated at the same height and supposing that the equidensity layers are horizontal and plane, the angle α_o (see Fig. 1) at the surface is obtained from the vertex curve and from Snell's formula, viz.

$$n_o \sin \alpha_o = n_v \sin \alpha_v \qquad (4)$$

Values at the vertex of the ray are here denoted with an index "v". As $\alpha_v = 90°$ we have

$$\sin \alpha_o = \frac{n_v}{n_o} \qquad (4')$$

In the more realistic case with spherical density layers we have to start from Eq. (3). This equation can, however, easily be transformed into one similar to Eq. (2), viz. by writing

$$r = r_o + z \quad \text{and} \quad N = n \left(1 + \frac{z}{r_o}\right),$$

z being the height above the ground (with $r = r_o$).

When the conditions are ideal, i.e. with strictly horizontal density layers, it is possible to study atmospheric refraction in detail. However, the conditions are seldom ideal. In the highly stratified atmospheric layers waves or even more irregular displacements occur, which give rise to changes in the angle of incidence of the light-rays and also to local increases and decreases of the vertical gradient of the refractive index n, compare above. A certain ray will then undergo fairly rapid changes in direction with time, and the general appearance of the mirages may also change markedly with time. Besides, the conditions in that respect may be different in different directions depending upon the character of the underlying surface. The undulations of the surface is here an important decisive factor. The appearance of the mirages and the magnitude of the refraction is then more or less unpredictable in its details.

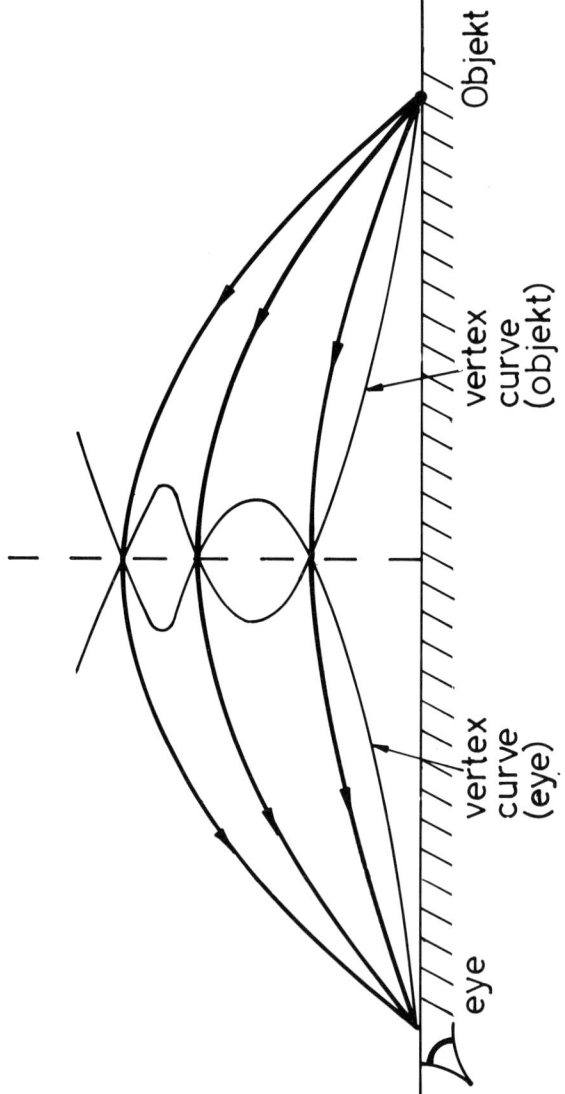

Fig. 2 From the variation of the refractive index n with height it is possible to derive the vertex curve, i.e. the locus of vertex points of rays leaving an object or reaching the eye of an observer. With object and eye at the same height the two vertex curves are the same and the vertex is situated at their intersection, i.e. halfway between object and eye.

The atmospheric refraction may also show short-periodic fluctuations. There exist small wavelets within the stratified layers and also irregular, small-scaled turbulent fluctuations. As a consequence a ray will pass through small volumes of air possessing densities which vary repeatedly along its path. As a result the image of an object is seen to change its position in an irregular way. This wellknown phenomenon - scintillation - may in extreme cases blot out the contours of surrounding objects; the latter may in extreme cases appear as a mere blur, even at distances less than 500 m. The wavelets within the stratified layers also give rise to another typical phenomenon. The upper rim of objects near the horizon, e.g. hills or hillocks, show small-scaled waves which are seen to move in the direction of the wind. Terrestrial scintillation becomes most pronounced in situations, when there exists a marked thermocline at a height of a few meters.

An abnormally great density gradient in the low layer of the atmosphere leads to an apparent displacement upwards of surrounding objects. Waves or other disturbances within these layers will affect this displacement, as both the density gradient and the angle of incidence of light are then changed. Even horizontal displacements may then occur. As mentioned previously, to an observer these vertical and horizontal displacements may seem to be almost stochastic. It should, however, be pointed out that marked displacements are found only near the horizon.

When a heavenly body is seen at the horizon, it is in reality well below it, viz. about 35' or roughly $0,6°$. However, observations do exist, where the sun has been seen when in reality it was about $4°$ below the horizon. This phenomenon was first described by the Dutch polar expedition under W. Barentz in 1597. From the winter quarters of the expedition in northernmost Novaya Zemlya the sun was observed 14 days before it was due to appear at the end of the polar night. This observation was considered as a "good story" until in 1915 a similar observation was made in the Weddell Sea pack-ice by the "Endurance" expedition under Sir Ernest Shackleton (Visser, 1956). Later - in 1950 and 1951 - the sun was observed at Maudheim (lat. $71°S$) when in reality it was $3,7°$ and $4,3°$ respectively below the horizon (Liljequist, 1964). On another occasion the sun was seen when it was in fact $1,0°$ below the horizon.

From Fig. 3 is seen that such abnormal refraction phenomena can be met with, when over long distances there exists a fairly undisturbed inversion, in which the solar rays are total-reflected. Such undisturbed conditions can best develop over an ice-covered sea, situated in the direction (from the observer) towards the sun at about local noon during the polar night. Such conditions existed for all the observations cited above. Over an ice-covered sea a marked inversion can develop in the low layer and this inversion is not affected by any underlying topography - the surface is more or less flat.

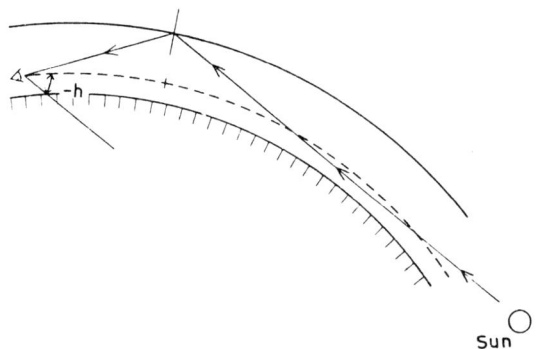

Fig. 3 The Novaya Zemlya phenomenon. Total reflection - one or more times - can make the sun appear above the horizon when in reality it should be one or more degrees below it. Note: the ray in the figure has for the sake of simplicity been drawn linear.

In conclusion we may note that temperature inversions at the surface or at height of a few hundred meters may introduce unexpected errors in geodetic measurements. Atmospheric refraction in situations with a marked inversion is generally such that reliable corrections cannot be introduced, at least not when the measurements are made close to the horizon. Inversions in the low layers are especially marked in the polar regions during the winter, but they are also found during fine-weather situations in the winter in the temperate regions. Marked inversions also develop in high and middle latitudes over the sea in spring and early summer, i.e. when the water is still cold, but the air over land is warm. In some situations this warm air may be driven out over the sea.

The geodesist is familiar with the adverse refraction effects met with over the land in the warm season, and generally he then avoids doing any measurements in the middle of the day. The atmospheric refraction, which occurs in the winter and also often in other seasons during the night due to cooling from below, is generally not considered in the same way. The atmospheric refraction - and the errors due to it - may on such occasions sometimes become even greater than in situations with a strong heating from below.

REFERENCES

Liljequist, G.H.: 1964, Refraction pehnomena in the polar atmosphere. Norewg.-Brit.-Swed. Antarctic Exp., 1949-52, Scientific Results, Vol II Part 2 Special studies. Oslo Un. Press, Oslo.

Pernter, J.M. and Exner, F.M.: 1922, Meteorologische Optik. Wien.

Visser, S.W.: 1956, The Novaya Zemlya phenomenon. Koninkl. Nederl. Akad. van Wetersch., Proc., Ser. B, 59, No 4. Amsterdam.

COOPERATION BETWEEN ASTRONOMERS, METEOROLOGISTS AND GEODESISTS FOR SOLVING REFRACTIONAL PROBLEMS

Chairman: G. Teleki

DISCUSSION

G. Teleki: After professor Liljequist's lecture I am more pessimistic than before. I am sorry that Dr Remmer is not here, because this morning he was practically against any measurements. But in any case, professor Liljequist documented very well the importance of a cooperation between the astronomers and the geodesists, as regards the refractional investigations. I suppose that you agree with me. This session is devoted to discussions about the cooperation between astronomers, meteorologists and geodesists in the refractional area. I hope that special questions concerning ice covered surfaces and refraction etc. might be dealt with during informal talks between professor Liljequist and especially professor Angus-Leppan. I suppose we have finished - in the pervious session - the essentials of a future improvement of the cooperation between astronomers and geodesists. Professor Liljequist's contribution today has emphasized the difficulties in treating the refractional problem without careful meteorological information. I agree with him, but I hope that neither astronomers nor geodesists are intending to try to solve the refractional problem without meteorological information, especially in such cases when the atmosphere behaves in a very odd way. Both have always to select observational conditions which are not too far from so-called normal conditions.

B. Garfinkel: What I have in mind, specifically, is a systematic check of the polytropic theory of astronomical refraction, the one which I published in the Astronomical Journal 1967. This theory has not been in much use, because it involves IBM cards and large electronic machines. Now, what I propose to have done, is that one systematically calculates the O minus C residuals for a range of zenith distance from $45°$ to $90°$. The reason for this choice of the range is that below $45°$ most theories will give the same results, no matter how bad they might be. This has been pointed out by professor Saastamoinen this morning. He said, that all we had to know is the boundary conditions at the observing station and also at infinity, where the refractive index is 1. This is all very true, if the zenith distance does not exceed $45°$. But what I am interested in is large zenith distances, in particular those going down to $90°$. One can get $90°$ under certain conditions from ground station. And I am concerned about the observations of cosmic refraction made by astronauts. It is possible to use their results for the purpose as well, because the range there actually extends all the way to $180°$, and these results might be used for a check of the consistency of my theory. Now, some details of my proposal. First step: observations. What I propose here is direct measurements of vertical refraction. I am not concerned here with anomalous refraction, it is not possible to fit it with any existing theory. The more I listen to the discussions here about anomalous refraction, the more I have to regard it as a stochastic process, which this kind of theory does not cover. However, the vertical component of refraction can be measured directly, let us say, by taking a star, the declination of which is approximately equal to the latitude of the station. So it goes to the zenith, where the refraction is negligible. So one knows its true coordinates, and then, when it descends to the horizon, one can measure its apparent zenith distance and calculate what the refraction is. That's one approach. Another one is the dispersion method, which was developed or at least proposed here by Dr Currie, as soon as this method becomes available. That is step number one, the observation. Now incidentally, before we go on to step number two, this observation must be correlated with the polytropic index. This parameter of the theory is not a constant, but it varies with the season of the year and possible also with the geographical location. That is the polytropic index, which is related to the temperature gradient. Now, there are no other temperature derivatives invloved, because we assume that temperature is linear in the geopotential. We have to know what the polytropic index is, if we want to have a good calculation for a large zenith distance. Now the sources of information as regards the temperature gradients are several. One can have resort to direct meteorological measurements of vertical temperature gradients by means of balloons or, if that is not available, one can use publications of, let us say, the U S standard atmosphere, that gives the distribution of the gradients at different locations on the earth and different seasons of the year. Or, if that is not easily available, one can use the mean value of this polytropic index, which is taken as 4.256 in the U S standard atmosphere. Now, point number three. With regard to calculations, of course, the electronic machines are necessary, and in the published paper we have a flow chart, but that has been superseded by an improved version, and we also have new sets of the IBM cards available. Those would be furnished on request to anybody who has a sincere interest in checking the theory. Also we will furnish corrected reprints of the paper, with removal of all misprints found so far. Now, this is my proposal. I also have prepared some supporting arguments. Shall I read them too?

G. Teleki: I propose you to discuss these in our working group on astronomical refraction in detail. Have you any proposition how to promote the joint research, the cooperation between the geodesists and the astronomers? I myself, have proposed the organization of joint meetings in the future too. It is one step. As next step I propose the organization of an interunion commission for studies of refractional and atmospherical influences. Why atmospherical? Because that is connected with lunar ranging. Therefore it is a wider problem. The name of the commission is not so important, but the content of the activity. Have somebody some other proposition or comments connected with this proposition?

E.G. Anderson: Just a comment on the actual terminology: in geodesy there is also work progressing on the gravitational effect of the atmosphere, so if you use "atmospheric influences" you would need to carefully distinguish between refraction and other effects such as gravitation.

G. Teleki: In that case we choose the word refraction.

T.J. Kukkamäki: It would be fine if this proposal could be ready for the Executive Meeting of the International Association of Geodesy, which will be held at September 19-20, so that the Executive Committee can consider this question already there. And then the final establishing from geodesists' side will be ready next year in Canberra.

G. Teleki: Some other propositions or comments, ideas?

K. Poder: We discussed in the resolution committee that such an interunion commission might be supported to take care of all things, and it might be some trouble. I might be too careful or too pessimistic about it. I ask professor Kukkamäki directly, is there any reason for not having this commission? I fear that people will say: now, there is a commission to take care of all these things, and anyone who wants to supply must be backed up by the commission.

T.J. Kukkamäki: From our side it will be fine if it will be possible to consider that next September. This Executive Committee cannot establish this kind of cooperation or commission. That will be made in Canberra next year. But it must be prepared carefully beforehand. This might be a good start.

K. Poder: What I fear is, that if somebody is going to work on these problems later on, he might be unsupported because one would refer to the commission to take care of such research. I don't know, maybe I am too cautious here.

T.J. Kukkamäki: So we are not ready to make this proposition?

K. Poder: Yes, by all means. I was just explaining the reason for my hesitation.

J.A. Hughes: In a brief note which I had from Dr Teleki he made a very good point; that refraction is something which has been investigated piecemeal in the past with everyone going off in different directions. Sometimes, I think with artificial distinctions, certainly not always. In any event, the opportunity here for a synthesis of ideas would be valuable.

E. Tengström: There exist interunion commissions before, which take wide areas into account, for instance in geodynamics. And there they are working in so-called working groups. Each working group has its special interest, but the common task is in the commission. So I don't think there will be any difficulty to cover all areas and include all interested people in this total research of the commission. Its work also comprises the use of all types of earth-bound observations, and this proposed commission on refraction will be able to carry out research in a much closer collaboration with all interested scientists in the field than has ever been possible before. We will have the opportunity to give one another good advice. So I don't think there will exist any difficulties from any personal directions in this work.

J. Dommanget: I would like to mention that there has been already some collaboration proposed between astronomers and meteorologists in the case of image quality which is directly connected with irregularities in refractive index. The first official such collaboration to my knowledge may be found in the proceedings of the IAU Symposium No 19, held at Rome in October 1962 (Bulletin Astronomique, Paris, 34, 1964, pp. 85-160). See also Transactions of the IAU, XII B, 1964, Commission 9, pp. 133-135. Before writing any proposal, one should especially consider the resolutions given on p. 135 of this reference.

WORKSHOP ON WAVE PROPAGATION

Chairman: H. Kahmen

H. Kahmen: (introduction)
When we are measuring distances with electronic equipments, the recordings must be corrected by velocity and also the curvatory corrections must be used. During the symposium of Wageningen, Mr Hugget presented his first results of his three-colour instrument, and these results showed us that these instruments can achieve a short periodic accuracy of about 0.1 ppm. But when we consider the long periodic accuracy we see that also these instruments only achieve an accuracy of 1 ppm. So there is no fundamental correction for the velocity til now, which gives us better results than 1 ppm. Yesterday we saw that the curvatures of the electromagnetic waves follow quite different ways. When we are correcting in geodesy the results of electronic distance measurements, the corrections are based on the assumptions that the curvature is constant. For longer distances the uncertainty of these corrections will still more limit the accuracy. During the large workshop in Wageningen 1977, professor Tengström started new activities for finding better corrections of angles and for electromagnetic measured distances. He made two proposals. At first, he wished that the turbulence tapes of the atmosphere should be taken into account, and on the second hand, the problems of the fore-runners should be studied. Professor Angus-Leppan has been so kind to give us here a short report about Brunner's studies into the special problem of the fore-runners.

ON PRECURSORS OF ELECTROMAGNETIC WAVE PROPAGATION

F.K. Brunner
University of New South Wales,
Kensington, 2033, Australia.

ABSTRACT. The precursor theory of an electromagnetic wavetrain propagating through a dispersive medium is briefly reviewed. The few successful experimental determinations of these precursors, which have been reported in the literature, are limited to transients in waveguides. The determination of precursors in the optical frequency range appears to be achievable in the laboratory. However, in the geodetic context of long wavepaths through the troposphere the utilisation of precursors for measuring distances and directions free of atmospheric effects is judged as not feasible at the present time.

1. INTRODUCTION

1.1 Preamble

An investigation concerning the state-of-the-art of precursors of electromagnetic wave propagation was instigated by Professor E.Tengström, President of the Special Study Group I:42 of IAG, during a meeting of this group at Wageningen, The Netherlands, in May 1977. The writer was invited to take charge of this investigation. The following note is essentially a compilation of material and information on this subject which the writer has been able to gather since May 1977.

1.2 Precursors in the geodetic context

In 1914, A. Sommerfeld and L. Brillouin showed in two papers that the *front* of an electromagnetic wavetrain propagates in a *dispersive medium* with *velocity* c_0, the velocity of light in vacuum [a translation of both papers is contained in the book by Brillouin (1960)]. It is important here to distinguish clearly between phase-, group-, signal- and front-velocity (Sommerfeld, 1954). The arriving front is immediately followed by some very weak oscillations, called precursor or forerunner. Unfortunately the precursor has very small amplitudes and their experimental determination seems to be extremely difficult. However, the utilisation of precursors could be envisaged as an elegant method of measuring distances and directions free of atmospheric effects for geodetic purposes.

It appears that in the geodetic literature the first reference to precursors was given by Henriksen (1969), and the relevant paragraph from his paper is quoted here in full length: "A second question, first raised by Dr.H.W. Straub of ESSA, is whether the refraction problem can be eliminated entirely by using ultra-short pulses. A considerable amount of work has been done recently in synthesizing light pulses of less than pico second duration. A pulse 10^{-14} second long in time occupies only 1 micron of length; it contains less than two complete wavelengths of green light. Now it is easy to find from Maxwell's equations, that any electromagnetic wave consists of a precursor at a very low level of intensity, and a later packet of high-energy content. The precursor passes through without refraction, but the high-energy packet (which is what we usually observe) does get refracted. Straub's suggestion was that the ultra-short pulses may force enough energy into the precursor to allow detection of this portion and therefore do away with refraction problems."

From then on the Presidents of the Special Study Groups I-19 and I-23 of IAG encouraged in their reports further investigations of the precursor phenomenon (Denison, 1971; Tengström, 1971; Poder, 1975). However, at least to the writer's knowledge, investigations related to the precursor problem have not been carried out at geodetic organisations.

2. THEORY OF PRECURSORS

The precursor phenomenon occurs when a limited wavetrain or a wave packet propagates through a dispersive medium. In a dispersive medium the phase velocity, v, is a function of the wave's frquency, ω. If the dispersion is normal then an increase in ω leads to a decrease in v, and it can also be shown that for this case the group velocity v_g is less than the phase velocity v. For anomalous dispersion an increase in ω leads to an increase of the phase velocity v.

The original Sommerfeld-Brillouin problem is that of a limited sinusoidal wavetrain, which at the time t=0 is incident upon a boundary plane x=0 of a dispersive medium. The wavetrain propagates into the dispersive medium which extends between x=0 and x=∞. Interest now focuses on the propagation of this transient signal, especially the behaviour of the amplitude and frequency of the wave with time at a certain depth x.

The mathematical treatment of this problem which is essentially the solution of the appropriate Fourier integral is rather demanding. A detailed discussion of the original derivations is given in the books by Sommerfeld (1954), Brillouin (1960) and Stratton (1941). Various asymtotic expansions (saddle-point method and stationary-phase method) have been used for the solution of the Fourier integral (Brillouin, 1960). In connection with these mathematical solutions the papers by Elices and

Garcia-Moliner (1968) and Felson (1969) should also prove useful. The propagation of transient signals in lossless waveguides has received considerable attention in recent years, and Schulz-DuBois (1970) has given a review of previous work. He also has discussed a rigorous theory of waveguide transients (Schulz-DuBois, 1970), and Vogler (1970a) has derived an exact expression for transients in an ideal waveguide using the Fourier integral convolution theorem.

In the following the result of the original development by A. Sommerfeld and L. Brillouin is briefly reviewed in a rather descriptive way. The very front of the sinusoidal wavetrain propagates with velocity c_0 into the dispersive medium and reaches x at the time $t=x/c_0$, see Figure 1. This is understandable in that the electrons in the molecules of the medium are to be considered at rest when the wavefront reaches them. The following passage from the book by Sommerfeld (1954, p.117) may further help to explain this point: "This also is made clear by the following consideration: the dispersion electrons are originally at rest (their thermal agitation which is in no way related to the rhythm of the light wave can obviously be disregarded). But according to our theory, refraction and dispersion are due to the induced periodic oscillations of the electrons or ions. Thus, to begin with, the medium is *optically void* like vacuum. The propagation velocity is equal to c_0 and the index of refraction, if one still cares to speak of one, is equal to 1."

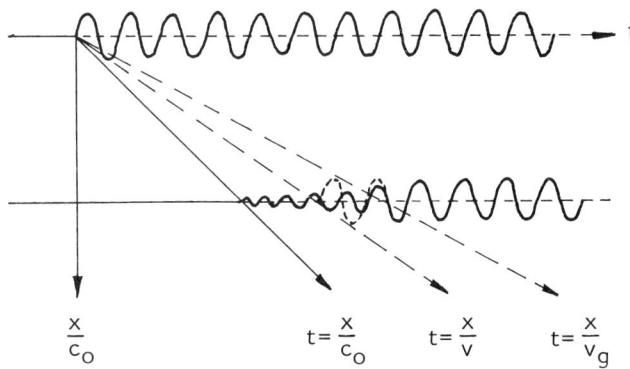

Figure 1: Scheme of propagation of a wave to a depth x in a dispersive medium. Transition of the precursor to the steady state situation. After Sommerfeld (1954).

Immediately behind the wavefront a disturbance arrives which is termed a precursor or first forerunner. The first forerunner arrives with zero amplitude which increases afterwards steadily. The low amplitude of the precursor may be explained by the withdrawal of energy from the incident wave to build up the oscillation of the electrons. The frequency of the precursor is initially very high (but finite) in comparison to that of the signal, and decreases continously. A second forerunner with an

initially very low frequency follows the first forerunner and from then
on the frequency and amplitude increase toward the steady-state values.
An illustration of the forerunners is given in Figure 2.

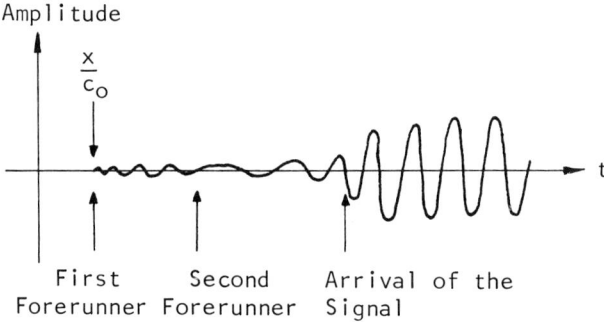

Figure 2: Variation in amplitude and frequency of the
precursors, after Brillouin (1960).

The transition between precursor and the steady-state sinusoidal wave
occurs rather suddenly but still continously. The time when the amplitude reaches half of the value of the steady-state amplitude defines
the signal velocity. The signal velocity and the group velocity are
equal but less than the phase velocity for normal dispersion. The
incident wavetrain at x=0 reproduces itself identically at the depth x,
its phase being merely shifted by (x/v), see Figure 1.

It appears from the literature on electromagnetic wave propagation, that
the theory of precursors has been accepted without much doubt since the
pioneering work of A. Sommerfeld and L. Brillouin. The only exception
seems to be an unpublished paper by Vogler (1970b). He found that in
an exact solution for the propagation of a step-modulated sine wave
through a lossless waveguide no precursor occurs. However, the precursor
appears when the method of stationary phase is used for the solution of
this wave propagation problem.

3. EXPERIMENTAL DETERMINATION OF PRECURSORS

Precursors of sound waves and ultrasonic stress waves seem to have been
detected experimentally. Brief notes about these experiments have been
given by Elices and Garcia-Moliner (1968, p.171) and by Ito (1965).

Very few experimental determinations of transient signals in waveguides
have been reported. Saxton and Schmitt (1963) have investigated transient oscillations in a large waveguide, and Ito (1965) has reported
about the behaviour of very short pulses in a waveguide. Pleshko and
Palócz (1969) have experimentally detected the Sommerfeld-Brillouin
precursors in the microwave frequency range. The precursor phenomenon

has also been studied experimentally using a travelling-wave maser (E.O. Schulz-DuBois, personal communication, 1978).

No experiments on the determination of precursors in the optical range of electromagnetic wave propagation seem to have been reported in the literature, at least to the writer's knowledge. The suggestion by H.W. Straub to use ultra short pulses of laser light for such experiments (see Section 1.2), does not appear to be associated with actual research in this direction by him or his institution. However, it seems that the recent advent of highly controllable and very short-pulse laser beams should raise hope for the experimental demonstration of precursors at optical wavelengths (A.E. Siegman, personal communication, 1978).

4. TEMPORARY CONCLUSION

The front of an electromagnetic wavetrain propagates through a dispersive medium with light velocity c_o (vacuum). It has been shown in numerous theoretical papers that very weak oscillations, so called precursors, should follow this wave front. The precursor has initially a very small amplitude but a high frequency, and both vary until the steady-state situation is reached. Geodesists have been interested in the precursor concept because its possible technical utilisation would provide an elegant solution to the atmospheric refraction problem of electromagnetic distance measurements and direction observations.

The theoretical prediction of precursors, which started with the pioneering work of A. Sommerfeld and L. Brillouin in the year 1914, has been followed up by a small number of laboratory experiments in which precursors have been successfully detected in waveguides at microwave frequencies. Therefore mathematical methods which do not predict the existence of precursors (Vogler, 1970b) should be carefully examined to resolve that discrepancy between theory and experiment.

It appears that detection of precursors should also be possible with presently available laser instrumentation in the optical frequency range. In designing such experiments the fact that the initial frequency of the precursor is very high (in comparison with the signal frequency), will become important for the selection of an optimal frequency for the transient. Furthermore, the choice of the depth x at which the determination of the precursor phenomenon is attempted, will play a crucial role, because Brillouin (1960) has shown that the amplitude of the precursor decreases exponentially with depth x. This supports the writer's opinion that for *geodetic field measurements* along wavepaths of up to several kilometres through an inhomogeneous and mostly turbulent atmosphere the chances are to be considered rather remote of successfully using precursors of electromagnetic wave propagation for the solution of the geodetic refraction problem.

ACKNOWLEDGEMENTS

The writer is grateful to R.H. Ott, J.R. Pollard, E.O. Schulz-DuBois, A.E. Siegman, M.C. Thompson Jr., L.E. Vogler and D.C. Williams for personal discussion and correspondence on this subject.

REFERENCES

Brillouin, L.: 1960, *Wave Propagation and Group Velocity*, Academic Press, New York.
Denison, E.W.: 1971, Report Special Study Group I-19 of IAG, *XV General Assembly of IAG*, Moscow.
Elices, M. and Garcia-Moliner, F.: 1968, in W.P. Mason (ed.) *Physical Acoustics*, Academic Press, New York, 5, pp. 163-219.
Felson, L.B.: 1969, *IEEE Trans. Antennas and Propagation* AP-17, pp.191-200.
Henriksen, S.W.: 1969, Ultra-Precise Measurement of Base Lines, *Symposium EDM and Atmos. Refraction*, Boulder, Colorado.
Ito, M.: 1965, *IEEE Trans. Microwave Theory and Techniques* MTT-13, pp. 357-364.
Pleshko, P. and Palócz, I.: 1969, *Phys. Rev. Letters* 22, pp. 1201-1204.
Poder, K.: 1975, Report Special Study Group I-19 of IAG, *XVI General Assembly of IAG*, Grenoble.
Saxton, W.A. and Schmitt, H.J.: 1963, *Proc. IEEE* 51, pp. 405-406.
Schulz-DuBois, E.O.: 1970, *IEEE Trans. Microwave Theory and Techniques* MTT-18, pp. 455-460.
Sommerfeld, A.: 1954, *Optics*, Academic Press, New York.
Stratton, J.A.: 1941, *Electromagnetic Theory*, McGraw-Hill, New York.
Tengström, E.: 1971, Report Special Study Group I-23 of IAG, *XV General Assembly of IAG*, Moscow.
Vogler, L.E.: 1970a, *Radio Science* 5, pp. 1469-1474.
Vogler, L.E.: 1970b, A Note on "Precursors", *Unpublished Report*, (Institute of Telecommunication Sciences, Boulder, Colorado).

DISCUSSION

H. Kahmen: Thank you for this report. Is there anybody who can tell us if there has been any success in using these fore-runners?

J.C. de Munck: I have quite the same ideas as the conclusions by Brunner, more from intuition than from real knowledge. We cannot use them, because they are too small.

H. Kahmen: In Karlsruhe we discussed these problems with some physicists, and they also came to this result.

P.V. Angus-Leppan: There has been interest in the subject for a long time. I think we should be grateful to Dr Brunner, that at last someone has taken the trouble to investigate the matter and to give firm conclusions about its usefulness.

H. Kahmen: I think it would be very interesting to remember the report we got yesterday by the professor of meteorology, Gösta Liljequist. He showed us that the curvatures of the electromagnetic waves follow very difficult ways, and he told us that endpoint measurements will not be able to give medium values of refractive indices. Now, we have done much work in improving the measurements at the endpoints. And I think we should discuss about these methods, if there are some possibilities to improve the measurements by observing more meteorological parameters at the endpoint.

T.J. Kukkamäki: I can tell you about some efforts we have made to get some information about the field of the atmosphere between these endpoints. It has been rather expensive, and not possible in the field work. At our 22 km long baseline, we have built two observation towers, 40 m high, at both ends. In these towers we are measuring the temperature and humidity at three different elevations, recording all the time when the measurements are going on. So this kind of measurements are not possible in field work. But we understand, that from these observations we get some knowledge, at least some experience. And we hope, that the extrapolation of our experience from this test field will have some value, at least in a qualitative sense, so that we can avoid some dangerous situations. However, this is for geodimeter measurements. When we have levelled across water, we have tried to determine the temperature field at both ends to see how the light beam is bending just at both ends. That is, of course, the most dangerous areas. So we have put there a set of termistors and determined the temperature field at these short areas. And we have found that, when the temperature on land is higher than on the sea surface, as is usually the case, and when the wind is in the direction from the land, these isotermes are going in one way at one point and in another way at the other, so the situations are different. We have at least been able to estimate, and even to compute, corrections for that. We also tried by running with a boat back and forth, to determine the integrated gradient along the whole length. These are some examples of our attempts to get more information.

H. Kahmen: Thank you, professor Kukkamäki. I think the Australian geodesists tried to find new assumptions by which the mean values of refractive index can be corrected. Professor Angus-Leppan, can you tell us how much the values might be improved by using your additional parameters?

P.V. Angus-Leppan: Yes, though our testing is somewhat limited. In one test we had a number of EDM lines, which were measured throughout 24 hours' period by microwave instruments. They were reduced in the normal way, and then with our particular atmospheric model, the TTM. On most of the lines the standard deviation was improved spectacularly down to less than 50% of the ordinary values. The standard method of reduction gives you a distance which is correlated with the temperature readings, whereas with the model the errors appear to be random. But on one line the improvement was not very great. It came down to about 80% of the previous values. Similarly, we used the model for trigonometrical heighting over a fairly short line, but one which was grazing all the way, so it was a difficult case. The coefficient of refraction, which has a standard value of 0.13, varied by several units! But the prediction of that coefficient of refraction by our model was down to about 0.1 unit. So that it was able to model a very big variation rather accurately. What really impressed us, was the fact that the model was able to take into account some rapid changes, when shadow came over the line, and the agreement between the observed angles and the predicted ones was very good, even with the sudden changes.

Could I ask a question? One of the very easy things one can do in taking the endpoint measurements in EDM is to put the thermometers high, say 5 m above the surface. I get the impression that the results are not really improved conclusive. I think, that Mayer and Feletschin at Karlsruhe have tried this technique, and it doesn't appear that it improves the results.

H. Kahmen: In Karlsruhe we did many experiments concerning endpoint measurements. There was one other experiment where we measured vertical angles simultaneously with distances, and tried to correct the distances by the vertical angles. We found, that these further measurements cannot improve the results, but they can make the measurements more sure. If you are measuring vertical angles simultaneously, you can find out if the conditions are quite rough or if they are good. But there is no scientifically certain correction to the distances. I think that when we try to find a new method to correct our distances, it must give an error which is not much greater than 0.1 ppm. Only then we can call it a new method. And I think that now such a method could not be found. In England you have got experiences with some two-colour instruments. Are there any results til now?

D.C. Williams: Most of our two-colour angular results have been obtained during cloudy afternoons or evenings, and on every occasion we have found a k-value close to the standard adiabatic value. So the results are not much of interest for the present discussion. We are planning to make some measurements in the early morning, when one expects the k-values to be higher.

H. Kahmen: In Wageningen Mr Hugget showed us for the first time long period measurements with three-colour instruments over lines of 4 km, 6 km, 7 km and 10 km. Every day he made measurements for nearly one year, and he found out that the accuracy of the measurements was 0.1 ppm, when the mean square errors were calculated only for one day. But when he considered the results he got for many days, one could clearly see that the long time accuracy was only 1 ppm. We tried to discuss about the reason for this, but I think there was no quite clear answer then. There may be two reasons. The first reason could be that the atmospheric model is too idealistic, and the second reason may be that the curvature corrections are too rough here.

T.J. Kukkamäki: On this 22 km line I told you about earlier, where we have these meteorological towers, we were able to determine the gradients along the whole line, and also to compute the curvature. Then we observed simultaneously with a theodolite the vertical angles, and they were in very good agreement, not of course to 100%, but somewhat. Unfortunately I cannot remember any exact values to mention here. It was quite clear a diurnal period of this situation. So we have used this experience in our field work, when we do not have these direct gradient measurements. We think that the isothermic layers are following the form of the topography, and so we were able to determine the real temperature along the whole line through some extrapolation.

J. Milewski: I think that the main question of the inaccuracy of 1 ppm is in the geometrical reduction. The data, e.g. illustrated by Mårtensson, give us short periodic variations of the atmosphere, and from his concrete data we see that the short period variations can change the dispersion with very great magnitudes. Mr Mårtensson has detected variations in 20 minutes of 1000 centesimalseconds in refraction. This means, that in a turbulent atmosphere the paths of the rays are stochastic, and short periodic variations for them cannot be accounted for. So, I think that the nature of the atmosphere limits our achievements in this domain, the very short variations of the atmosphere. When they can give such great differences in dispersion, it means that they give great differences in the real geometrical path over the distance. I see this as the main limit of the possibility to obtain the accuracy of distance measurements. And I think that the two-colour method is limited because of the geometrical reduction, which changes rapidly with time as a consequence of the atmospherical turbulence.

H. Kahmen: What do you think of the limit of the scale factor b?

J. Milewski: I cannot answer now. I think, that when we investigate the true data of the real atmosphere, and when we have from the dispersometer such records as these of Williams' instrument, and records of short exposure times by Mårtensson, we can calculate the limit of possibility to reduce geometrically the distance measurements. But the differences in geometrical paths, I think, means the limit of accuracy when determining the true distance. And I think it is possible to calculate approximately this limit.

J. Saastamoinen: In these experiments, was the line very close to the ground?

H. Kahmen: About 10 m high.

J. Saastamoinen: There might be a possibility, when raising the line by about 10 m, to improve the measurements, because then the geometric corrections would smooth out.

J. Milewski: Mårtensson's height above the ground is 30 m. And he has such great variations.

H. Kahmen: There is another question concerning dispersive measurements. Are we quite sure today that the dispersive angle, that we can measure with these new instruments, is an integral over the distance?

J. Milewski: I think that we can be sure that we measure dispersion. The accuracy of the measurements, this is another question. It might be investigated, because we have many sources of possibilities, that the accuracies are not very high, in Williams' device and in Tengström's device, but I think we measure dispersion, that is the integral of the curvatures over the path, assumed identical.

D.C. Williams: When we ask whether we are measuring the dispersion correctly, we are really asking whether the long term averages of the direction of the red light and the direction of the blue light are reliable. They could only be unreliable if there were asymmetry in the turbulence distribution, which is conceivable in the vertical direction. But as far as one can tell, our telescope images and electronic signals appear to be symmetrical, so the average dispersion ought to be correct.

H. Kahmen: Did you find a clear correlation between the refractive angle and the dispersive angle?

D.C. Williams: During a particular set of measurements, we have not seen a significant change either in the refraction angle or in the dispersion angle. The changes between different occasions are also rather small, so we are hoping to make further measurements at times which will give larger refraction, or over a longer distance. Any lack of correlation that we are observing at present is probably due to instrumental imperfections rather than to real effects of the atmosphere.

S-G. Mårtensson: I think that these dispersion methods have to be checked with absolute tests. At the absolute tests we have done in Uppsala so far, only a few of our measurements show that the refraction and the dispersion are correlated. But that might be due to the fact that we are using very short exposure times. If we integrate over long periods we may have a very good value. But that is something we can't say today, because we can't measure with long exposure times because then the accuracy will be very low.

H. Kahmen: Are there any connections between three-colour distance measurements and dispersive angle measurements? I mean, is it possible by these two methods to find out something about the turbulent activities of the atmosphere?

J. Milewski: There are two sources of two colours, the red and the blue. The first problem is that the waves in the atmosphere are not exactly the same. And when one has a turbulence in the atmosphere, the dispersion depends not only on the refraction along the paths, but there are two waves, and we can make such an investigation. But we can also let the waves from two sources agree in the observation point. Then we can investigate, if these waves are closer in the part near the observer, but more separated in the source part. I think it would be good to suggest to the physicists to make laboratory experiments to explain the main question of the correlation between refraction and dispersion. Then we will be able to answer, what is the part of influence of turbulence of the atmosphere, and what is the part of wave propagation of the two colours.

H. Kahmen: But to make such experiments in the laboratory, where we have very short distances, will probably fail.

J. Milewski: No matter. It will be in the medium of controlled air, and means exist to study even extremely small effects, e.g. by the Michelson device introduced by professor Tengström.

H. Kahmen: So you think the atmospherical effects will be detectable in the laboratory, despite their extremely small values?

J. Milewski: Yes, with the great resolving power of the Michelson device, it will certainly be possible. The accuracy of our laboratory measurements of dispersion is something like 0".0005. The base is 30 m long, which means a normal refraction in the laboratory air of 0".1. Using UV and red, which means a magnification factor of about 25, dispersion toward refraction, we can study the non-turbulent situation to within 10% about. We have all of us some difficulties to treat questions about turbulence effects, method of registration of recordings, the accuracies in our instruments, the integration intervals, the quality of recordings etc. I think it would be very good to suggest to the physicists to help us solving these problems for our two- or three-colour methods.

H. Kahmen: I think, at this time we should leave the towers of our instruments and go back into the laboratory. But I think we have to do much work in analyzing the best time averaging procedures in the field. Are there some final results until now about the way such an averaging procedure must be done? It is an open question, and I think we can discuss the results when we have long time series of measurements. Apparently there is a great difference between the short periodic accuracy and the long periodic one. Are there any more results now about long periodic accuracy? Did you measure some long time series of refraction angles? Did you do some measurements during different periods of the year? And did you find there some significant differences in accuracy?

S-G. Mårtensson: Most of the experiments we have done in Uppsala have been done during summer time, for practical reasons. But I can tell that we have never been in the neighbourhood of what we call normal refraction. It seems more like we are measuring in periods where we have two times or even three times the normal refraction. With normal refraction I mean a k-value of 0.13 or 0.14. We have never been in that region. We have maybe selected a very bad base for our refraction studies. I think we have to investigate the base, as well, because most of the times we had two to three times the normal refraction, and even just one week ago we had a k-factor equal to 1.04 at 2 o'clock in the night. For practical reasons we only measure during night-time, and our extreme results might be explained by our existing inversion layers.

H. Kahmen: I heard something about altitude measurements which gave great differences. Is there any explanation for this error?

S-G. Mårtensson: You mean systematic errors between IDM and theodolite?

H. Kahmen: No. I have heard of great systematical effects in your measurements which could not be explained.

S-G. Mårtensson: Yes. Variations by IDM of 1150 centiseconds in twenty minutes of time.

H. Kahmen: I think that all your corrections are based on values which must be quite uncertain at a certain moment. And I have another question regarding the dispersive method. When do you think you will get the best result, when there is a large refraction, or when there is a normal refraction? Are there any results?

S-G. Mårtensson: I think we should get good results whatever the refraction is, hopefully. But maybe it is so, that due to the fact that the two rays, blue and red, travel through different atmospheric media, they are separated by some cm in space at maximum. Maybe the turbulence affects them differently. They are not affected with some correlation, for sure. In that case, of course, small refraction values should give better results, because then the rays are closer to each other and hopefully travel through about the same atmospheric layers. But when we have big refraction, the separation is big and the beams travel in different turbulent media, if looking at micro-turbulence media, not macro.

H. Kahmen: By experiences we have made in Karlsruhe, using both lasers and microwaves for distance measurements, we found that the results with microwaves were usually too small, and the laser results too long. We use formulas for integrating the correct distances. And we have found that these formulae of interpolation are very good if the distances are great, but quite unsure for small distances. Why is it so? All formulas are based upon assumptions and approximations, and therefore it is very important to consider the error propagation with various numbers of legs.

T.J. Kukkamäki: When measuring our 900 km traverse with 25 legs, 30 km long, with geodimeter, we have found that there is some slight systematical difference between the observations before noon and after noon. And we have not been able to find out the reason for that. So, it means there are some systematical errors in these reduction formulas and reduction observations which we have used to reduce our observations to one and the same consistent system. This geodimeter has been calibrated

on that 22 km absolute baseline. Observations have been made before
noon and after noon, with the same procedure as in the fieldwork,
together ten observations in the morning and twenty in the afternoon.
In this way we hope that this systematical error is eliminated in the
fieldwork and in the results. But there is some slight unexplained
difference between morning and afternoon measurements.

H. Kahmen: I think that in Finland you have very good field conditions.

B. Gächter: Is there any dependence of the dispersion from the polarization state of the light, from theoretical point of view or from experiment?

D.C. Williams: It can be said with certainty that the effect of the atmosphere on the state of polarization of light is very small. Some EDM instruments such as the Mekometer exploit this feature. Whether the effect of the state of polarization on the angle of bending is negligible, I am not sure. One guesses that it will be negligible, because the angles involved are so small.

H. Kahmen: I think that we may now conclude from the discussion that there is much fundamental work to be done concerning the dispersion method. One very important thing should be to find out, maybe by theoretical considerations, where there will be a limit for geodetic accuracy, for direction measurements and for distance measurements. What methods do you think could be used, e.g. the one by Tatarski in his "Wave propagation in a turbulent medium"? Have you made some investigations in this respect? I think there should be made some further investigations, so that also one can find the limit of accuracy on the method, theoretically. And I think this may be successful. On the other hand I think we must do much laboratory work concerning two- and three-colour distance measurements and two-colour direction measurements. Will there be some more success in correcting atmospherical models, geometric models? As noone seems to have any further comment about this, I think we can close the workshop now.

WORKSHOP ON LOCAL GEODYNAMICS FROM REFRACTION CORRECTED MEASUREMENTS

Chairman: A.H. Dodson

A.H. Dodson: (introduction)
This last session means really a slight movement away from what we have been talking about during the rest of the meeting. We are coming down to much smaller scale measurements. In the paper I am going to present as an introduction to the workshop, you may notice already existing successful attempts in engineering surveys to study local dynamics with high precision. I shall mainly talk about studies of dam deformations in England.

THE ROLE OF REFRACTION IN THE MEASUREMENT OF THREE-DIMENSIONAL MOVEMENTS BY GEODETIC METHODS

A.H. Dodson,
Department of Civil Engineering,
University of Nottingham,
ENGLAND.

1. INTRODUCTION

The measurement of three-dimensional movements by geodetic techniques involves the determination of spatial co-ordinates by the observation of horizontal theodolite angles and EDM distances, with either vertical angles or spirit levelling providing the height control. For correct interpretation of the results it is essential that estimates of the accuracy of the position determinations are also provided.

Recent legislation, in the United Kingdom (1975 Reservoirs Act) has necessitated the continuous monitoring of water retaining structures. Geodetic techniques have already been used successfully [Ashkenazi, 1975] and further applications of the technique are at present being undertaken. However, one of the major problems in the analysis of the results is the determination of the à priori observational accuracies. Field experiments, aimed at providing estimates of the accuracies which can be expected, with various combinations of instruments, have been conducted at Nottingham over several seasons [Ashkenazi and Dodson, 1975] and [Dodson, 1977]. Both the field tests and the monitoring schemes have shown atmospheric refraction to be a major problem, not only because of its effect on the accuracy which can be achieved but also since it makes the estimation of à priori accuracies difficult.

This paper briefly describes the field experiments and monitoring schemes (§2 and §3) giving the accuracies which have been achieved. The effect of refraction is discussed in §4 and the conclusions drawn are summarized in §5.

2. FIELD EXPERIMENTS

The basic experiment consisted of observing a movable target from two base stations, the target and base stations forming an approximate isosceles triangle. Observations were made to the target before and after it had been moved to simulate a deformation. This experiment has been conducted on several sites over different ranges using various combinations of instruments.

For all the tests the base stations have consisted of concrete pillars with forced centring for the instruments. The target consists of a cube corner retro-reflector and separate optical target positioned in an accurately machined mount. This mount provides movement in three mutually perpendicular directions and incorporates vernier scales capable of measuring the movement of the target to 0.1 mm. Since a 'deformation' measurement was conducted in a matter of hours the long term stability of the base pillars was unimportant.

Various combinations of observations were used to test the significance of the inclusion of EDM distances at the varying ranges Horizontal angles were the mean of three rounds whereas the vertical angles were averaged from two rounds.

The accuracies with which the simulated deformations were monitored, at the various ranges, are given in Table 1 and Table 2. These results are given for movements calculated from two observed horizontal angles, two vertical angles and two distances.

Table 1 shows that with a single-second theodolite the component movements were detected to an accuracy of approximately 1 mm at 50 m range and 3 mm at 350 m range. Over the shorter distance vertical refraction would play little part in the accuracies achieved, however, at 350 m it might be expected that it would be significant. The results do not confirm this expectation, since the δz term is of the same order as the δx and δy terms. This may be due to the particular site where the observations were made, and the fact that only a single-second theodolite was used.

Over the much larger range of 900 m there is a significant difference between the error in z movement determination and the errors in x and y deformation measurement. This difference was almost certainly due to vertical refraction and its effect has been high-lighted by the use of higher precision theodolites. It should be noted that the longer range tests were conducted on a different site from the shorter range tests, where the line of sight was closer to the ground. Because of the much larger errors in the determination of the vertical movements some further experiments were conducted at this longer range.

Firstly the vertical angle to the target pillar was monitored continuously throughout the day. A variation of up to 47 seconds of arc was measured.

Secondly an auxiliary, fixed, target was set up adjacent to the movable target and the vertical angle to this second target was

RANGE (m) and No. of sets	INSTRUMENTS	MOVEMENT: OBSERVED - ACTUAL		
		δx (mm)	δy (mm)	δz (mm)
50 :- 12 sets	WILD T2 GEODIMETER 6A	1.1	0.8	0.7
350 :- 7 sets	WILD T2 KERN MEKOMETER	4.3	3.2	3.3
350 :- 3 sets	WILD T2 TELLUROMETER MA 100	6.2	3.4	2.8

Short Range Deformation Tests
Mean Experimental Accuracies
Table 1

RANGE (m) and No. of sets	INSTRUMENTS	MOVEMENT: OBSERVED - ACTUAL		
		δx (mm)	δy (mm)	δz (mm)
870 :- 4 sets	WILD T3 HEWLETT-PACKARD 3800B	3.1	2.1	16.4
870 :- 6 sets	KERN DKM3 HEWLETT-PACKARD 3800B	3.4	3.7	10.1
950 :- 13 sets	KERN DKM3 TELLUROMETER MA 100	2.7	1.8	29.6

Long Range Deformation Tests
Mean Experimental Accuracies
Table 2

measured both before and after the movement of the deformation target. Thus the refraction effects, on the vertical angle to the deformation target, could be eliminated by referring all the vertical angle measurements to the fixed auxiliary target. Table 3 gives the results obtained by correcting the observations for vertical refraction in the manner described. It can be seen that a significant improvement has been obtained, and that the accuracy of the determination of vertical movement has been brought nearer to that achieved for horizontal movement.

RANGE	INSTRUMENTS	MOVEMENT: OBSERVED-ACTUAL					
		BEFORE CORRECTION			AFTER CORRECTION		
		δx	δy	δz	δx	δy	δz
950 m :- 5 sets	KERN DKM3 TELLUROMETER MA 100	3.6	2.1	36.3	3.5	1.8	9.4

<u>Long Range Deformation Tests</u>
<u>Refraction Correction</u>
Table 3

3. DEFORMATION MONITORING SCHEMES

A deformation monitoring scheme has already been successfully employed [Ashkenazi, 1975] and two further schemes, involving contracts with Nottingham University, are being undertaken at present. These schemes consist of the establishment of a control network, observed by theodolite and EDM instruments, which is to be adjusted by a rigorous least squares procedure. Vertical control is provided by reciprocal (though not simultaneous) vertical angle measurement, with the possibility of some spirit levelling. Numerous detail points on the structures in question can be monitored from any two control stations. Some detail points will not be easily accessible and hence spirit levelling will be impracticable. The accuracies required from these schemes vary from 1-2 mm to 10 mm and the ranges over which the observations are made are between 100 m and 2 km.

The accuracies achieved in the first scheme, where the instruments used were a WILD T3 theodolite and a TELLUROMETER MA 100, were of the order of 3 mm for absolute horizontal position [Ashkenazi, 1975]. However, tests have indicated that an à priori accuracy, for height control, of approximately 10 mm per km can be expected for a line levelled by reciprocal vertical angle measurement. This would lead to à posteriori accuracies of absolute vertical position of the order of 5 mm. Nevertheless even this accuracy can be drastically affected if there exist either unusually poor conditions for observation or lines which are particularly susceptible to refraction errors. Vertical positional errors of the order of 50 mm have been experienced on certain occasions. The possible accuracy of the vertical control on these schemes is thus much less than that of the horizontal control.

The accuracy of the determination of the detail point positions obviously depends largely on the accuracy of the control stations used. Provided the range from the control stations to the detail points is short (< 200 m) then, unless very unusual conditions exist, the additional error due to the detail point observations will be of the order of 1-3 mm (see §2).

4. REFRACTION

The major influence of refraction on the measurement of three dimensional movements is its effect on the determination of heights by vertical angle observations. §2 described the results obtained in field experiments and it is clear that large errors can arise from one-way vertical angle measurements. §3 explains how, even with reciprocal (though not simultaneous) observations the accuracy of height control is less than that which can be reasonably easily obtained for horizontal position.

The errors in vertical angles described in §2 are of the order of 5 - 10 seconds of arc, in gernal, and experience has shown that this is also the magnitude of error to be expected from one-way observations in actual monitoring schemes. The use of reciprocal observations, where possible, reduces this error but at a considerable cost in time and manpower. Simultaneous observations would almost certainly reduce the error still further but again at increased cost. The obvious alternative of spirit levelling would provide better precision but involves greatly increased expenditure and, because of the terrain and accessibility of stations, is not always feasible. However, for the more precise surveys, unless vertical refraction errors can be reduced to the order of 1 - 2 seconds of arc, spirit levelling remains essential.

The second, less significant, effect of atmosphere refraction is on the distances measured using EDM instruments. The uncertainties in the refractive index can cause random errors of the order of 5 ppm. However, our experience at Nottingham [Ashkenazi, 1975] , [Dodson, 1977] , [Ashkenazi and Dodson, 1977] has shown that the random proportional errors, when using instruments such as the TELLUROMETER MA 100 and KERN Mekometer ME 3000, may be estimated at approximately 2 ppm (an average value throughout a network) and that, assuming the instruments used have been properly calibrated, this error does not significantly affect the accuracy of horizontal position determination provided sufficient horizontal angles are measured. Obviously any systematic refraction effects (due maybe to the determination of the atmospheric parameters) will not be included in this accuracy estimate. Reduction of atmospheric refraction errors in EDM distances to the order of 1 ppm (including any systematic effects) would obviously increase the accuracy of movement determination but more significantly would allow fewer observations to be made (particularly time consuming angular measurements) whilst still maintaining the precision now available.

Whether the reduction of atmospheric refraction errors is achieved through instrument development [Huggett and Slater, 1977], [Tengström, 1977] and [Williams, 1977] or by an improvement in atmospheric modelling techniques [Brunner, 1977], [Brunner and Fraser, 1977], [Felletschin, 1977] and [Maier, 1977] is a matter for discussion but, in either case the improvement would be valuable for the accurate determination of three-dimensional movements.

5. CONCLUSIONS

(a) Field experiments have shown atmospheric refraction to have a significant effect on the determination of the vertical movement of a target at 950 m range.

(b) If corrections are applied for vertical refraction three-dimensional movements can be monitored to about 3 mm in the horizontal plane and 9 mm vertically, at a range of 950 m.

(c) Horizontal position accuracies of 2 - 3 mm have been achieved for a deformation monitoring scheme but atmospheric refraction limits the height control (determined by reciprocal vertical angle measurements) to between 5 and 10 mm generally, with some observations being in error by 50 mm.

(d) Vertical position accuracies could be of a comparable magnitude to the horizontal accuracies if vertical refraction effects could be reduced by means of either dispersometer type instrumentation or improved atmospheric modelling techniques.

(e) At present high accuracy vertical control can only be provided (where possible) by time consuming and thus expensive spirit levelling techniques.

(f) The effect of uncertainties in the refractive index of the atmosphere introduces errors into EDM observations. A reduction in these errors, either by use of multi-wavelength EDM or by atmospheric modelling, would enable higher accuracies to be achieved in horizontal movement determination.

(g) Furthermore, improvements in observational accuracy would enable economies to be made in the number of observations required to produce a network of a given strength.

(h) High precision monitoring of three-dimensional movements would benefit greatly from the reduction of the errors in observations due to atmospheric refraction.

REFERENCES

Ashkenazi, V. (1975) *Deformation measurements at the Empingham Dam site*, Proc. F.I.G. Symp., Cracow.

Ashkenazi, V. and Dodson, A.H. (1975) *Experimental accuracies in the remote measurement of spatial deformations by geodetic methods*, Proc. F.I.G. Symp., Cracow.

Ashkenazi, V. and Dodson, A.H. (1977) *The calibration and evaluation of EDM instruments*, Proc. XV F.I.G. Congress, Stockholm.

Brunner, F.K. (1977) *Experimental determination of the co-efficients of refraction from heat flux measurements*, Proc. I.A.G. Symp., Wageningen.

Brunner, F.K. and Fraser, C.S. (1977) *An atmospheric turbulent transfer model for EDM reduction*, Proc. I.A.G. Symp., Wageningen.

Dodson, A.H. (1977) *The measurement of spatial displacements by geodetic methods*, Ph.D. Thesis, University of Nottingham.

Felletschin, V. (1977) *Possibilities for raising the accuracy of light and microwave instruments*, Proc. I.A.G. Symp., Wageningen.

Huggett, G.R. and Slater, L.E. (1977) *Recent advances in multi-wavelength distance measurement*, Proc. I.A.G. Symp., Wageningen.

Maier, U. (1977) *Distance measurements over the Rhine valley*, Proc. I.A.G. Symp., Wageningen.

Tengström, E. (1977) *Some absolute tests of the results of IDM measurements in the field with a description of the formulae used in the tests*, Proc. I.A.G. Symp., Wageningen.

Williams, D.C. (1977) *First field tests of an angular dual wavelength instrument*, Proc. I.A.G. Symp., Wageningen.

DISCUSSION

This was mainly a dialogue between Dr Dodson and professor Kukkamäki about the possibility of using levelling instead of vertical angles to determine accurate heights. Dodson agreed with Kukkamäki's statement that levelling is desirable, but pointed out that the costs for such a survey, and the difficulties to carry it out, were so great that he had been forced to rely upon trigonometrical levelling, the accuracy of which usually satisfied his customers' requirements. Moreover, by correcting for refraction by means of methods which are under development, the accuracy will probably be increased to satisfy all needs in the future. Dr Holdahl interfered, as regards the costs of levelling;

S.R. Holdahl: Another advantage of leveling, in locations where you can use it, is that a detailed profile of elevation can be established at low cost. When leveling 1 km, it is customary to determine a height at each of 10 or more turning points, one for each setup of the instrument. If the turning points (points on which the rods are set) are solid monuments, the observational cost of determining heights at the ten points is no different from the cost of determining the height difference between the first and last points. The cost of additional monumentation is an expenditure made only once, at the time of the first survey.

CLOSING SESSION

>T.J. Kukkamäki, President of the International Association of
>Geodesy (IAG)
>G. Teleki, Chairman of the Scientific Organizing Committee
>E. Tengström, Convener and Chairman of the Local Organizing
>Committee

T.J. Kukkamäki: Dear colleagues! I have been asked to preside this
closing session, and it is a great pleasure for me to to this, because
of your fine work here. Now I ask the Chairman of the Scientific Organizing Committee, Dr Teleki, to give us his opinion about the work of
this symposium and its results.

G. Teleki: Mr Chairman, dear colleagues! First I would like to stress
that I am full of a great number of impressions. Therefore, it is
practically impossible for me to give a critics, complete and real,
concerning this symposium. Please, allow me, therefore, and professor
Tengström to summarize the results of this symposium for the proceedings (see encl. addendum). I believe that the symposium covered all
parts of refractional influences. We had several very interesting review papers, which are very useful for everybody I suppose. We obtained
a good information about the present state and the future about this
kind of investigations. I suppose that our proceedings will be a very
useful handbook for every researcher in refractional problems. I must
say that I am completely satisfied with this symposium. I expect now
critics and your reactions. The symposium was very intimate. If I remember, at least 40 participants, but 40 very good participants were
here. Discussions and contributions were very effective. Therefore,
I can conclude that this symposium was very active. We permanently
discussed about the cooperation between geodesists and astrometrists.
Why must we cooperate? Allow me to stress the reasons. We need some
recapitulation, connected with this document we must send to the
Executive Committees of our unions. First of all I can say, that the
atmospheric effects on astrometric and geodetic observations put a
limit on the real accuracy. Therefore, it is a common problem, which
I think we must solve together. The second common field is the latitude -
longitude and azimuth observations in geodetical astronomy, or earth
rotation in astrometry. Therefore, it is the same problem we investigate. The third question is the determination, or rather the elimination,
of refractional influences at a star observation. We have this kind of
needs in both geodetical astronomy and astrometry. The fourth common
field is the modelling of the atmosphere, not only the whole atmosphere.

In astrometry we also need a modelling of the atmosphere very near to our instruments. Therefore, it is really a common problem. And I can notice that the prevention is also a problem, common for you in geodesy and for us in astrometry. No discussion that we really need meteorological help. Therefore, I am completely satisfied with the resolutions, because the resolutions no 5 and no 8 recommend the establishment of a commission, which includes not only the astronomers and the geodesists, but also meteorologists and physicists. So I expect that this commission will be very useful for us. At the end of my speech, I would like to express our thanks to all speakers and participants in the discussions. And also allow me to express your and my own thanks to the Local Organizing Committee, especially to professor Tengström.

T.J. Kukkamäki: Thank you, Dr Teleki, for this summarizing of our work. We have now the next point in our agenda. That is accepting of our resolutions. I ask if you at this plenary session of our symposium accept the resolutions. I can see, from your raised hands, a quite clear majority. So, the resolutions are accepted by the symposium.
I think we have made a valuable work here, and we see that black on white when we read our resolutions. I would like to mention one of the resolutions, namely no 8, where we say that this joint work of astronomers and geodesists must be continued. That will be made at a more official level when we recommend the IAU and IUGG (IAG) to establish a special joint commission to take care of this kind of research. On behalf of the International Association of Geodesy, I thank again the Scientific Organizing Committee, especially its Chairman, Dr Teleki. On behalf of my association and all the participants, I thank the organizers of this meeting, which is one merit among other great merits of professor Tengström. We have to thank him and his staff, not least miss Ohlsson, the secretary. She has made a wonderful job in arranging our business, travelling, accomodation etc. With these words I close this formal session. There is, however, still one point, which is very pleasant for us, namely Dr Angus-Leppan will tell us something which will happen next year.

P.V. Angus-Leppan: informed, as chairman of the Organizing Committee for the General Assembly, about the meeting of the XVII General Assembly of the International Union of Geodesy and Geophysics, to be held in Canberra, Australia, December 2-15, 1979.

CLOSING SESSION

ADDENDUM

Dr Teleki asked, during his speech at the closing session, for a demonstration of the common opinion of the chairmen conducting the IAU working group WGAR and the IAG special study group I.42, as regards the results of the symposium work.

After this symposium, Dr Teleki and professor Tengström have concluded:

1) That all papers and discussions were of great value for planning the future work.

2) That all the scientific resolutions passed, adequately describe various important areas of research to be considered by adequate bodies.

3) That specifically most discussions have reflected the opinion that the ultimate goal of all investigations of refractional effects in astrometry and geodesy, should be to find methods of deriving accurate corrections, due to these effects, involving a minimum number of meteorological parameters, derivable from simple and feasible observations at - or close to - the actual time of measurement.

4) That the participants agreed that, in order to minimize the influences of refractional effects in astrometry, the prevention - the best possible location of instruments, most adequate pavilion and observing methods, etc. - is recommendable.

5) That the predominant part of the participants envisaged the dispersion method - although not yet sufficiently tested in the field for absolute significance - as the only way of reaching the goal in 3), at least at present

6) That, furthermore, the contributions to the discussions around astronomic, parallactic, photogrammetric (cosmic) and terrestrial refraction, have demonstrated
a) that improved atmospherical models might - more or less - satisfy long time researches, e.g. the stellar catalogue work and also the needs for certain investigations in satellite geodesy, but
b) that such models should be avoided and replaced by other methods - especially in astrogeodetic work, e.g. in low elevation stellar triangulation in photogrammetry and in terrestrial geodesy (including precise levelling) - namely such methods with which the effect of the actual instantaneous local atmospherical state can be accurately derived.

7) That interesting papers and interferences from various participants have shown that, when the effects of micro-turbulence on refraction for short observation times and the behaviour of electromagnetic wave propagation under such conditions are better known for various atmospheric conditions, three ways of eliminating such effects should be

considered, namely
a) at first, by measuring during clearly settled non-turbulent conditions,
b) second, by means of appropriate spatial and/or time averaging of signal observations,
c) third, to use statistically computed corrections to eventually minimize the effect of turbulence in the results.

8) That the participants, noting the fact that micro-turbulent atmospherical effects on refraction are also of interest to meteorology, recommend steps to be taken for storing astronomic and geodetic short-time observational data concerning refraction, so that they could be used by the meteorologists in their studies of atmospherical turbulence under various thermodynamical conditions.

9) That the role of air humidity for refraction and wave propagation was considered as being important to study further for different regions of the electromagnetic spectrum.

10) That the participants agreed on the needs for improving methods and techniques for determining refraction-free vertical angles in local geodynamical investigations, to be used as supplementary information beside ranging and levelling.

11) That the participants, though expressing certain doubts concerning the future possibility of detecting the so-called fore-runners of electromagnetic waves, still realized the importance of further studies of such means to observe directly refraction-free quantities in astronomy and geodesy.

RESOLUTIONS

Prof. K. Ramsayer, chairman of the resolution committee, read the proposed resolutions.

After a long discussion, the details of which are not reported here, the following resolutions were passed with these definitive texts:

Resolution No 1

The Symposium,

considering the fact that atmospheric refraction is still a severe limitation on the attainable accuracy of measurements in astrometry and geodesy,

recommends that the existing astronomical refraction tables or computation methods should be improved so as to bring them nearer to reality.

Resolution No 2

The Symposium,

recommends that for higher accuracy demands, the real state of the atmosphere should be taken into account. Especially, means should be found to determine the influence of the deviations of the atmospheric layers from the hypothetical shape.

Resolution No 3

The Symposium,

recommends that all possible preventive measures be taken by astrometric observers to minimize the effects of refraction. Careful site selections, using the appropriate scientific criteria is of utmost importance. Attention must also be given to, e.g. the design of buildings and the organization of the observing programme.

Resolution No 4

The Symposium,

recommends that established observing sites be investigated by all means possible, in order to determine the local refraction effects not corrected by the application of the general theories in use.

Resolution No 5

The Symposium,

recommends that since investigations to determine the influence of refraction on directions and distances, using two or more wavelengths, showed encouraging results, all efforts should be made to test these methods in the field and to bring them to a state, where they can be applied in astrometric and geodetic practice.

Resolution No 6

The Symposium,

recommends that since after extensive researches into refraction effects in geometric and trigonometric levelling there are still problems with refraction in relation to the current demands for higher accuracy, more investigations of the refraction be encouraged either using classical means or new methods and techniques.

Resolution No 7

The Symposium,

recommends that since the meteorological approach to refraction problems has led to encouraging results, further investigations should be undertaken along these lines, including the testing, development and application of improved atmospheric models, and improved methods for probing and remote sensing of atmospheric parameters.

Resolution No 8

The Symposium,

recommends, since both astronomers and geodesists are concerned with refraction problems, that this first cooperation between the IAU and the IAG in the field be continued, and that therefore a joint commission of the IAU and the IAG should be established with the participation of interested meteorologists and physicists, and that this recommendation be transmitted to the Executive Committees of the IAU and the IAG.

Resolution No 9

The Symposium,

recommends that in view of the possibilities of using different space laboratory facilities to investigate refractional effects in the atmosphere, steps should be taken by the IAU and the IAG to request various space agencies to consider such research.

Resolution No 10

The Symposium,

expresses its most cordial thanks to the IAU, the IUGG, the IAG, the Swedish Natural Science Research Council, the Wallenberg Foundation and the University of Uppsala for their sponsoring the meeting, to the Organizing Committees for the excellent arrangements made for a successful joint meeting, and to the University of Uppsala and its Geodetic Institute at Hällby for their kind hospitality.

JOINT MEETING OF THE WORKING GROUP ON ASTRONOMICAL REFRACTION, IAU COMMISSION 8, AND SSG I.42 ON "ELECTROMAGNETIC WAVE PROPAGATION AND REFRACTION IN THE EARTH'S ATMOSPHERE" OF IAG.

Chairman: G. Teleki

Participants: W. Altenhoff, B. Garfinkel, L. Hradilek, J.A. Hughes, J.C. de Munck, J. Saastamoinen, C. Sugawa, G. Teleki and E. Tengström.

E. Tengström summed up the results of the symposium. Those present pointed out the success of the meeting and the necessity of the realization of the resolutions. They also rendered full homage to the organizers of the symposium.

The members thought that it would be very useful to organize a joint meeting of the groups, with a small number of topics, in 1980 or 1981, maybe in Yugoslavia. It is indispensable that the organizer publishes the proceedings of that meeting.

Both groups need detailed plans of the future work and therefore the chairmen of the groups asked the members for proposals, which should be in a written form.

The participants underlined the importance of collaboration with meteorologists and physicists, which has to be considered in the preparation of the future plans for the groups.

At the end of the joint meeting, the members expressed their satisfaction with the successful cooperation between the groups, and asked the chairmen of the groups to try to strengthen this collaboration in the coming years.

The members supported the symposium resolution no 8, on the establishment of the special joint commission of IAU and IAG for research into refraction.

INDEX

Aerial photography 322
aerological data 28
aerology
  refractional calculation
    from 6
angle of arrival 227-238
astrographic images
  reduction of 40,54ff
astrometrists
  cooperation with geodesists
    381
astrometry
  calculation of refraction
    in 119-124
  geodesic, refraction
    corrections in 129-130
  refraction calculation
    for each observation
    120
  tangential coordinates,
    refraction effect on
    39-65
  terrestrial and extra-
    terrestrial, survey
    2ff
astronomers
  cooperation with meteorol-
    ogists and geodesists
    351-354, 387
atmosphere
  boundary layer 234ff
  density variation 32
  models 169ff,281-283,381,
    382
    astronomical refraction
      in 35-37
    refraction calculation
      in 73-94
    temperature and humidity
      gradients in 166
  real shape of 385
  two-layer model 80

atmosphere
  two-spheres model 104
  turbulent 241-247
    high-frequency spectrum 242ff
atmospheric correction 285-292
atmospheric density
  refraction anomaly and 27-33
atmospheric dispersion 165
atmospheric integrals 79ff
atmospheric stability parameter
  228ff

Boundary layer meteorology 6
boundary layer
  statistics of 234ff

Calculation
  continued fraction 95-101
  coefficient of refraction 5,27
    43,103,165,168,196,203-212,213-
    225,268ff,277,344
  gradient 228ff
  time series of 214,218ff
  variation of 166
conduction effects of heat island,
  model of 14ff
convection of heat island, model
  of 14ff
coordinates
  pseudoequatorial 41
  tangential, expressions for 41
    general expression of 49ff
    refraction effect on 39-65

Damping factor 73,83ff
dams, continuous monitoring of
  373-380
declination, absolute 125-128
deep space tracking 157-162
deformation measurement 374
differential delay, water effect on
  11

391

dispersion methods 6,383
dispersometer, absolute test
 of 239-240
Doppler shift, microwave 157-162

Electronic Distance Measurement
 (EDM) xvii,377ff
 atmospheric models in 171ff
 correction to, eikonal equation
  167ff
 curvature correction to 167
 instruments 285-292
 refraction corrections to
  conventional 167
 temperature error in 166
 velocity correction to 167
environment
 influence on observation 179-189

Flux-profile relationships 235ff
frequency calibration 291

Geodesists
 cooperation with astrometrists
  381, 387
 and meteorologists 351-354
geodesy
 astrometric xviii
 instrumental accuracy in xvii
 levelling errors in 305-319
 refraction elimination in 327-330
 refraction evaluation by 191-193
 satellite,
  range correction in 73
  refraction calculation in
   119-124
 three-dimensional movement
  measurement 373-380
geodetic framework
 trilateration and 267-274
geodimeter 172
geodynamics 371
gravitational radiation 157-162

Humidity gradient 329,363
 in model atmospheres 166
hygrometer
 infrared 11

IDM measurements 241-247
 refraction angle determination
  by 249-266
 systematic errors in 368
image dancing 233
instrumental accuracy 286ff,368ff,
 376ff
 in levelling 298-299
 in photogrammetry 324
instrumental housing 7
interferometer
 single aperture amplitude 154
inversion layers 343-350
ionosphere 11
isopycnic tilts 13-25
 model of 16ff

Levelling 293-299,386
 errors in, detection of 301-303
 geodetic, errors in 305-319
location of instruments 7
location of observatories 14
lower atmosphere 6

Meridian observations
 fundamental 14
meteorologists
 cooperation with astronomers and
  geodesists 351-354
meteorology 352,363ff,386
 and levelling 293ff
 in compuation of refractional
  effects 165-178
 refractional study and xxiii
model
 temperature perturbation 14ff

Northern hemisphere
 astronomical refraction in 103-117
Novaya Zemlya phenomenon 349
numerical methods 213-225

Observational methods 7
observation
 systematic errors in 180ff

Parallel measurement 165-178

INDEX 393

path length variation 157-162, 163-164
photogrammetry 321-325
photographic vertical circle 125-128
photon detection 134, 135
photon noise 136ff

Quadrant photosensor 134
quadrant sensor system 133, 151ff

Radioastronomy, refraction in 11, 12
radiometers, water vapor 158
refraction
  abnormal 343-350
  angle 249-266
    vertical 227-238
  anomalous 13-25, 131-155
    atmospheric density and 27-33
    determination of 4
  astrometric 385
  astronomical 83ff
    asymptotic expansion of 95-101
    in northern hemisphere 103-117
    model calculations 73-94
    precise calculation of 119-124
    survey of 1-10
    three-dimensional model atmosphere 35-37
    variations of 27-33
  atmospheric
    effect on tangential astrometric coordinates 39-65
    in geodesy 373-380
  chromatic 125-128
  coefficient see coefficient of refraction
  cosmic, astronaut observation of 352
  error, in geodetic levelling 305-319
  extraterrestrial observation of 387
  formulae 42ff
  geodetic, elimination of 327-330

refraction
  geodetic estimation of 191-193
  importance of knowledge of xx
  integral 95-101, 277
  lateral xvii
  local 386
  longitudinal xvii
  meteorological influences on xxiii
  meteorology and 165-178
  mitigation of effects 7ff
  nivellitic xxii
  parallactic 73, 275-279, 281-283
  photogrammetric 73
  pure, tables of 3ff
  radio, diurnal changes of 163-164
  tables 125, 385
    correction of, analytical model for 27-33
    empirical 28
    primitive calculation compared with 119-124
  terrestrial 203-212
    in Iraq 269ff
  total effect, determination of 4ff
  tropospheric 157-162, 163-164
  vertical xvii
refractometer
  Mekometer 286ff
  two-color 131-155, 286, 364, 386
    calibration of 291ff

Satellites, artificial 7
scale height 111
scale height variation 32
solar radiation 309ff
  temperature gradient and 306
spacecraft tracking 157-162
spectral transparency 126
star catalogs
  precision 131-155
  refraction corrections in 129-130
statistical methods 179-189, 234ff
  random error 180
stellar image models 136ff
stellar triangulation 275-279
stochastic models 214ff
stratosphere in model atmosphere 80
symposium conclusions 383ff

Target blurring 233
Telescope, Goddard 48-in. 154

tellurometer 172
temperature
  error in EDM 166
  function 296
  gradient 293ff,306,343-350,363
    direct measurement 171
    in model atmospheres 166
  measurement 346
  perturbation field model 14ff
  profile, vertical 315ff
  scaling 228ff
  variation 303
trigonometric height 168
  refraction in 170
trigonometric levelling 184
  refraction in 195-201
  traverse adjustment 198
trigonometry
  astronomical measurements and 44ff

trilateration 267-274
troposphere
  in model atmosphere 80
turbulent atmospheres 228ff,383,384

Upper atmosphere 6
urban heat islands 13-25
  temperature perturbation field model 14ff

Vertical angle measurements 213,376
vertical circle observation 125-128

Water vapor density 163-164
  refraction by 11
waveguides 359
wave propagation 355
  in turbulent atmospheres xxiii
  precursors 357-369

Zenith angles, reciprocal 203-212